Lecture Notes in Physics

Springer-Verlag Berlin Heidelberg GmbH

The Editorial Policy for Proceedings

The series Lecture Notes in Physics reports new developments in physical research and teaching – quickly, informally, and at a high level. The proceedings to be considered for publication in this series should be limited to only a few areas of research, and these should be closely related to each other. The contributions should be of a high standard and should avoid lengthy redraftings of papers already published or about to be published elsewhere. As a whole, the proceedings should aim for a balanced presentation of the theme of the conference including a description of the techniques used and enough motivation for a broad readership. It should not be assumed that the published proceedings must reflect the conference in its entirety. (A listing or abstracts of papers presented at the meeting but not included in the proceedings could be added as an appendix.)
When applying for publication in the series Lecture Notes in Physics the volume's editor(s) should submit sufficient material to enable the series editors and their referees to make a fairly accurate evaluation (e.g. a complete list of speakers and titles of papers to be presented and abstracts). If, based on this information, the proceedings are (tentatively) accepted, the volume's editor(s), whose name(s) will appear on the title pages, should select the papers suitable for publication and have them refereed (as for a journal) when appropriate. As a rule discussions will not be accepted. The series editors and Springer-Verlag will normally not interfere with the detailed editing except in fairly obvious cases or on technical matters.
Final acceptance is expressed by the series editor in charge, in consultation with Springer-Verlag only after receiving the complete manuscript. It might help to send a copy of the authors' manuscripts in advance to the editor in charge to discuss possible revisions with him. As a general rule, the series editor will confirm his tentative acceptance if the final manuscript corresponds to the original concept discussed, if the quality of the contribution meets the requirements of the series, and if the final size of the manuscript does not greatly exceed the number of pages originally agreed upon. The manuscript should be forwarded to Springer-Verlag shortly after the meeting. In cases of extreme delay (more than six months after the conference) the series editors will check once more the timeliness of the papers. Therefore, the volume's editor(s) should establish strict deadlines, or collect the articles during the conference and have them revised on the spot. If a delay is unavoidable, one should encourage the authors to update their contributions if appropriate. The editors of proceedings are strongly advised to inform contributors about these points at an early stage.
The final manuscript should contain a table of contents and an informative introduction accessible also to readers not particularly familiar with the topic of the conference. The contributions should be in English. The volume's editor(s) should check the contributions for the correct use of language. At Springer-Verlag only the prefaces will be checked by a copy-editor for language and style. Grave linguistic or technical shortcomings may lead to the rejection of contributions by the series editors. A conference report should not exceed a total of 500 pages. Keeping the size within this bound should be achieved by a stricter selection of articles and not by imposing an upper limit to the length of the individual papers. Editors receive jointly 30 complimentary copies of their book. They are entitled to purchase further copies of their book at a reduced rate. As a rule no reprints of individual contributions can be supplied. No royalty is paid on Lecture Notes in Physics volumes. Commitment to publish is made by letter of interest rather than by signing a formal contract. Springer-Verlag secures the copyright for each volume.

The Production Process

The books are hardbound, and the publisher will select quality paper appropriate to the needs of the author(s). Publication time is about ten weeks. More than twenty years of experience guarantee authors the best possible service. To reach the goal of rapid publication at a low price the technique of photographic reproduction from a camera-ready manuscript was chosen. This process shifts the main responsibility for the technical quality considerably from the publisher to the authors. We therefore urge all authors and editors of proceedings to observe very carefully the essentials for the preparation of camera-ready manuscripts, which we will supply on request. This applies especially to the quality of figures and halftones submitted for publication. In addition, it might be useful to look at some of the volumes already published. As a special service, we offer free of charge LaTeX and TeX macro packages to format the text according to Springer-Verlag's quality requirements. We strongly recommend that you make use of this offer, since the result will be a book of considerably improved technical quality. To avoid mistakes and time-consuming correspondence during the production period the conference editors should request special instructions from the publisher well before the beginning of the conference. Manuscripts not meeting the technical standard of the series will have to be returned for improvement.

For further information please contact Springer-Verlag, Physics Editorial Department II, Tiergartenstrasse 17, D-69121 Heidelberg, Germany

B. Wolf O. Stahl A.W. Fullerton (Eds.)

Variable and Non-spherical Stellar Winds in Luminous Hot Stars

Proceedings of the IAU Colloquium No. 169
Held in Heidelberg, Germany, 15-19 June 1998

Springer

Editors

Bernhard Wolf
Otmar Stahl
Landessternwarte Heidelberg
Königstuhl
D-69117 Heidelberg, Germany

Alex W. Fullerton
Department of Physics and Astronomy
University of Victoria
P.O. Box 3055
Victoria, BC
Canada V8W 3P6

Library of Congress Cataloging-in-Publication Data.

Die Deutsche Bibliothek - CIP-Einheitsaufnahme

Variable and non-spherical stellar winds in luminous hot stars :
held in Heidelberg, Germany, 15 - 19 June 1998 / B. Wolf ... (ed.). -
(Proceedings of the IAU colloquium ; No. 169) (Lecture notes in physics ; Vol. 523)
ISBN 978-3-662-14210-3 ISBN 978-3-540-49062-3 (eBook)
DOI 10.1007/978-3-540-49062-3

ISSN 0075-8450
ISBN 978-3-662-14210-3

Originally published by Springer-Verlag Berlin Heidelberg New York in 1999
Softcover reprint of the hardcover 1st edition 1999

Typesetting: Camera-ready by the authors/editors
Cover design: *design & production*, Heidelberg

SPIN: 10644327 55/3144 - 5 4 3 2 1 0 – Printed on acid-free paper

Preface

Since the discovery of strong P Cygni-type profiles in their satellite-UV spectra more than 25 years ago, the mass loss of luminous hot stars by stellar winds has been a subject of considerable interest. This discovery not only changed our physical concept of the atmospheres of hot stars dramatically, but also had important consequences for the theory of the evolution of massive stars. The presence of winds from these stars influences many other astrophysical situations: in particular, they are an important source of chemically enriched material for the interstellar medium, and provide a substantial contribution to the interstellar energy budget. With large telescopes, individual luminous blue stars can be studied with high spectroscopic resolution, even in extragalactic systems. They are, therefore, important probes of the physical conditions in extragalactic systems of different Hubble types. Luminous blue stars have always been considered to be potentially important distance indicators.

Due to the importance of this ubiquitous mass loss, various IAU symposia (83, 116, 162, and 163) and IAU colloquia (59 and 113) have been held during the past twenty years, each of which emphasized different astrophysical aspects of the mass loss from hot stars.

Recently, however, there have been important discoveries and developments that have resulted in a new view of the properties of hot-star winds and the processes that affect them. Spectroscopic time series in the satellite UV (with IUE) and in ground-based wavelength ranges have shown that time variability is a general characteristic of the winds of luminous hot stars and that steady-state flows do not provide a good description of their observed properties. Rotation and pulsation most likely play an important role in triggering this variability, and some observational facts can be explained only by the presence of magnetic fields. Likewise much evidence has been accumulated that the winds are often non-spherical. In some spectacular cases the asymmetric outflows can be directly imaged with the HST. There have also been many new theoretical advances concerning pulsational effects and the formation of the disks observed around luminous hot stars.

These new results made it desirable to have an IAU colloquium that brought together the specialists in this field from all continents, to further develop the emergent concepts of wind variability and asphericity and to try to find answers to these most urgent questions, with the aim of developing a better physical understanding of the process of mass loss from hot stars and its evolutionary consequences.

There were several good reasons to hold IAU colloquium No. 169 on "Variable and Non-spherical Stellar Winds in Luminous Hot Stars" in Heidelberg during June 15 – 19, 1998. First and foremost is that these dates coincided almost exactly with the hundredth anniversary of the inauguration of the Landessternwarte Königstuhl on June 20, 1898. The investigation of variable, luminous hot stars has a long tradition at the Landessternwarte Königstuhl. Indeed, the founder of the Landessternwarte – Max Wolf, one of the pioneers of using photographic plates for astronomy – discovered the LBV Var2 of M33 even before the extragalactic nature of this system was recognized. His first photographic plate of this object dates from October 1902. In 1998, LBVs and other variable luminous hot stars constituted one of the main fields of research at the Landessternwarte. Hence, the Landessternwarte's centennial also marks nearly 100 years of investigating variable luminous hot stars, which is further cause for celebration.

The colloquium was opened by the Oberbürgermeisterin of Heidelberg, Frau Beate Weber, who welcomed 150 participants from 30 countries. During the course of the meeting, 62 oral papers and approximately 70 poster papers were presented. The social program included a reception in the Max-Planck-Haus on June 14, a BBQ at the Landessternwarte Königstuhl in the evening of June 16, and a boat trip on the Neckar in the afternoon of June 18. A special problem during the colloquium was an influenza outbreak caused by the so-called "Heidelbug" which affected a number of participants. A proven remedy – beer – was distributed prophylacticly every evening after the last session and helped to limit the pain of the participants. A dive into the Neckar was also shown to be an effective way to kill the bug.

The scientific program of the colloquium was supported by Division IV and Commissions 25, 27, 28, 35, and 36 of the IAU. Financial support from the IAU and the German Research Council (DFG) is gratefully acknowledged.

The Editors, on behalf of the SOC (E. Chentsov (Russia), P.S. Conti (USA), P. Mc Gregor (Australia), R.M. Humphreys (USA), G. Koenigsberger (Mexico), R.P. Kudritzki (Germany), H.J.G.L.M. Lamers (The Netherlands), C. Leitherer (USA), A. Maeder (Switzerland), C. Sterken (Belgium), B. Wolf (*Chair*, Germany)) and LOC (I. Appenzeller, B. Baschek, A. Kaufer, G. Klare, D. Lemke, H. Mandel, T. Rivinius, O. Stahl (*Chair*), T. Szeifert, B. Wolf), thank everyone associated with organizing the colloquium. We thank Holger Mandel for ensuring that the colloquium budget was balanced and Andreas Korn for transcribing the discussions. We would particularly like to thank the graduate and postgraduate students of the Landessternwarte for their help both before and during the meeting. Last but not least, we thank all participants for their contributions to IAU Colloquium No. 169, the variety and quality of which ensured its success.

Heidelberg, December of 1998

B. Wolf
O. Stahl
A. Fullerton

Contents

Session I Observations of Non-spherical Winds

chair: J.M. Marlborough and P. Williams

Session II Theory of Non-spherical Winds

chair: B. Baschek

Session III Variable Winds

chair: A.W. Fullerton, N. Markova and D. Massa

Session VI Evolutionary Aspects

chair: C. Sterken

List of Participants

Appenzeller, Immo I.Appenzeller@lsw.uni-heidelberg.de
Landessternwarte, Heidelberg, Germany,

Baade, Dietrich dbaade@eso.org
ESO, Garching, Germany,

Babel, Jacques babeljr@bluewin.ch
Geneva, Switzerland,

Baschek, Bodo baschek@ita.uni-heidelberg.de
ITA, Heidelberg, Germany,

Bastian, Ulrich s01@ix.urz.uni-heidelberg.de
ARI, Heidelberg, Germany,

Berghöfer, Thomas thb@ssl.berkeley.edu
SSL, Berkeley, USA,

Bisikalo, Dmitry bisikalo@inasan.rssi.ru
Institute of Astronomy, Moscow, Russia,

Bjorkman, Jon jon@physics.utoledo.edu
University of Toledo, USA,

Bjorkman, Karen karen@astro.utoledo.edu
University of Toledo, USA,

Blomme, Ronny Ronny.Blomme@oma.be
Royal Observatory, Brussels, Belgium,

Bohlender, David david.bohlender@hia.nrc.ca
Herzberg Institute, Victoria, Canada,

Bomans, Dominik J. bomans@astro.uiuc.edu
University of Illinois, Urbana, USA,

Bono, Giuseppe bono@oat.ts.astro.it
Trieste Observatory, Italy,

Boonyarak, Chayan chayanb@nu.ac.th
Naresuan University Phitsanulok, Thailand,

Breysacher, Jacques jbreysac@eso.org
ESO, Garching, Germany,

Brillant, Stephane stephan@imspark.obspm.fr
Observatoire de Paris-Meudon, France,

Briot, Danielle danielle.briot@obspm.fr
Observatoire de Paris, France,

Brown, John C. john@astro.gla.ac.uk
University of Glasgow, Scotland,
Busfield, Graeme g.busfield@ic.ac.uk
Imperial College, London, United Kingdom,
Cakirli, Omur cakirli@bornova.ege.edu.tr
EU Observatory, Izmir, Turkey,
Cassinelli, Joseph cassinelli@madraf.astro.wisc.edu
University of Wisconsin, Madison, USA,
Chalabaev, Almas chalabae@laog.obs.ujf-grenoble.fr
Observatoire de Grenoble, France,
Chentsov, Eugene echen@sao.stavropol.su
SAO, Karachai-Circessia, Russia,
Chesneau, Olivier chesneau@obs-azur.fr
Observatoire de la Cote d'Azur, France,
Clampin, Mark clampin@stsci.edu
STScI, Baltimore, USA,
Clark, James jsc@star.cpes.susx.ac.uk
University of Sussex, Brighton, United Kingdom,
Dachs, Joachim
Tübingen, Germany,
de Jong, Jeroen A. jdj@astro.uva.nl
Astronomical Institute, Amsterdam, The Netherlands,
Dorodnitsyn, Anton dora@mx.iki.rssi.ru
Space Research Institute, Moscow, Russia,
Drew, Janet j.drew@ic.ac.uk
Imperial College, London, United Kingdom,
Echevarria, Juan jer@astroscu.unam.mx
UNAM, Mexico City, Mexico,
Eversberg, Thomas eversberg@astro.umontreal.ca
Universite de Montreal, Canada,
Fabrika, Sergei fabrika@sao.ru
SAO, Karachai-Cherkess, Russia,
Feldmeier, Achim achim@pa.uky.edu
University of Kentucky, Lexington, USA,
Friedjung, Michael friedjung@iap.fr
Institut D'Astrophysique, Paris, France,
Fullerton, Alexander awf@pha.jhu.edu
Johns Hopkins University, Baltimore, USA,
Gayley, Kenneth kenneth-gayley@uiowa.edu
University of Iowa, Iowa City, USA,
Georgiev, Leonid georgiev@astroscu.unam.mx
UNAM, Mexico City, Mexico,
Gittings, Michael gittings@lanl.gov
Los Alamos, New Mexico, USA,

Glatzel, Wolfgang wglatze@eden.uni-sw.gwdg.de
Uni-Sternwarte, Göttingen, Germany,

Guzik, Joyce Ann joy@lanl.gov
Los Alamos, New Mexico, USA,

Gvaramadze, Vasilii vgvaram@mx.iki.rssi.ru
Astrophysical Observatory, Tbilisi, Georgia,

Hamann, Wolf-Rainer wrh@astro.physik.uni-potsdam.de
Universiät Potsdam, Germany,

Hambaryan, Valeri V.Hambaryan@lsw.uni-heidelberg.de
Byurakan Astrophysical Institute, Armenia,

Henrichs, Huib huib@astro.uva.nl
University of Amsterdam, The Netherlands,

Hujeirat, Ahmad hujeirat@astro.uni-wuerzburg.de
Astronomisches Institut, Würzburg, Germany,

Hulbert, Stephen hulbert@stsci.edu
STScI, Baltimore, USA,

Humphreys, Roberta roberta@isis.spa.umn.edu
University of Minnesota, Minneapolis, USA,

Ignace, Richard rico@astro.gla.ac.uk
University of Glasgow, Scotland,

Jankovics, Istvan ijankovi@gothard.hu
Eotvos University, Szombathely, Hungary,

Janot Pacheco, Eduardo janot@sismo.iagusp.usp.br
Universidade de Sao Paulo, Brazil,

Kanschat, Guido kanschat@iwr.uni-heidelberg.de
IWR, Heidelberg, Germany,

Kaper, Lex lkaper@eso.org
ESO, Garching, Germany,

Kaufer, Andreas A.Kaufer@lsw.uni-heidelberg.de
Landessternwarte, Heidelberg, Germany,

Keskin, Varol keskinv@alpha.sci.ege.edu.tr
Ege University Observatory, Izmir, Turkey,

Khalack, Victor khalack@mao.kiev.ua
Astronomical Observatory, Kyiv, Ukraine,

Klare, Gerhard
Landessternwarte, Heidelberg, Germany,

Koenigsberger, Gloria gloria@astroscu.unam.mx
UNAM, Mexico City, Mexico,

Koesterke, Lars lars@astro.physik.uni-potsdam.de
Institut fuer Physik, Potsdam, Germany,

Kolka, Indrek indrek@aai.ee
Tartu Observatory, Toravere, Estonia,

Korn, Andreas A.Korn@lsw.uni-heidelberg.de
Landessternwarte, Heidelberg, Germany,

Koubský, Pavel koubsky@sunstel.asu.cas.cz
Astronomical Institute, Ondrejov, Czech Republic,
Krautter, Joachim jkrautte@lsw.uni-heidelberg.de
Landessternwarte, Heidelberg, Germany,
Kubat, Jiri kubat@sunstel.asu.cas.cz
Academy of Sciences, Ondrejov, Czech Republic,
Kudritzki, Rolf kudritzki@usm.uni-muenchen.de
Uni-Sternwarte, München, Germany,
Kuznecov, Oleg kuznecov@spp.keldysh.ru
Keldysh Institute, Moscow, Russia,
Lafon, Jean-Pierre J. jpj.lafon@obspm.fr
Observatoire de Paris, France,
Lamers, Henny h.lamers@sron.ruu.nl
Utrecht University, The Netherlands,
Langer, Norbert ntl@astro.physik.uni-potsdam.de
Inst. f. Physik, Potsdam, Germany,
Le Mignant, David dlemigna@eso.org
ESO, Santiago, Chile,
Ludke, Everton eludke@sm.conex.com.br
Universidade Federal de Santa Maria, Brazil,
MacFarlane, Joseph jjm@icf.neep.wisc.edu
University of Wisconsin, Madison, USA,
MacLow, Mordecai-Mark mordecai@mpia-hd.mpg.de
MPIA, Heidelberg, Germany,
Maeder, Andre andre.maeder@obs.unige.ch
Geneva Observatory, Switzerland,
Magalhães, Mário mario@argus.iagusp.usp.br
Instituto Astronomico, Sao Paulo, Brazil,
Mandel, Holger hmandel@lsw.uni-heidelberg.de
Landessternwarte, Heidelberg, Germany,
Markova, Nevena rozhen@sm.unacs.bg
Institute of Astronomy, Smolian, Bulgaria,
Marlborough, J. Michael marlboro@astro.uwo.ca
University of Western Ontario, London, Canada,
Marston, Anthony tm9991r@acad.drake.edu
Drake University, Des Moines, USA,
Massa, Derck massa@xfiles.gsfc.nasa.gov
Raytheon STX, Greenbelt, USA,
Mathys, Gautier gmathys@eso.org
ESO, Santiago de Chile, Chile,
Mellema, Garrelt garrelt@astro.su.se
Stockholm Observatory, Saltsjoebaden, Sweden,
Messegier, Claude Claude.Megessier@obspm.fr
Observatoire de Paris-Meudon, France,

Meynet, Georges georges.meynet@obs.unige.ch
Geneva Observatory, Sauverny, Switzerland,

Millar, Carol cmillar@astro.uwo.ca
University of Western Ontario, London, Canada,

Miroshnichenko, Anatoly anatoly@physics.utoledo.edu
University of Toledo, USA,

Moffat, Anthony F.J. moffat@astro.umontreal.ca
Universite de Montreal, Canada,

Morrison, Nancy nmorris2@uoft02.utoledo.edu
University of Toledo, USA,

Müller, Patrick pmueller@ita.uni-heidelberg.de
ITA, Heidelberg, Germany,

Muratorio, Gerard muratorio@observatoire.cnrs-mrs.fr
Marseille Observatory, France,

Negueruela, Ignacio ind@astro.livjm.ac.uk
John Moores University, Liverpool, Great Britain,

Niedzielski, Andrze aniedzi@astri.uni.torun.pl
Torun Centre for Astronomy, Poland,

Nota, Antonella nota@stsci.edu
STScI, Baltimore, USA,

Oedegaard, Knut J. R. knutjo@astro.uio.no
Theoretical Astrophysics, Oslo, Norway,

Okazaki, Atsuo okazaki@elsa.hokkai-s-u.ac.jp
Faculty of Engineering, Sapporo, Japan,

Osaki, Yoji osaki@dept.astron.s.u-tokyo.ac.jp
Department of Astronomy, Tokyo, Japan,

Oudmaijer, Rene r.oudmaijer@ic.ac.uk
Astrophysics Group, London, United Kingdom,

Owocki, Stan owocki@bartol.udel.edu
University of Delaware, Newark, USA,

Pamyatnykh, Alosza alosza@camk.edu.pl
Copernicus Astronomical Center, Warsaw, Poland,

Pasquali, Anna apasqual@eso.org
ESO, Garching, Germany,

Pavlovski, Kresimir kresimir@geof.hr
University of Zagreb, Croatia,

Perinotto, Mario mariop@arcetri.astro.it
Astronomia e Scienza dello Spazio, Firenze, Italy,

Peters, Geraldine gjpeters@mucen.usc.edu
University of Southern California, Los Angeles, USA,

Petrenz, Peter uh101bv@usm.uni-muenchen.de
Uni-Sternwarte, München, Germany,

Porter, John jmp@astro.livjm.ac.uk
John Moores University, Liverpool, Great Britain,

Proga, Daniel d.proga@ic.ac.uk
Imperial College, London, United Kingdom,
Puls, Joachim uh101aw@usmu01.usm.uni-muenchen.de
Uni-Sternwarte, München,
Ragland, Sam sam@arcetri.astro.it
Arcetri Astrophysical Observatory, Firenze, Italy,
Rauw, Gregor rauw@astro.ulg.ac.be
Institut d'Astrophysique, Liège, Belgium,
Rivinius, Thomas T.Rivinius@lsw.uni-heidelberg.de
Landessternwarte, Heidelberg, Germany,
Rodrigues, Cláudia Vilega claudia@das.inpe.br
INPE, Sao Jose dos Campos - Sao Paulo, Brazil,
Röser, Siegfried s19@ix.urz.uni-heidelberg.de
ARI, Heidelberg, Germany,
Ružić, Željko zruzic@geof.hr
University of Zagreb, Croatia,
Sapar, Lili lilli@aai.ee
Tartu Observatory, Tartumaa, Estonia,
Sapar, Arved-Ervin sapar@aai.ee
Tartu Observatory, Tartumaa, Estonia,
Schäfer, Dominik D.Schaefer@lsw.uni-heidelberg.de
Landessternwarte, Heidelberg, Germany,
Schenker, Klaus schenker@astro.unibas.ch
Astronomisches Institut, Binningen, Switzerland,
Schmid, Hans Martin hschmid@lsw.uni-heidelberg.de
Landessternwarte, Heidelberg, Germany,
Schneider, Hartmut hschnei@gwdg.de
Uni-Sternwarte, Göttingen, Germany,
Schulte-Ladbeck, Regina E. rsl@phyast.pitt.edu
University of Pittsburgh, USA,
Schweickhardt, Jörg jschweic@lsw.uni-heidelberg.de
Landessternwarte, Heidelberg, Germany,
Shaviv, Nir nir@tapir.caltech.edu
California Inst. of Technology, Pasadena, USA,
Sholukhova, Olga olga@sao.ru
SAO, Karachai-Cherkess, Russia,
Shore, Steve sshore@paladin.iusb.edu
Indiana University South Bend, USA,
Singh, Mahendra msingh@upso.ernet.in
India,
Škoda, Petr skoda@sunstel.asu.cas.cz
Academy of Sciences, Ondrejov, Czech Republic,
Smith, Nathan nathans@bu.edu
Boston University, USA,

Smith, Linda ljs@star.ucl.ac.uk
University College, London, United Kingdom,
Stahl, Otmar O.Stahl@lsw.uni-heidelberg.de
Landessternwarte, Heidelberg, Germany,
Stee, Philippe stee@altair.obs-azur.fr
Observatoire de la Cote d'Azur, France,
Štefl, Stanislav sstefl@sunstel.asu.cas.cz
Academy of Sciences, Ondrejov, Czech Republic,
Sterken, Christiaan csterken@vub.ac.be
Univ. of Brussels, Belgium,
Szeifert, Thomas tszeifer@lsw.uni-heidelberg.de
Landessternwarte, Heidelberg, Germany,
Trams, Norman ntrams@iso.vilspa.esa.es
ESA, Madrid, Spain,
Tubbesing, Sascha stubbesi@lsw.uni-heidelberg.de
Landessternwarte, Heidelberg, Germany,
Vakili, Farrokh vakili@obs-azur.fr
Observatoire de la Cote d'Azur, France,
Valchanov, Tashko rozhen@sm.unacs.bg
Institute of Astronomy, Smolian, Bulgaria,
van der Hucht, Karel A. k.vanderhucht@sron.ruu.nl
SRON, Utrecht, The Netherlands,
van Genderen, Arnout genderen@strw.LeidenUniv.nl
Leiden Observatory, The Netherlands,
Veen, Pieter M. veen@strw.leidenuniv.nl
Leiden Observatory, The Netherlands,
Verdugo, Eva ev@vilspa.esa.es
ISO Observatory, Madrid, Spain,
Vincze, Ildiko ivincze@gothard.hu
Eotvos University, Szombathely, Hungary,
Vink, Jorick J.s.Vink@fys.ruu.nl
University of Utrecht, The Netherlands,
Voors, Robert voors@fys.ruu.nl
Utrecht University, The Netherlands,
Vreux, Jean-Marie vreux@astro.ulg.ac.be
Institut d'Astrophysique, Liège , Belgium,
Wang, Jun-Jie Wangjj@class1.bao.ac.cn
Beijing Astronomical Observatory, P.R. China,
Waters, Rens rensw@astro.uva.nl
Astronomical Institute, Amsterdam, The Netherlands,
Weaver, Robert rpw@lanl.gov
Los Alamos, New Mexico, USA,
Wehrse, Rainer wehrse@ita.uni-heidelberg.de
ITA, Heidelberg, Germany,

Weis, Kerstin kweis@etacar.ita.uni-heidelberg.de
ITA, Heidelberg, Germany,
White, Stephen white@astro.umd.edu
University of Maryland, USA,
Williams, Peredur pmw@roe.ac.uk
Royal Observatory Edinburgh, United Kingdom,
Wolf, Bernhard B.Wolf@lsw.uni-heidelberg.de
Landessternwarte, Heidelberg, Germany,
Yudin, Ruslan ruslan@pulkovo.spb.su
Astronomical Observatory, St.Petersburg, Russia,
Zickgraf, Franz-Josef zickgraf@cdsxb7.u-strasbg.fr
Observatoire de Strasbourg, France

The Landessternwarte Königstuhl in its hundredth anniversary year as seen by the participants.

Session I

Observations of Non-spherical Winds

chair: J.M. Marlborough and P. Williams

Rotationally Modulated Winds of O Stars

Alex W. Fullerton

Dept. of Physics and Astronomy, University of Victoria, Victoria, BC, Canada; and FUSE Science Operations Center, Dept. of Physics and Astronomy, The Johns Hopkins University, Baltimore, MD 21218, USA

Abstract. The stellar wind diagnostics of some well-studied O stars exhibit cyclical variations with periods that are probably related to the rotational period of the underlying star. This rotational modulation is usually attributed to large scale, persistent structures in the wind, which are thought to be generated and maintained by photospheric processes that alter the emergence of the wind from different regions of the stellar surface. In this review, three case studies are used to illustrate the patterns of variability that are attributed to rotational modulation and to highlight some open issues connected with this hypothesis. The problems associated with establishing the occurrence of rotational modulation rigorously are also discussed.

1 Introduction

Intensive ground- and spaced-based spectroscopic monitoring programs have shown that the stellar winds of luminous OB stars vary systematically on time scales that are longer than their estimated flow times; see, e.g., the contributions to this Colloquium by Kaufer, Kaper, and Massa. Often these variations have a cyclical component; in a few well-studied cases, the time scales associated with these cycles can be related to the estimated rotational period of the underlying star. Consequently, rotation is now believed to be one of the main processes controlling the variability of hot-star winds. Much of the observed variability is attributed to the presence of large-scale structures in the stellar wind, which persist for many rotational cycles and alter the wind diagnostics seen by a distant observer as they are carried around the star. These structures are presumably maintained by photospheric processes that affect the emergence of the stellar wind from localized regions of the stellar surface in some way.

The hypothesis of "rotational modulation" is the focus of many of the contributions to this Colloquium. The purpose of this review is to emphasize some of the difficulties associated with establishing this hypothesis rigorously, and to illustrate the diverse phenomenology currently associated with rotationally modulated stellar winds.

2 Can Rotational Modulation Be Demonstrated?

Rotational modulation can be demonstrated by supplying evidence of cyclical or quasi-cyclical variations in a stellar wind diagnostic (e.g., a P Cygni profile), and showing that the period associated with these variations is related to the rotational period of the star to within some tolerance. Acquiring evidence for modulation is subject to the constraints imposed by the sampling theorem, weather, or time allocation committees, but is otherwise straightforward. However, demonstrating that an observed period is consistent with the rotational period (or a meaningful fraction of the rotational period) is much more difficult, for two reasons.

First, there are large uncertainties associated with the radius of an individual star. If $R_\star$ is the equatorial radius of the star in $R_\odot$ and $v_{\rm rot}$ is the equatorial rotational velocity in $\rm km\,s^{-1}$, then the rotational period in days is

$$P_{\rm rot} = 50.61\, R_\star \,/\, v_{\rm rot} \ . \tag{1}$$

The uncertainty in estimates of $P_{\rm rot}$ is dominated by the uncertainty in measurements of the stellar radius: even in the best cases (e.g., ζ Puppis), this amounts to a fractional uncertainty of ~25%. Consequently, the fractional uncertainty in estimates of $P_{\rm rot}$ will be at least this much for single early-type stars.

Second, this large uncertainty is compounded by the fundamental difficulty in determining the inclination of the rotation axis to the observer's line of sight. Since only the projected rotational velocity, $v_{\rm rot} \sin i$, can be measured, only a upper limit on the true rotational limit can be obtained:

$$P_{\rm rot}^{\rm max} \equiv 50.61\, R_\star \,/\, (v_{\rm rot}\ \sin i) = P_{\rm rot} / \sin i \ . \tag{2}$$

The traditional strategy for overcoming this problem is to limit the investigation to a sample of stars with very similar radii. If the observed periods, $P_{\rm obs}$, are due to rotational modulation, then (2) shows that they are inversely proportional to $v_{\rm rot} \sin i$. Consequently, the generally accepted "proof" that $P_{\rm obs}$ for the whole sample can be attributed to rotational modulation is the demonstration that the observed periods occupy a region of the Period–$v_{\rm rot} \sin i$ plane defined by $P_{\rm obs} \le P_{\rm rot}^{\rm max} \propto (v_{\rm rot} \sin i)^{-1}$. However, this approach is not very useful for O-type stars as a class, since they exhibit a large range in radii. The only alternative strategy, which in fact has guided much of the work for the O-type stars, is to select targets with the largest values of $v_{\rm rot} \sin i$ for their spectral classification. This implies that, in some broad statistical sense, the inclination must be close to 90°.

Since the rotational periods of single O stars cannot be determined precisely, the basic prediction of the rotational modulation hypothesis cannot be confirmed observationally. Instead, ancillary observational evidence must generally be introduced to argue for or against the hypothesis: e.g., the universality of the phenomenon; the constancy and phase stability of the observed period; the presence or absence of multiple periods; the prevalence of related physical processes that can be observed directly. These ancillary arguments often represent model-dependent inferences, which may not be correct. Thus, we are left with the unsatisfying recognition that the hypothesis of rotational

Table 1. Physical parameters for case studies

Quantity	HD 66811		HD 37022		HD 64760	
Spectral Type	O4 I(n)f	(1)	O7 V	(2)	B0.5 Ib	(3)
$\dot{M}$ [$\times 10^{-6}$ $M_\odot$/yr]	5.9	(4)	...		0.1	(5)
$R_\star$ / $R_\odot$	19	(4)	8	(6)	22	(7)
v_∞ [km s^{-1}]	2250	(4)	1600–3600	(8)	1500	(9)
$v_{\rm rot} \sin i$ [km s^{-1}]	219	(10)	53	(10)	216	(10)
$P_{\rm rot}/ \sin i$ [days]	4.4		7.6		5.2	
$P_{\rm obs}$ [days]	5.075	(11)	15.422	(2)	1.202	(12)

(1) Walborn 1972 (2) Stahl et al. 1996 (3) Hiltner et al. 1969 (4) Puls et al. 1996 (5) Fullerton & Puls, unpublished (6) Howarth & Prinja 1989 (7) Humphreys & McElroy 1984 (8) Walborn & Nichols 1994 (9) Massa et al. 1995b (10) Howarth et al. 1997 (11) Moffat & Michaud 1981 (12) Prinja et al. 1995

modulation is equally difficult to prove or disprove. If an observed period is within 20 or 30% of the "best-guess" rotation period, then rotational modulation must at least be considered to be a viable explanation for the variations.

3 Case Studies

Despite the difficulties discussed above, there is a growing consensus that the winds of at least some OB stars are rotationally modulated. Here, three well-studied stars that span the range of O-star temperatures – ζ Puppis (HD 66811), θ^1 Orionis C (HD 37022), and HD 64760 – are presented as "case studies" to illustrate the patterns of wind variability attributed to rotational modulation, and to highlight some open issues connected with the hypothesis. Salient parameters for these objects are listed in Table 1.

3.1 HD 66811

The bright, early-type O supergiant HD 66811 = ζ Puppis offers favourable circumstances to search directly for rotationally modulated stellar wind diagnostics because its radius is comparatively well determined from measurements of its angular diameter and distance, and from spectroscopic analyses. Its unusually large value of $v_{\rm rot} \sin i$ suggests that $i \approx 90°$, which therefore implies that $P_{\rm rot} = 4.4 \pm 1.3$ days.

Cyclical variations with period 5.075 ± 0.003 days have been observed in the morphology of both the Hα emission feature (Moffat & Michaud 1981; Berghöfer et al. 1996) and the absorption trough of the Si IV resonance lines (Howarth et al. 1995). Since $P_{\rm obs}$ is within the large uncertainty associated with the measured value of $P_{\rm rot}$, it is attributed to a disturbance in the wind that recurs once per rotation cycle. Both Moffat & Michaud (1981) and

Howarth et al. (1995) suggested that the disturbance is due to the effects of a weak, low-order magnetic field whose axis of symmetry is inclined with respect to the rotation axis (i.e., an oblique magnetic rotator). Such a field will tend to suppress the emergence of the wind from the region near the magnetic equator. However, in order to exhibit only one modulation per rotation, the magnetic geometry cannot be symmetric about its axis, either because of the presence of a quadrapole component or because the dipolar field is offset from the center of the star along the magnetic axis (i.e., a decentered oblique rotator).

The photosphere and wind of ζ Pup also display a host of other periodic variations. Discrete absorption components (DACs) are the dominant component of wind variability, but these recur with a period of 19.2 hours (Howarth et al. 1995), which seems to be unrelated to $P_{\rm rot}$. A soft X-ray period of 16.7 hours (in 1991; Berghöfer et al. 1996) or 15 hours (in 1996; Berghöfer, this Colloquium) is attributed to shock structures in the wind, but apparently these structures are not directly related to $P_{\rm rot}$ or the DACs, though Hα seems to vary with the 16.7-hour period. Deep-seated variations with a period of 8.54 hours have been attributed to nonradial pulsations (NRP; Baade 1988; Reid & Howarth 1996), but this period cannot be linked to the others (except possibly the 16.7-hour X-ray period), so it is not clear that NRP are plausible triggers for the formation of wind structures. Finally, Eversberg et al. (1998) have presented evidence to suggest that the wind of ζ Pup consists in part of stochastically evolving clumps, in addition to large-scale, coherent structures.

Thus, although ζ Pup provides direct evidence for rotational modulation, its wind appears to be subject to a variety of perturbations. Both NRP and magnetic structures are believed to be present, but cannot be directly linked to the large-scale wind structures represented by DACs, the periodic shock phenomena seen in X-rays, or the stochastic evolution of smaller-scale clumps. At the very least, the bewildering array of apparently unrelated variations exhibited by this exceptionally well-studied star cautions against interpreting stellar wind variability in terms of a single process.

3.2 HD 37022

HD 37022 = θ^1 Orionis C is a youthful object situated in the heart of the Trapezium in Orion. It has a history of spectroscopic peculiarities, which include reports of variable inverse P Cygni profiles (Conti 1972) and progressive changes in spectral type from O6–O4 over an interval of about a week (Walborn 1981).

Stahl et al. (1993) detected cyclical changes in the shape and strength of the Hα wind profile of θ^1 Ori C. The period associated with these changes is known very accurately: 15.422 ± 0.002 days (Stahl et al. 1996). The same period has been recovered from the wind profiles of the C IV and Si IV UV resonance lines (Walborn & Nichols 1994; Stahl et al. 1996), soft X-ray emission (Gagné et al. 1997), and photospheric line profiles (Stahl et al. 1996). *IUE* observations obtained over a baseline of ~15 years indicate that the variations have been phase-locked for at least this long. In contrast to most other O stars, the UV P Cygni profile variations are not dominated by progressive DACs, but consist of absorption modulations that affect a large range of velocities nearly simultaneously. Maximum emission in Hα corresponds to minimum absorption in the C IV resonance line.

The regularity of the period, coupled with the phase relationship between Hα and the UV wind lines, can be explained qualitatively by an oblique magnetic rotator model (Stahl et al. 1996; Gagné et al. 1997). However, as with ζ Pup, some form of asymmetry in the field appears to be required to explain the single episode of modulation per rotation. Babel & Montmerle (1997; see also the contributions by Babel and Shore to this Colloquium) have developed a detailed description of the magnetosphere of θ^1 Ori C in terms of their "magnetically confined wind shock" model, which seems to be able to reproduce the observed X-ray variability with a surface magnetic field strength of ~300 G.

However, in the context of the oblique magnetic rotator model, the 15.422-day period must be associated with the rotation period of the star. If the radius of θ^1 Ori C is similar to other O7 dwarfs, then Table 1 shows that the expected rotational period is in fact much shorter. Either the radius of this star is bigger than expected, or the photospheric lines are substantially broadened by a mechanism other than rotation. There is some evidence to support that the latter possibility since Stahl et al. (1996) found that the photospheric variations (which presumably arise from surface abundance anomalies caused by the magnetic field) are confined to a velocity range of $\pm 12\,\mathrm{km\,s^{-1}}$ centered on the systemic velocity of the star. This is much less than the measured breadth of photospheric lines (~50 $\mathrm{km\,s^{-1}}$; Table 1), which also appear to be very symmetric. The origin of the excess broadening is unknown.

Despite the discrepancy between P_{obs} and the estimated P_{rot}, there is nearly universal agreement that θ^1 Ori C is a hot analog of the chemically peculiar, magnetic Bp stars. As such, it is probably the best candidate for the detection of a magnetic field in an O-type star. Even though fields that are well below the current detection thresholds could strongly influence the emergence of its stellar wind, it is nonetheless disappointing that the magnetic field of θ^1 Ori C remains undetected (Mathys; this Colloquium).

3.3 HD 64760

HD 64760 has an extraordinarily large $v_{\mathrm{rot}} \sin i$ for its spectral type and luminosity class (Table 1), which implies that it is an intrinsically rapid rotator with an inclination close to 90°. Unfortunately, its radius can only be estimated from coarse calibrations based on spectral classification. Consequently, its rotational period is not known reliably: the best guess is $\sim 5 \pm 1$ days, though values outside this range are not excluded.

Prinja et al. (1995) detected periodic variations in the UV wind lines of HD 64760 in the long time series obtained during the *IUE* MEGA Campaign (Massa et al. 1995a). Fullerton et al. (1997) showed that this periodic component results from two quasi-sinusoidal fluctuations with periods of 1.202 ± 0.004 and 2.44 ± 0.04 days. The 1.2-day period has been detected weakly in UV photospheric lines (Howarth et al. 1998), while the 2.4-day period has been seen in Hα (Kaufer, private communication) and in UV data from a previous *IUE* campaign (Fullerton et al. 1997). It is not completely clear which of these periods is the more fundamental. Since they are broadly consistent with a quarter and a half of the estimated $P_{\rm rot}$, they have been attributed to rotational modulation by wind features that recur 4 times around the circumference of the star, with every second feature being different for some reason. However, as Howarth et al. (1998) discuss, this interpretation is not very secure because of the large uncertainties in $P_{\rm rot}$.

The periodic variations coexist with DACs, but do not appear to be linked to their recurrence or propagation. In contrast to DACs, the periodic component of the wind variability consists of modulations of the line flux that affect a large range of velocities at any given time, particularly in the absorption trough of a P Cygni profile. However, since the modulations can also be traced through the emission lobe, they must be caused by structures in the wind that are longitudinally extended. Within a P Cygni absorption trough, the modulations exhibit the curious property of simultaneously evolving toward larger and smaller line-of-sight velocities, a phenomenon that is known as “phase bowing”. Owocki et al. (1995) showed that phase bowing can be explained quite naturally by corotating, spiral-shaped wind structures, which exit the column of absorbing material projected against the stellar disk simultaneously at two different line-of-sight velocities.

Thus, phase bowing is an important diagnostic of the geometry of the stellar wind structures responsible for the observed modulations. Fullerton et al. (1997) used this phenomenon to show that higher ions are concentrated along the inner, trailing edge of spiral-shaped perturbations in the wind of HD 64760. By using a simplified kinematic model, they inferred that the velocity law governing the radial flow of material is normal. They suggested that the periodic variations in the wind of HD 64760 are due to corotating interaction regions (CIRs), which are spiral-shaped perturbations caused by the collision of fast and slow wind streams that emerge nearly radially from different longitudinal sectors of the stellar surface; see, e.g., the contribution by Owocki to the Colloquium. Howarth et al. (1998) concluded that the 1.2-day photospheric period is due to NRP, which could serve as the source for these longitudinally-spaced fast/slow wind streams. Although there are difficulties associated with this conjecture (e.g., the pulsations may not corotate with the stellar surface), the detection of the same period in the photospheric and wind lines of HD 64760 provides strong evidence that wind structure is generated by deep-seated, photospheric processes.

4 Concluding Remarks

Until the radii of individual stars can be measured precisely, there seems to be little hope of showing definitively that the periodic component of stellar wind variability is due to rotational modulation. Nevertheless, the case studies presented here provide circumstantial evidence that such modulation does occur. However, they also caution against trying to fit all observations with one model: *none* of the case studies look very similar to each other, which implies that a variety of processes may be at work or that there are strong observational selection effects. For example, the visibility of phase bowing is expected to depend strongly on the value of $v_{\rm rot}/v_{\infty}$, which might explain why it is rarely detected in OB stars. Observer aspect could also play a large rôle in determining the characteristics exhibited by a given class of wind structure. These difficulties can only be addressed by enlarging the sample of stars that have been monitored extensively. Acquiring such time series data remains a formidable observational challenge.

Several theoretical challenges also need to be addressed. One such challenge is to understand the circumstances that are required to ensure that large-scale wind structures can survive the ravages of the line-driven instability (see, e.g., the contributions by Feldmeier and Owocki to this Colloquium). The effect of these structures on spectroscopic estimates of the global mass-loss rate also needs to be clarified. Although the resolution of these observational and theoretical issues will require sustained effort, it will also provide deeper insights into the physics of radiatively driven winds and the photospheres from which they flow.

Acknowledgements

I am grateful to the SOC for their invitation to deliver this review, the LOC for financial support to attend the Colloquium, and to many colleagues world-wide for stimulating discussions on hot-star winds.

References

Baade D. (1988): *Ground-Based Observations of Intrinsic Variations in O, Of, and Wolf-Rayet Stars,* in O Stars and Wolf-Rayet Stars, ed. P.S. Conti & A. B. Underhill, (NASA SP-497), p. 137

Babel J., Montmerle T. (1997): ApJ **485**, L29

Berghöfer T. W., Baade D., Schmitt J. H. M. M., et al. (1996): A&A **306**, 899

Conti P.S. (1972): ApJ **174**, L79

Eversberg T., Lépine S., Moffat A. F. J. (1998): ApJ **494**, 799
Fullerton A. W., Massa D. L., Prinja R. K., et al. (1997): A&A **327**, 699
Gagné M., Caillault J.-P., Stauffer J. R., et al. (1997): ApJ **478**, L87
Hiltner W.A., Garrison R.F., Schild R. E. (1969): ApJ **157**, 313
Howarth I.D., Prinja R.K. (1989): ApJS **69**, 527
Howarth I.D., Prinja R.K., Massa D. (1995): ApJ **452**, L65
Howarth I. D., Siebert K. W., Hussain G. A., et al. (1997): MNRAS **284**, 265
Howarth I. D., Townsend R. H. D., Clayton M. J., et al. (1998): MNRAS **296**, 949
Humphreys R.M, McElroy D.B. (1984): ApJ **284**, 565
Massa D., Fullerton A. W., Nichols J. S., et al. (1995a): ApJ **452**, L53
Massa D., Prinja R. K., Fullerton A. W. (1995b): ApJ **452**, 842
Moffat A.F.J., Michaud G. (1981): ApJ **251**, 133
Owocki S. P., Cranmer S. R., Fullerton A. W. (1995): ApJ **453**, L37
Prinja R.K., Massa D., Fullerton A.W. (1995): ApJ **452**, L61
Puls J., Kudritzki R.-P., Herrero A., et al. (1996): A&A **305**, 171
Reid A.H.N., Howarth I.D. (1996): A&A **311**, 616
Stahl O., Kaufer A., Rivinius Th., et al. (1996): A&A **312**, 539
Stahl O., Wolf B., Gäng Th., et al. (1993): A&A **274**, L29
Walborn N.R. (1972): AJ **77**, 312
Walborn N.R. (1981): ApJ **243**, L37
Walborn N.R., Nichols J.S. (1994): ApJ **425**, L29

Discussion

H. Henrichs: θ^1 Ori C is problematically placed on the pre-main sequence in your HRD. Wouldn't you rather like to see a much larger value for its radius to solve the rotation problem as well?
A. Fullerton: There often seem to be problems placing early O-type dwarfs in the theoretical HRD. The fundamental parameters I have adopted for θ^1 Ori C are solely based on calibrations with spectral type and may not be entirely appropriate for such a peculiar object. It would be very interesting to try to constrain these parameters (in particular the radius) more precisely by comparison with model atmosphere calculations.

A. Moffat: Can you say that all O stars show rotationally modulated winds?
A. Fullerton: At this point, I think it would be premature to say that all O stars show rotationally modulated winds. There are a handful of well-studied cases for which the evidence of rotational modulation is strong; but even the objects I have emphasized here suggest that there might be several different "flavours" of rotational modulation, which might be caused by different photospheric processes. There may also be some strong selection effects (e.g., observer aspect, ratio of $v_{\rm rot}/v_\infty$) that determine whether rotational modulation will be observable even if structures are present. So, I'm reluctant to generalize too widely, even though the necessary ingredients – rotation, a wind, and wind structures as manifested by DACs – do seem to be universal among the O stars.

Rotationally Modulated Winds of BA-Type Supergiants

Andreas Kaufer

Landessternwarte Heidelberg, Königstuhl 12, D-69117 Heidelberg, Germany

Abstract. Extended spectroscopic monitoring programs with high resolution and coverage in wavelength and time have revealed a new picture of the winds and the circumstellar environments of late B- and early A-type supergiants. Dramatic line-profile variations (LPV) of the wind-sensitive Hα line with characteristic cyclical V/R variations indicate the presence of deviations of the envelopes from spherical symmetry. Time-series analysis of these LPVs suggest that the wind variations are caused by rotating surface structures which modulate the lower wind region. Occasionally observed high-velocity absorptions (HVA) indicate the presence of rotating extended and dense streakline or loop structures in the envelopes. The potential use of these circumstellar features to determine the true stellar rotation periods is discussed.

1 Introduction

Based on extended spectroscopic monitoring programs with high resolution and coverage in wavelength and time a new picture of the winds and the circumstellar environments of late B- and early A-type supergiants (in the following BA supergiants) has emerged in the last years (cf. e.g. Kaufer 1998b).

In this contribution the observational evidence for rotational modulation of the winds of BA supergiants is discussed.

2 Signature and time scales of wind variability

The Heidelberg group has carried out extended spectroscopic monitoring campaigns on BA supergiants to examine their photosphere and envelope variability (cf. Kaufer 1998a for a description of the campaigns).

Their crucial finding about the complex wind variability patterns in *all* of the six examined program stars is that – independent of the timely average appearance of the most wind sensitive Hα profile – the variability pattern is localized symmetrically about the system velocity with the maximum power of the variations just beyond the $\pm v \sin i$ velocities of the star (Kaufer et al. 1996a). The variations are mainly due to additional violet (V) and red (R) shifted emission components superimposed on the otherwise constant underlying wind or even photospheric profiles. The amplitudes of the modulations were measured for all program stars in the corresponding 'temporal variance spectra' (TVS) and give equal amplitudes for the V and R peaks with values between 5% and 20% which are characteristic for the individual object. This characteristic V/R variability is highly indicative for deviations of the circumstellar envelopes of BA supergiants from spherical symmetry: a presumably equatorial concentration of the emitting circumstellar material is favored by Kaufer et al.

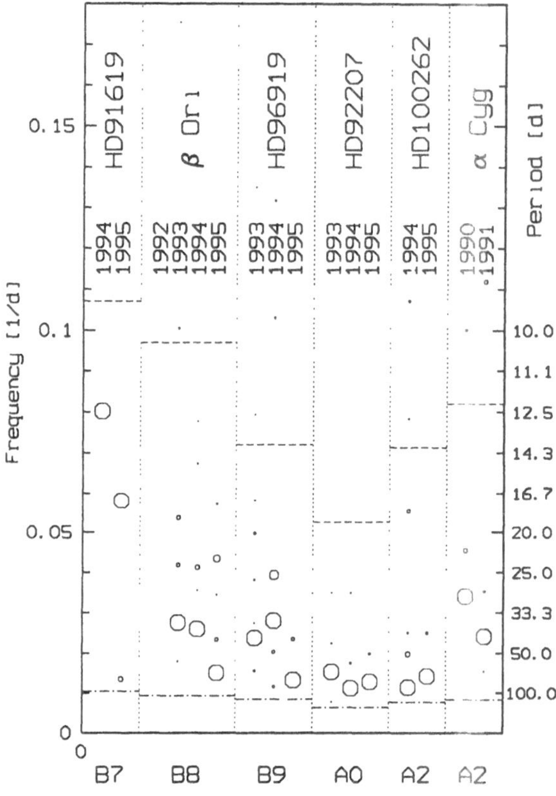

Fig. 1. CLEANed frequencies from Hα equivalent-width curves of the program stars of Kaufer et al. Horizontal dashed lines indicate the frequencies of expected photospheric radial fundamental pulsation modes $P_{\mathrm{rad,fund}}$ for $\log Q = -1.4$. Dot-dashed lines indicate the estimated lower limits for the rotational frequencies given by $P_{\mathrm{rot}}/\sin i$. Note that always one dominant period is found.

The measured equivalent-width curves of the variable 'excess' emission display best the cyclical variability of the wind profiles. Figure 1 shows the annual CLEANed frequencies from Hα equivalent-width curves of the six program stars in comparison with the frequencies of the expected photospheric radial fundamental pulsation modes and the lower limits for the rotational frequencies derived from $P_{\mathrm{rot}}/\sin i$. These time-series analyses of the equivalent-width curves reveal for all examined stars basically *one* dominant period, which is in all cases close to the rotational period as estimated from $v \sin i$

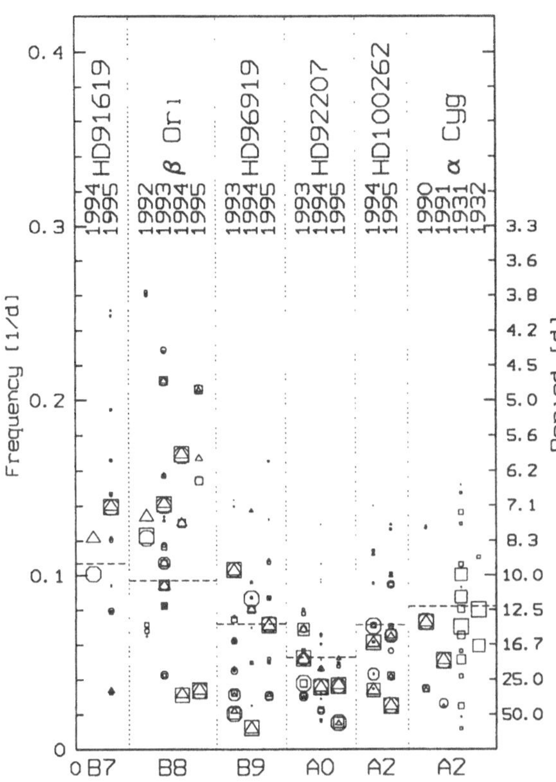

Fig. 2. CLEANed frequencies from the radial-velocities curves, which were measured with a cross-correlation method. The different symbols refer to the different groups of lines (triangles = weak, squares = medium, circles = strong); the size of the symbols represent the power (the significance) of the respective peaks in the periodograms. The horizontal dashed lines give $P_{\text{rad,fund}}$ for $\log Q = -1.4$. Note the multiperiodicity and the variability from year to year of the frequency spectrum and the concentration around $P_{\text{rad,fund}}$.

and the stellar radius. At this point it is worth to note that the values for $v \sin i$ of BA supergiants cannot be derived directly from the width of the photospheric line profiles since BA supergiants (as most of the evolved hot stars in the upper HRD) are known to show significant additional broadening mechanisms apart from rotation. The approach used by Kaufer et al. is to measure the width of the variable part of the profile by the use of 'temporal variance spectra' as suggested by Reid et al. (1993). The idea is that possible surface structures like spots or pulsation patterns which probably cause the LPVs are Doppler-mapped by the stellar rotation to a maximum velocity, which is to be identified with $v \sin i$. This method assumes that an azimuthal velocity component of the surface structures is small compared to the value of $v \sin i$. Apart from this, the even larger uncertainty for the estimate of upper limits for the rotation periods $P_{\text{rot}}/\sin i$ of the program stars comes from the difficulties in the determination of their radii due to badly known distances and luminosities.

Even with keeping in mind these uncertainties and difficulties, the wind modulation frequencies shown in Fig. 1 give strong evidence for a rotational modulation of the lower circumstellar envelope of BA supergiants by stellar surface structures. Kaufer et al. suggest that an efficient coupling of the stellar rotation into the lower wind region could be provided by weak magnetic surface structures. From their detailed examination of the simultaneously recorded photospheric line-profile variability (Kaufer et al. 1997) they could rule out in first order e.g. non-radial pulsation patterns as modulating surface structures. Figure 2 shows the corresponding CLEANed photospheric frequencies to Fig. 1 as derived form radial-velocity curves. Multi-periodic NRPs are present in their data sets in form of low-order g-modes but occur on time scales distinct from the wind variability and are more coincident with the expected photospheric radial fundamental pulsation periods as indicated as dashed horizontal lines in Figs. 1 and 2.

Note that the pulsation periods in BA supergiants are clearly distinct from the rotation periods. However, in the case of the proposed non-radial g-modes, which are expected to display longer periods than the radial fundamental modes, this separation of time scales might shrink. On the other hand, Kaufer et al. report that the photospheric line-profile variations display prograde traveling pseudo-emission and pseudo-absorption features with crossing times from $-v \sin i \rightarrow +v \sin i$ of the order of the break-up rotation period and therefore, identify these features as NRP modes and not as rotating surface features.

Also the clear presence of multi-periodicity of the photospheric pulsations (cf. Fig. 2) could systematically shift the photospheric time scales towards longer time scales of the order of the estimated rotation periods by the beating of multiple periods. Recently, Rivinius et al. (1998) have presented strong observational evidence for mass-loss events triggered by the beating of closely spaced non-radial pulsation periods for the Be star μ Centauri. In addition to the multiplicity of the photospheric period spectra in BA supergiants, the variability from year to year of the period spectrum itself (see the periods from different years for the individual objects in Fig. 2) so far inhibits any precise determination of the pulsation modes, which would be required in order to test the tempting hypothesis of pulsation-driven mass-loss events in BA supergiants.

3 Repetition times of high-velocity absorptions

Temporarily the envelopes of BA supergiants display extraordinarily large and extended circumstellar structures which become observable as suddenly appearing, highly blue-shifted, and unusually strong absorption features; the so-called high-velocity absorptions (HVA) (cf. Kaufer et al. 1996b, Israelian et al. 1997). Both groups suggest localized regions of enhanced mass loss on the stellar surface to build up these extended circumstellar structures; Kaufer et al. favor rotating streak lines in the equatorial plane, Israelian et al. rotating loops, the latter supported by the observation of red-shifted absorption as indication of strong infall of material during an HVA event. In their picture the observed HVAs are the result of the *rotation* of these structures through the line of sight.

In two cases (HD 96919 and β Ori) it seems plausible that two extreme rotating circumstellar structures were strong enough to survive for several months and have been observed over several rotational cycles. This would allow for the first time a *direct determination of the (circum)stellar rotation period.*

In the case of the extremely strong and accidentally well-observed HVA event in HD 96919 in 1995, Kaufer et al. were able to derive directly a repetition time of 93 days from the reappearance of double-peaked emission components within one contiguous time series. With this 'period', the reappearance after four cycles of this very strong and therefore presumably quite time persistent HVA was predicted for March 22, 1996 (MJD 50164 = time of the maximum blue absorption) and indeed observed in a follow-up campaign at La Silla exact on the predicted date. The repetition time of 93 days is consistent with the estimated $P_{\rm rot}/\sin i = 119$ days for HD 96919 and therefore was identified as the true (circum)stellar rotation period.

In the second case of β Ori, the maximum blue absorptions of two comparatively strong HVAs in autumn 1993 and spring 1994 were observed 108 days apart. This interval is in excellent agreement with the estimated $P_{\rm rot}/\sin i = 108$ days for this star.

The crucial point about this direct determination of true (circum)stellar rotation periods is obviously the observational distinction between HVA events caused by the *same* circumstellar structure and HVA events caused by two individual, timely and spatially independent circumstellar structures. Basically, the latter can never be completely ruled out even not with a continuous (spectroscopic) monitoring of the object. From the observational material obtained so far, i.e., the two events described above, criteria for the identification of physically related HVAs might defined as (i) an integer cycle numbers between two events, (ii) similar characteristics of the line-profile and velocity signatures, i.e., the double-emission peaks, the depth and velocity of the maximum absorption, the slope of the rising branch of the equivalent-width curve of the event, (iii) a continuous weakening of the structure from cycle to cycle since the HVAs are observed to fade over several months until they disappear.

4 Discussion

The observational evidence for rotational modulation of the winds of BA supergiants has been presented in this paper.

The analysis of the line-profile variability of the most wind-sensitive line Hα has provided new insight in (i) the 'general' and always present wind variability of BA supergiants, which produces quite dramatic LPVs in Hα but a moderate 5 – 20% modulation of the integrated equivalent width, and (ii) the 'exceptional' variations as observed during the HVA events. Figure 3 illustrates the two intensity scales of the two types of observed variability.

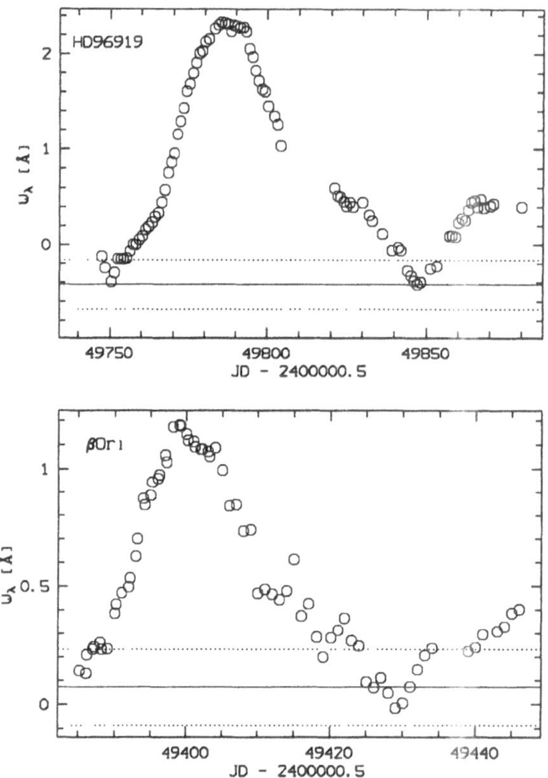

Fig. 3. Time development of the total Hα equivalent width during the HVA events of β Ori in 1994 (bottom) and HD 96919 in 1995 (top). For comparison the mean equivalent widths of the respectively preceding years (without HVAs) are indicated by full lines. The broken lines indicate the corresponding standard deviations representing the 'general' Hα variability. These statistics were computed from 86 spectra of β Ori and 83 spectra of HD 96919 obtained in 1993. Note the striking similarity in the run of the equivalent widths of two HVAs in two different stars.

Interestingly, both could be *independently* related the stellar rotation, (i) to rotating surface structures modulating the lower circumstellar envelope (based on time scale arguments), and (ii) to rotating extended circumstellar streaklines or loop structures also rooted to the stellar surface in localized regions of enhanced mass loss (based on the observation of the reappearance of the bona fide identical circumstellar structure).

For the so far two objects where both types of wind variability could be observed quantitatively, i.e., HD 96919 and β Ori, the corresponding dominant periods as found from (i) are *not* strict integer fractions of the directly determined periods from (ii). But in all cases the former are in the range of 0.5 – 0.3 times $P_{\rm rot}/\sin i$ whereas the direct periods are very close to the estimated rotation periods $P_{\rm rot}/\sin i$.

This leads to the conclusion that the 'general' rotational modulation of the winds of BA supergiants is caused by several ($\sim 2-3$) simultaneously present surface features, which are probably not located in equidistant stellar longitudinal sectors but are distributed irregularly. This is consistent with the on the first look definitely non-periodic appearance of the Hα LPVs. On the other hand, the 'exceptional' structures observed as HVAs are comparatively rare and have to be attributed to singular but therefore more extended and dense structures reaching up to several stellar radii (Israelian et al. 1997) into the circumstellar envelope. It seems plausible to speculate that individual of the always present surfaces structures could develop under certain – so far unknown – circumstances into strong and extended streaklines or magnetic loops.

References

Israelian G., Chentsov E., Musaev F., 1997, MNRAS **290**, 521

Kaufer A., Stahl O., Wolf B., Gäng Th., Gummersbach C.A., Kovács J., Mandel H., Szeifert Th., 1996, A&A **305**, 887

Kaufer A., Stahl O., Wolf B., Gäng Th., Gummersbach C.A., Jankovics I., Kovács J., Mandel H., Peitz J., Rivinius Th., Szeifert Th., 1996, A&A **314**, 599

Kaufer A., Stahl O., Wolf B., Fullerton A.W., Gäng Th., Gummersbach C.A., Jankovics I., Kovács J., Mandel H., Peitz J., Rivinius Th., Szeifert Th., 1997, A&A **320**, 273

Kaufer A., 1998, Variable circumstellar structure of luminous hot stars: the impact of spectroscopic long-term campaigns. In: *Reviews in Modern Astronomy* **11**

Kaufer A., 1998, Cyclic variability in BA-type Supergiants. In: *Cyclic variability in stellar winds*, eds. L. Kaper, A.W. Fullerton, ESO Astrophys. Symp., Springer, p. 114

Reid A.H.N., Bolton C.T., Crowe R.A., Fieldus M.S., Fullerton A.W., Gies D.R., Howarth I.D., McDavid D., Prinja R.K., Smith K.C., 1993, ApJ **417**, 320

Rivinius Th., Baade D., Stefl S., Stahl O., Wolf B., Kaufer A., 1998, Predicting the outbursts of the Be star μ Cen. In: *Cyclic variability in stellar winds*, eds. L. Kaper, A.W. Fullerton, ESO Astrophys. Symp., Springer, p. 207

Discussion

H. Henrichs: I am puzzled by your remark that many of your stars have rotationally modulated winds and at the same time show different periods from season to season. Could differential rotation account for this ?

A. Kaufer: We do not find an obvious connection between rotationally modulated winds and photospheric pulsation. And it is the frequency spectrum of the photospheric variations which is found to vary from year to year. It looks to me like we have objects with rich period spectra, but only selected pulsation modes are excited; and the same modes are not necessarily excited each time we look at them.

T. Eversberg: Maybe I can add a reminder: such NRP mode variations are already known for δ Scuti stars observed by the Breger group in Vienna. They explained these quick changes by changes in the inner structure of the star.

S. Owocki: Could you clarify again why you say that the "High-Velocity Absorptions" (HVAs) you see are different from the "Discrete Absorption Components" (DACs) seen in UV wind lines? In particular, do you have the time resolution to rule out acceleration on the wind acceleration time scale? Also, a comment: the azimuthal speed declines so quickly outward, that it is difficult for material to be brought into the disk line-of-sight anywhere away from the surface. On the other hand, this does occur much more readily for rigidly rotating structures, as from, say, a magnetic structure extending to a stellar radius or so.
A. Kaufer: For a beta-type velocity law and typical stellar-wind parameters with $\beta \approx 0.8 \ldots 1.5$ and $v_{\rm inf} \approx -200 \ldots -300\,{\rm km\,s^{-1}}$, the flow time scales (defined as the time span for acceleration from $0.2\,v_\infty$ to $0.8\,v_\infty$) are on the order of 20 to 60 days; during a HVA event the blue absorption over the same velocity range grows to its maximum depth within some 1 to 5 days without any clear acceleration. This is why we favor the rotation of an existing structure into the line-of-sight as the model for the HVA. Concerning your comment: in our simple streakline model we had to include rigid rotation of the inner envelope up to about 1 stellar radius to account for the quasi-instantaneous deepening of the line-of-sight absorption over the full velocity range (cf. Kaufer et al. 1996b).

I. Appenzeller: Concerning the temporary presence and absence of pulsations in your stars, in spite of the fact that these stars seem to be in a stable region of the H-R diagram: the absence of instabilities does not mean that a star cannot pulsate. It only means that the pulsations are damped. But if the star gets a "kick" from some nonstationary process, it will pulsate for some time. The mode (temporarily) excited will depend on the symmetry (or asymmetry) of the "kick".

Using Spectropolarimetry to Determine Envelope Geometry and Test Variability Models for Hot Star Circumstellar Envelopes

Karen S. Bjorkman

Ritter Observatory, Dept. of Physics & Astronomy, University of Toledo, Toledo, OH 43606-3390, USA

Abstract. A survey and monitoring of the spectropolarimetric characteristics of hot stars over the entire visible wavelength range has been carried out over the past 8 years using the HPOL instrument at the Pine Bluff Observatory. Data from these projects is being used to derive physical characteristics of circumstellar envelopes. Quantitative modeling of the polarization, in combination with optical interferometry, has shown that the circumstellar disks of classical Be stars are geometrically thin, consistent with either the wind-compressed disk model or with hydrostatically supported Keplerian disks. Furthermore, spectropolarimetric variability, which is significant in a large fraction of the hot stars observed, provides information about changes occuring in the circumstellar envelope. For example, polarimetric changes provide a critical test of the one-armed density wave models proposed to explain observed V/R variations.

1 Introduction

One of the difficult aspects of trying to study the nature of stellar winds and circumstellar envelopes is that the radiation from the star itself dominates the observed spectrum. Furthermore, the circumstellar environment is rarely directly resolvable with current instruments and techniques. However, polarimetry, and particularly spectropolarimetry, provides a means of probing the disk directly. The polarization in hot stars is produced primarily by electron scattering in the circumstellar environment. This polarized flux is affected by passage through the circumstellar material, via pre- and post-scattering attenuation of the flux. The measured polarization is the ratio of the polarized flux to the total flux (which is dominated by the direct starlight). Because of this, in the ratio the wavelength dependence of the stellar spectrum cancels out. Thus, any residual wavelength dependence of the polarization level is determined by the opacity within the circumstellar envelope. So observations of the polarization (as opposed to the total flux) give insights into the physical conditions of the circumstellar material. Furthermore, the combination of spectropolarimetric observations with other techniques, such as high-dispersion spectroscopy, optical interferometry, and infrared photometry, can provide strong constraints on models of the circumstellar environment.

2 Spectropolarimetric Observations

We are continuing an optical spectropolarimetric survey and monitoring program for several types of early-type stars (as well as other kinds of objects). The data have been obtained using the Halfwave Polarimeter (HPOL) system (designed by K. Nordsieck) on the 1-m telescope at the University of Wisconsin's Pine Bluff Observatory (PBO) and on the 3.5-m WIYN telescope at Kitt Peak National Observatory. Types of hot stars included in the program include Oe/Be stars, Herbig Ae/Be stars, Wolf-Rayet stars, LBV's, and OB supergiants. An atlas of the Be star observations from 1989-94 is in preparation (Bjorkman, Meade & Babler 1997), and the 1995-98 observations will be published in a follow-up paper. Using these data, we are now developing techniques to analyze the nature of the circumstellar envelopes of these types of stars. In our analysis we also include ultraviolet spectropolarimetry from the Wisconsin Ultraviolet Photo-Polarimeter Experiment (WUPPE), flown on Astro-1 (1990) and Astro-2 (1995). Polarimetric observations of hot stars are also discussed in other papers in these proceedings (c.f. contributions by Rodrigues, Magalhães, and Schulte-Ladbeck et al.)

3 Diagnostics of Circumstellar Disks

Recently we have used spectropolarimetry to diagnose the physical geometry and density of circumstellar disks around classical Be stars (Wood, Bjorkman, & Bjorkman 1997; Quirrenbach et al. 1997). As discussed by Wood et al. (1997), we find that measurement of the peak polarization level observed, together with the size of the polarimetric Balmer jump, provides a diagnostic of both the geometrical thickness of the disk (the opening angle) and the optical depth in the disk, which indicates the density of material. For the case of ζ Tau, our results indicate a very thin disk with an opening angle of only 2.5°, which agrees with predictions of either the wind-compressed disk model (Bjorkman & Cassinelli 1993) or a hydrostatically-supported Keplerian disk. Evidence for the thin nature of the disk is confirmed by a combination of optical interferometry and spectropolarimetry (Quirrenbach et al. 1997), and the good agreement between these two different techniques demonstrates the power of spectropolarimetry to diagnose disks even in cases that cannot be resolved at all.

4 Variability Issues

Combining contemporaneous spectropolarimetric observations with more traditional types of observation, such as spectroscopy, can also place much stronger constraints on proposed models for variability and asymmetries in hot star envelopes. For example, in collaboration with D. McDavid, we have just begun looking at the question of whether the combined Hα spectroscopic and polarimetric variations observed in Be stars can be adequately explained by the proposed "one-armed density wave" model (c.f. Okazaki 1997). While this model has been successful as a potential explanation for some cases of observed V/R variations of the Hα line profile, our preliminary analysis indicates that the polarimetric observations are 90° out of phase compared to theoretical predictions based on the one-armed density wave model. This work is still quite preliminary, and we are currently investigating whether a spiral geometry can remove the discrepancy. Other models make definite predictions of polarimetric variability as well, so this type of analysis demonstrates how polarimetric observations can provide a complementary and somewhat independent test of model predictions.

5 Future Work

We intend to continue our development of techniques for using spectropolarimetry as a diagnostic of the nature of the physical nature of circumstellar envelopes. Ultraviolet spectropolarimetric observations, which show pronounced polarization decreases in regions of strong line-blanketing (Bjorkman et al. 1991; 1993), can provide information about the temperature of the disk material. We are pursuing this possibility by developing temperature diagnostics based on reproducing this depolarization effects. We are also continuing to investigate specific models for circumstellar disks by combining spectropolarimetry with other kinds of data. These techniques are applicable to a number of different types of hot stars.

References

Bjorkman, J.E. and Cassinelli, J.P. (1993): *Ap.J.*, **409**, 429
Bjorkman, K.S., Meade, M.R., and Babler, B.L. (1997): *B. A. A. S.*, **29**, 1275
Bjorkman, K.S., et al. (1991): *Ap.J.*, **383**, L67
Bjorkman, K.S., et al. (1993): *Ap.J.*, **412**, 810
Okazaki, A.T. (1997): *A. & A.*, **318**, 548
Quirrenbach, A., Bjorkman, K.S., Bjorkman, J.E., Hummel, C., Buscher, D., Armstrong, J., Mozurkewich, D., Elias, N., & Babler, B.L. (1997): *Ap. J.*, **479**, 477
Wood, K., Bjorkman, K.S., and Bjorkman, J.E. (1997): *Ap. J.*, **477**, 926

Discussion

H. Henrichs: Is optical interferometry essential to constrain the disk parameters?

K. Bjorkman: Not essential: one can actually use other types of data instead (such as spectroscopy or IR photometry, in conjunction with spectropolarimetry) to constrain the disk parameters. But interferometry is certainly the best and most conclusive result to use. The combined interferometry-spectropolarimetry study gave us higher confidence in our ability to determine disk parameters from spectropolarimetric data and modelling. Even without the interferometry, the mid-IR data for ζ Tau also indicates that the thin disk solution is the preferred one.

A. Maeder: Can you get estimates of the disk mass from your observations?

K. Bjorkman: Yes, once we have the density parameters and opening angle, we can estimate the disk mass. For ζ Tau it was on the order of 10^{-9} or $10^{-10} M_{\odot}$.

G. Mathys: Your discussion dealt with the degree of polarisation. Can't you use the polarisation angle to enhance your diagnostics?

K. Bjorkman: Yes, and in fact we did use the position angle information in our analysis. We found that the polarisation position angle is exactly perpendicular to the disk as defined from interferometry. This had long been predicted, but this was the first test of the prediction from combined interferometry and spectropolarimetry. This held true for all four of the stars for which we were able to find an intrinsic position angle (see Quirrenbach et al. 1997), so this gives us confirmation that the polarisation position angle indeed does measure the intrinsic orientation of the disk on the sky for Be stars.

I. Appenzeller: What wavelength is most "helpful" in the case of polarimetry?

K. Bjorkman: For analyses such as the one that I have discussd here, the regions around the Balmer jump, Hα, and even the Paschen jump are most useful. However, higher resolution measurements (i.e., detailed spectropolarimetry) across Hα will also be useful; other groups are doing this sort of work.

M. Friedjung: Is the disk optically thick or thin in the continuum at optical wavelengths?

K. Bjorkman: It is optically thin, at least in the direction perpendicular to the disk. For a photon trying to travel radially outward within the disk, it would appear optically thick.

D. Baade: W. Hummel recently presented persuasive arguments that disks of Be stars may also be warped/tilted. Do you find supportive evidence in your polarimetry?

K. Bjorkman: Well, in some cases we do see a slight variability of the position angle about the intrinsic position angle of the disk. This might be interpreted as evidence for structure or warping in the disk. But the effects are very small and will require better S/N observations and more detailed analysis before we can say anything conclusive.

P. Stee: Are you able to fit the continuum energy flux with the very thin-disk model?

K. Bjorkman: Yes. In fact, in mid-IR the thin-disk solution fits the continuum flux slightly better than the thick-disk solution. Although I did not have time to discuss this here, the flux fitting is discussed in detail in our paper (Wood et al. 1997).

Bernhard Wolf, Andreas Kaufer and Dietrich Baade

Karen Bjorkman and Derck Massa

Disks of Classical Be Stars

Stanislav Štefl

Astronomical Institute, Academy of Sciences, CZ-25165 Ondřejov, Czech Republic

Abstract. Observational methods used in investigation of circumstellar disks of Be stars significantly changed during the last decade. While mostly the studies of line profiles in the optical and near-IR region were used in the 70th and 80th, the present progress is done also thanks to interferometric and polarimetric observations. These techniques enable to determine the geometrical parameters of the circumstellar disks directly for the first time. They are flattened and axially symmetric. Their H α emission region extends typically up to several tens of stellar radii. Systematic long-term spectroscopic monitoring helped to define and partly interpret the line-profile variability of emission lines on different time-scales. Variations on a time-scale of hours may appear after the matter is ejected during the outburst. The V/R variations with periods of the order of 10 years reflect the global disk oscillations.

1 Introduction

Jaschek et al. (1980) defined a Be star as a non-supergiant B-type star, whose spectrum has, or had at some time, one or more Balmer lines in emission. Because some other definitions include also supergiant stars, the adjective "classical" was later added to the stars that meet the Jaschek's narrower definition.

The emission in optical, mainly Balmer and Fe II, lines and in the IR continuum originates in circumstellar disks around Be stars. In this paper, I attempt to summarize the recent results of observations of the disks and their main geometrical and physical properties. For more general reviews on Be stars or those directed to other aspects of Be stars see the proceedings of IAU Coll. 92 and IAU Symp. 162.

2 Observations of Be star disks

2.1 Basic types of emission line profiles and their variations

Extensive observations of the Bochum group in the southern hemisphere and several other groups in the north lead to the classification of the emission profiles of Be stars. Classification of the emission line spectra and their variability was discussed e.g. by Dachs (1986), Hanuschik et al. (1988). In Fig. 1, examples of the prototype emission profiles are given.

The type of an emission line profile correlates with v sin i. The winebottle profile can be expected mainly for pole on stars, double-peak for medium

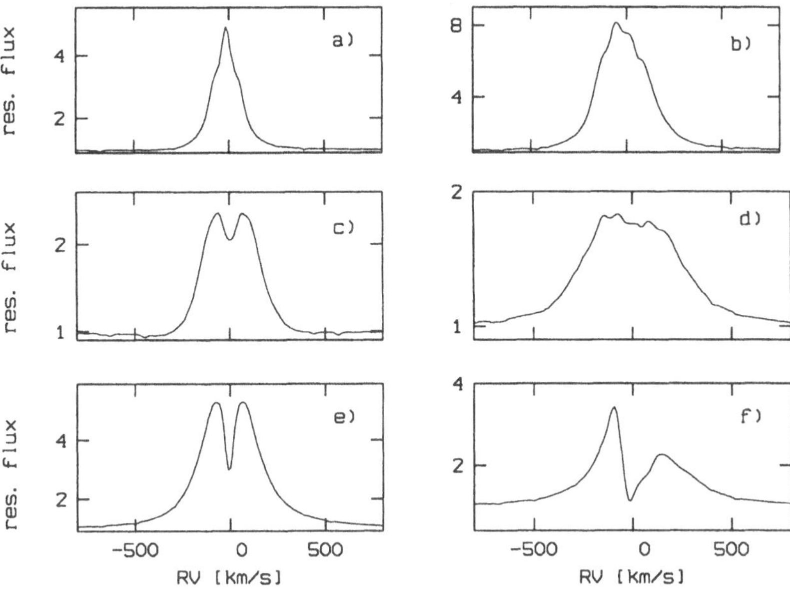

Fig. 1. The typical emission H α profiles of Be stars: a) wine-bottle symmetric (11 Cam on March 1, 1997), b) wine-bottle asymmetric (66 Oph on June 24,1994), c) double-peak symmetric (23 Tau on October 11, 1995), d) double-peak asymmetric (59 Cyg on November 1, 1994), e) shell symmetric (ψ Per on March 1, 1997) and shell asymmetric (ζ Tau on March 4, 1997). Profiles from the top downward demonstrate the effect of increasing inclination angle i. Asymmetric profiles (right column) are generally variable. The symmetric profiles are stable or can appear at a given phase of variable profiles. All spectra have been obtained with the 2m telescope of the Ondřejov observatory.

inclination and shell profile for equator-on stars. However, the inclination is not the only determining parameter. The type can vary on a time-scale of years. This can be documented by the history of the first known Be star γ Cas. During a short interval between 1930 and 1942, the type of the emission profile changed at least nine times and all types shown in Fig. 1 were observed in H α - H δ. In 1942, the emission even disappeared and a quasi-normal B type spectrum was observed for several months.

2.2 Long-term E/C and V/R variations

The E/C variations, defined as the maximum H α emission expressed in units of the continuum flux, take place on a time scale as long as 10-20 years. The time-scale is longer for later spectral types (Hubert-Delplace, 1981).

The V/R parameter is defined as the ratio of the flux in the violet and red peaks of the double peak emission line. Most frequently it is used for H α profiles. Their V/R variations appear in about three quarters of Be stars. The quasi-periods are in the range of 2 - 22 years and are strongly concentrated around 7 years. Lower amplitude variations on a time scale of tens of days - months may be superimposed, as in the case of 66 Oph (Štefl et al., in preparation). The V/R variability can cease for several years, as in γ Cas for 25 years. The quasi-periods do not depend on the spectral type or v sin i. They can change from cycle to cycle (e.g. 28 Tau: 22 and 6 years; Gulliver, 1977, Goraya and Tur, 1988). The shifts between V/R variations of different Balmer lines exclude the Huang's (1973) model of apsidal motion of material in an elliptical orbit. We suppose that the variations reflect the density waves in circumstellar disks (see Sect. 3).

2.3 Transient periodicities

The discovery and main properties of transient periodicities were reported by Štefl et al (1998). They were detected during the long-term monitoring of the Be stars 28 CMa, μ Cen, and η Cen. The transient periodicities co-exist with the photospheric line-profile variations (periods 0.5-2 days) and differ by $\approx$ 10 % from them. They are present for a few days (μ Cen) up to a few months (28 CMa). The transient periodicities are either not or only weakly detectable in photospheric lines, attain significant power only in lines formed (also) above the photosphere and are indicated mainly in the line wings. They do not re-appear with the same period as at the previous epoch. In the case of μ Cen, the transient periodicities excited during different outbursts fall in the interval ± 10 %. We suspect that transient periodicities reflect the processes in the region of star-disk interface, but this interpretation is more than preliminary.

2.4 Rapid V/R variations

Rapid V/R variations are observed on a time scale of hours to a few days. Rivinius et al. (1998a) published the study of μ Cen which contains probably the most detailed observations of the phenomenon. The V/R variations in the wings of He I lines have a temporary character. They appear within 1 - 3 days after the outburst, last for 5-10 days and then V/R converges to 1. Periods from different emission episodes differ up to 10 %. Very preliminarily, the authors interpret the variations as a cloud ejected during the outbursts and merging into the disk. Rapid V/R variations may reflect the same phenomenon as transient periods.

2.5 Infrared, interferrometric and polarimetric observations

The [12]-[60] vs. [12]-[25] colour diagram is consistent with free-free emission from the disk. IR excess decreases towards later spectral types (Gehrz et al.,

1974), but shows no clear correlation of IR excess vs. v sin*i*. Observations indicate correlations of V-[12] colour excess vs. spectral type and vs. H α net emission.

The first star, for which the radius of the circumstellar disk was directly measured in the interferometric observations, was γ Cas (Thom et al. 1986, Quirrenbach et al. 1993). The main results were achieved with the GI2T interferometer of the Observatoir de la Côte d'Azur and MKIII interferometer of the Mont Wilson Observatory. For more detailed and up-to-date new important results see the contributions by Stee and Vakili in these proceedings.

Be stars show a variable degree of polarization, but the polarization angle is constant. This indicates that the envelope is concentrated to a well defined plane and only negligible part of the polarization comes from the polar regions. Magnitude of polarization is wavelength dependent. It is lower in the lines and downward of Balmer discontinuity. It correlates with the 12 μm IR excess (Coté and Waters, 1987) and v sin i (Brown and McLean, 1977). The position angle depends only on the disk orientation. For a more complete summary of the polarimetric results see K. Bjorkman (this Coll.).

3 Geometrical and physical properties of the disks

Thanks to the recent interferometric and polarimetric observations, the long lasting discussion about sphericity of the Be envelopes was solved. The flattened disks are confirmed both by direct interferometric observations and by the correlation of interferometric and polarimetric positional angles. Their differences are very close to 90° in accordance with the model predicting scattering mostly in the equatorial plane of the disk.

The angular radii a and flattening r of the selected Be stars in H α were measured interferometrically by Quirrenbach et al. (1997). Their diameters derived with the help of Hipparcos distances are listed in Tab. 1. Although there are still only few measured stars for a statistics, a typical diameter of the circumstellar disk seems to be 20 - 30 stellar diameters. Thus, the interferometric observations confirmed typical diameters of the disks ($\approx 20\,R_\star$) found by Hanuschik (1986), who analyzed the H α emission profiles and applied the rotating disk model.

The diameter of the disk depends on the spectral line selected for the interferometric measurements. The VLA interferometry of ψ Per done at 15 GHz (Dougherty and Taylor, 1992) gives the angular diameter of 111±16 mas, it is 34 times the diameter measured in H α. The positional angles from H α and radio observations agree within the errors [(-33 ± 11) ° and (-32 ± 10) °, respectively]. This indicates that the disk plane is well defined up to a distance of several hundred of stellar diameters.

The inner radii of the disks were not measured so far. The spectroscopic observations indicate that they are very different for different stars, change

Table 1. Interferometrically measured geometrical parameters of Be star disks: $R_\star$- equatorial stellar radius (Harmanec, 1988), a - angular diameter, r - flattening (Quirrenbach et al., 1997), d_{Hip} - Hipparcos distance (Perryman et al., 1997), D - disk diameter in solar and in stellar radius.

Star	Spect. type	$R_\star$ [$R_\odot$]	a [mas]	r	d_{Hip} [pc]	D [$R_\odot$]	D [$R_\star$]
γ Cas	B0 IVe	5.21	3.47 ± .02	0.70 ± .02	188	140	27
ϕ Per	B1 Vep	4.63	2.67 ± .20	0.46 ± .04	219	126	27
ψ Per	B5 Ve	2.95	3.26 ± .23	0.47 ± .11	215	151	51
η Tau	B7 IIIe	8.2	2.65 ± .14	0.95 ± .22	113	64	8
48 Per	B3 Ve	3.56	2.77 ± .56	0.89 ± .13	170	101	28
ζ Tau	B4 IIIpe	≈ 5	4.53 ± .52	0.28 ± .02	128	125	25
β CMi	B8 Ve	2.43	2.65 ± .10		52	30	12

in the given star with its emission activity and may even approach the stellar photosphere.

The opening angel α cannot be derived unambiguously, because it is impossible to separate the inclination and thickness of the disks. The published results are inconsistent. Hanuschik (1996) derived $\alpha \approx 13°$ from the observed number of shell stars, Waters et al. (1987) $\alpha \approx 15°$ by modelling the IR excess. However, the wind compressed model of Bjorkman and Cassinelli (1993) requires $\alpha \leq 3°$.

The tilt angle ϵ between rotational axes of the star and the disk was assumed to be zero. Only recently Hummel (1998) published his study of the Be stars γ Cas and 59 Cyg, in which he breaks this taboo. Assuming a non-zero tilt angle increasing as the emission episodes develop, he could roughly model observed I) variations of the emission-line type, II) line width variations and III) long-term brightness variations. His model suggests a non-zero tilt angel only as a temporal effect in one emission episode for each of the two studied Be stars, but not as a general property. The model also does not explain the physical origin of the tilt. According to Porter (1998), the Be star disks may develop radiatively induced warps in their inner regions. It is possible that the effects explained by Hummel as variable tilt of the disk are in fact manifestations of the warps. The time-scale predicted for the warps is consistent with that derived by Hummel.

The densities of circumstellar disks were derived mainly from the IR excess (Waters et al. 1987) and by modeling the asymmetric UV lines (Snow, 1981). The former method indicates the relation $\rho(r) = \rho_0(r/R_\star)^{-n}$ with $\log\rho_0 = -(11.0 - 11.6)\, g/cm^3$ and $2.0 \leq n \leq 3.5$. The mass-loss $\dot{M}_{IR} \approx 10^{-8}\ M_\odot$/year and depends only weakly on luminosity. The latter method gives the mass-loss $\dot{M}_{UV} \approx 10^{-10}\ M_\odot$/year and it depends strongly on luminosity. Comparison of the both methods indicates that the mass-loss is strongly asymmetric and that the mechanisms responsible for the mass loss in the infrared and ultraviolet may be different.

4 Dynamical structure of the disks

The Be star disks are quasi-Keplerian. Okazaki (1991) found that global (so-called one-armed) oscillations can proceed in the disks in a retrograde way. Papaloizou et al. (1992) and Savonije and Heemskerk (1993) derived that prograde modes can be excited by perturbation from a rotationally flattened rapidly rotating star. In the Okazaki's (1997) hybrid scenario, the one armed oscillation can occur in the disk due to perturbation by radiation pressure and rotationally distorted star. The former effect is dominant for early Be stars (B0 - B3), the latter for the later Be stars.

The existence of one armed oscillations is supported by the following observational facts:

- predicted periods agree with the observations (Okazaki, 1997)
- line profiles calculated for $m = 1$ perturbation agree with observed line profile variability (Hummel and Hanuschik, 1997)
- appearance of UV absorption components in γ Cas is higher when $V/R \geq 1$ (Telting and Kaper, 1994)
- Observed correlation between V/R and long-term photometric variations (Mennickent et al., 1997) is consistent with the model of global oscillations and seems to support existence of prograde modes.
- in the hybrid scenario, the period decreases with the increasing value of radiative parameter for earlier spectral types and with larger deformation parameter towards later spectral types. As the result, the maximum in period vs. spectral type distribution should occur around B3. The statistics of the observed periods may agree with this prediction, although the scatter in observed periods is large and correlation by far not convincing (see Okazaki, 1997).

5 Formation of the Be star disks

The physical process responsible for the formation of disks is not known even after many decades of intensive studies of these objects and we cannot exclude that more than one mechanism is involved.

It seems that the disks are formed mainly by sudden emission outbursts. Rivinius et al. (1998b) showed that the emission outbursts in the Be star μ Cen coincide with times of beating of photospheric periods derived from mode and line profile variations. Mass flow into the disk is then concentrated in relatively short time intervals after the outbursts.

In some stars, the continuous mass outflow from the photosphere plays an important role in supplying material into disks. However, the formation of disks by the wind compressed model (Bjorkman and Cassinelli, 1993; Owocki et al., 1994) meets serious theoretical and observational problems.

Acknowledgement: The author thanks Drs. D. Baade, W. Hummel and Th. Rivinius for valuable comments on the manuscript.

References

Bjorkman J.E., Cassinelli J.P., 1993, ApJ 409, 429
Brown J.C., McLean I.S., 1977, A&A, 57, 141
Coté J., Waters L.B.F.M., 1987, A&A 176, 93
Dachs J., 1986, in IAU Coll. 92, p. 149
Dougherty S.M., Taylor A.R., 1992, Nature 359, 808
Gehrz R.D., Hackwell J.A., Jones T.W., 1974, ApJ 191, 675
Goraya P.S., Tur N.S., 1988, A&A 205, 164
Gulliver A.F., 1977, ApJS 35, 441
Harmanec P., 1988, Bull. Astron. Inst. Czechosl. 39, 329
Hanuschik R., 1986, A&A 166, 185
Hanuschik R.. 1996, A&A 308, 170
Hanuschik R., Kozok J.R., Kaiser D., 1988, A&A 189, 147
Huang S.S., 1973, ApJ 183, 541
Hubert-Delplace A.M., Jaschek M., Hubert H., Chambon M.Th., 1981, in IAU Symp. 98, p. 125
Hummel W., 1998, A&A 330, 243
Hummel W., Hanuschik R., 1997, A&A 320, 852
Jaschek M., Hubert-Delplace A.-M., Hubert H., Jaschek C., 1980, A&AS 42, 103
Mennickent R.E., Sterken C., Vogt N., 1997, A&A 326, 1167
Okazaki A.T., 1991, 1991, PASJ 42, 75
Okazaki A.T., 1997, A&A 318, 548
Owocki S., Cranmer S.R., Blondin J.M., 1994, ApJ 424, 887
Papaloizou J.C., Savonije G.J., Henrichs H.F., 1992, A&A 265, L45
Perryman M.A.C., Høg E., Kovalewski J., Lindegren L., Turon C., 1997, ESA SP-1200, *The Hipparcos and Tycho Catalogues*
Porter J.M., 1998, A&A 336, 966
Quirrenbach A., Hummel C.A., Buscher D.F., Amstrong J.T., Mozurkewich D., Elias II N.M., 1993, ApJ 416, L25
Quirrenbach A., Bjorkman K.S., Bjorkman J.E., Hummel C.A., Buscher D.F., Armstrong J.T., Mozurkewich D., Elias II N.M., Babler B.L., 1997, ApJ 479, 477
Rivinius Th., Baade D., Štefl S., Stahl O., Wolf B., Kaufer A.: 1998a, A&A, 333, 125
Rivinius Th., Baade D., Štefl S., Stahl O., Wolf B., Kaufer A., 1998b, in Proc. *A Half Century of Stellar Pulsation Interpretations: A Tribute to Arthur N. Cox*, eds. P.A. Bradley and J.A. Guzik, Astron. Soc. Pacific Conf. Series, 135, 341
Savonije G.J., Heemskerk M.H.M., 1993, A&A 276, 409
Snow T.P., 1981, ApJ 251, 139
Štefl S., Baade D., Rivinius Th., Kaufer A., Stahl O., Wolf B.: 1998, in Proc. *A Half Century of Stellar Pulsation Interpretations: A Tribute to Arthur N. Cox*, eds. P.A. Bradley and J.A. Guzik, Astron. Soc. Pacific Conf. Series, 135, 348
Telting J.H., Kaper L., 1994, A&A 284, 515
Thom C., Granes P., Vakili F., 1986, A&A 165, L13
Waters L.B.F.M., Coté J., Lamers H.J.G.L.M., 1987, A&A 185, 206

Discussion

H. Lamers: What is the observational evidence that the disk is tilted with respect to the equator?
S. Štefl: Hummel (1987) offers the hypothesis of non-zero and variable tilt angle and shows that it can explain quite well the long-term variations of the width of emission lines, brightness and the type of emission profiles. This makes the hypothesis attractive, but it is still not evidence that the disk is really tilted.

P. Stee: We have observed γ Cas in He I λ6678 and in the nearby continuum: we found that the emission comes from inner regions, e.g., from regions with $R \leq 4\,R_\star$ (see Stee et al. 1998, A&A 332, 268). Moreover, the He I λ6678 emission comes from smaller regions than the 0.6 μm continuum. These results are also confirmed by a recent paper by Moujtahid et al. (A&A, in press) from photometric observations. This is in contradiction with WCD models or concave models (see Hanuschick et al.) where you have no matter very close to the star.

S. Shore: Given the similarity of the disks you are discussing to those of some binaries, such as ω Ser, β Lyr, etc., what is known about the UV absorption in the systems you have discussed? In particular, models for ω Ser and other thick disk systems give strong Fe II/III, etc. absorption from the middle and outer disks. After all, whether binaries or single stars, one would expect that "disks are disks", right?
S. Štefl: I am not aware of any significant observational difference between disks of ω Ser stars and classical Be stars. My personal opinion is that in general the subordinate UV absorption lines of Fe II are weaker in the disks of Be stars.
J. Bjorkman: UV polarisation observations of ζ Tau (a star for which interferometric observations indicate a thin disk) show large line blankcting around 1900 Å and 2100 Å indicating the presence of Fe III and Fe II in the disk, close to the star. So indeed the physical conditions in Be star disks are similar to the systems you mentioned.

A. Sapar: Has the fact that tilted disks must have a changing (perturbed) plane of revolution, as in the case of artificial earth satellites, been taken into account? Can this effect destroy the unique plane of the disk?
S. Štefl: Yes, the effect of the perturbed plane of revolution was taken into account, but it is not necessary to explain the observations. This effect can really destroy the plane of the disk. This is indicated by the equation for the first order secular perturbation of the nodal position in Hummel (1998).

Evidence for Azimuthal Asymmetry in Be Star Winds

Geraldine J. Peters

Space Sciences Center, University of Southern California, University Park, Los Angeles, CA 90089-1341

Abstract. Extensive *IUE* observations from 1987-96 have provided compelling evidence for azimuthal asymmetry in the winds of Be stars. Two identified types of wind/FUV flux behavior are briefly discussed. Interpretations based upon both nonradial pulsations and rotation of one or more active regions appear plausible.

1 The *IUE* Campaigns

During the lifetime of the *IUE* satellite we carried through 12 multiwavelength campaigns to investigate the wind behavior in Be stars that have shown periodic variability in their optical spectra and light. A typical campaign consisted of 24–72 hours of uninterrupted, repeated *IUE* observations that were supported by simultaneous ground-based photometry, high resolution spectroscopy, and polarimetry. Two principal objectives were to ascertain if the wind strength correlates with the continuum flux and if the optical light variability is a result of a modulation of the photospheric temperature. A list of the 15 program stars and some of their properties can be found in Peters (1998). Three classes of behavior were identified: 1) correlated cyclic variability in wind strength and FUV flux, 2) cyclic wind variability only, and 3) cyclic FUV flux variability only. The first two groups are discussed here.

2 The NRP Case

Five of the Be stars for which we achieved good simultaneous ground-based spectroscopy (λ Eri, 28 Cyg, η Cen, ζ Tau, & 2 Vul) displayed nonradial pulsations (NRP) with sectorial, $\ell = -m = 2$ modes (Hahula & Gies 1994) and the periods derived from the NRP analysis agreed well with those from the FUV flux/wind variations (cf. Peters 1998, Table 1). The optical line profile data confirm that in 28 Cyg (B3 IVe) and η Cen (B2 IVe) a hot crest of the ℓ=2 NRP mode crossed our line-of-sight when the star was brightest in the FUV. The latter stars also showed the strongest correlation between the wind strength and the FUV flux. Typical wind/photospheric behavior for this class of periodic Be stars is shown in Fig. 1 where the temporal variations in the strengths of the C IV wind and photospheric C III (λ1247) and Si II

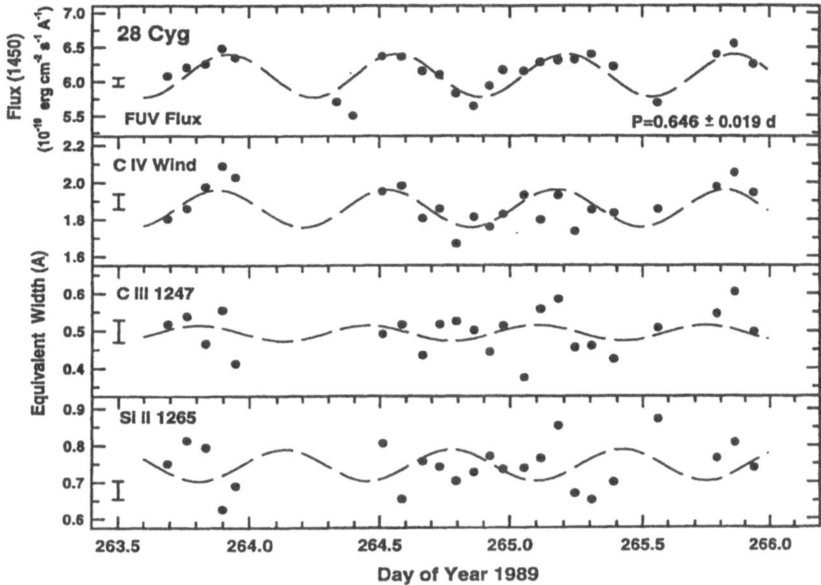

Fig. 1. The photometric, wind, and photospheric behavior of 28 Cyg during the 1989 September campaign. The *dashed* lines are sine curve fits to the 1450 Å flux and the EWs of the C IV wind and C III/Si II photospheric lines.

(λ1265) lines in 28 Cyg are plotted and compared with the FUV flux. A sine curve fit to the FUV light variations revealed a 10% modulation at 1450 Å with a period of $0\overset{d}{.}646$. The amplitude of the FUV light curve increases with decreasing wavelength, implying that the surface temperature is modulated by 500–750°. This interpretation is supported by the strength behavior of the temperature-sensitive photospheric C III and Si II lines (Fig. 1) that will be anticorrelated for early B stars (cf. results for η Cen in Peters 1998). For 28 Cyg and η Cen there is compelling evidence that ongoing (retrograde) NRP produces a modulation of the surface temperature and that mass loss is enhanced over the hot crests.

3 The Rotation Case

For a few of the program stars (ω Ori, ψ Per, DU Eri, and EW Lac) cyclic variability in the wind was apparent but there did not appear to be a strong correlation with the FUV flux. These Be stars typically show no evidence of NRP or cyclic variations in their photospheric temperatures and the period for the modulation of their wind lines is comparable to what one would expect from stellar rotation alone. The periods seen in the stars undergoing NRP are shorter because the NRP motion is retrograde and for an ℓ=2

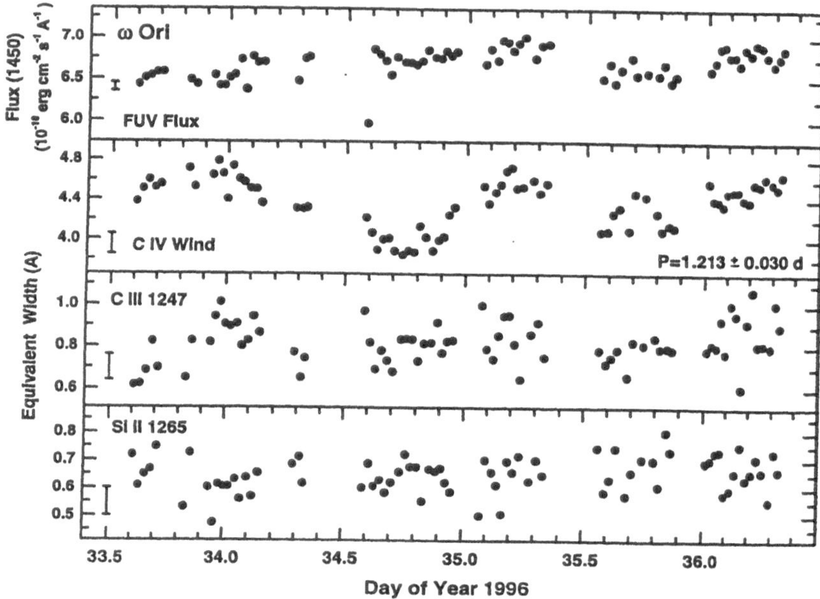

Fig. 2. Similar plot as in Fig. 1 for ω Ori. No variability was evident in the FUV flux or photospheric lines, but the wind was clearly modulated with a period of $\sim 1\overset{d}{.}2$ that is comparable to the star's expected rotational period.

mode there are two crests/troughs that we see each rotational cycle. The temporal flux/wind/photospheric behavior observed in ω Ori (Fig. 2) can be compared with the NRP objects 28 Cyg (Fig. 1) and η Cen (Peters 1998). The observations imply that enhanced mass loss occurs from an active region on the star that sweeps past our line-of-sight with each stellar rotation, but this region does not appear to differ substantially in temperature from the surrounding photosphere.

I extend my thanks to D. Gies, H. Henrichs, D. McDavid, and J. Percy who participated in most of the multiwavelength campaigns. This project was partially supported by NASA grants NSG-5422, NAG5-1296, & NAG5-2313.

References

Hahula, M. E., Gies, D. R. (1994): in *Pulsation, Rotation, and Mass Loss in Early-Type Stars*, ed. L. A. Balona, H. F. Henrichs, J. M. LeContel, (Kluwer,Dordrecht, Boston, London), 100–101

Peters, G. J. (1998): in *ESO Workshop on Cyclical Variability in Stellar Winds*, ed. L.Kaper & A. Fullerton (Berlin: Springer), 127–133

Discussion

H. Lamers: In most stars where the equivalent width of C IV changes in phase with the FUV flux, you have direct evidence that the ionisation is due to photoionisation rather than to Auger-ionisation. Do you see that this correlation depends on spectral type? (You expect it to disappear in later B stars.)
G. Peters: It has yet to be determined whether the modulation in the wind is entirely caused by the changing FUV flux due to the NRP. Detailed modelling will have to be undertaken to find out whether the wind is entirely radiatively driven or if the NRP imparts significant amounts of mechanical energy to assist the acceleration of the wind. There is no apparent correlation of the degree of FUV-wind variability with spectral type (nor $v \sin i$). The object with the earliest spectral type (2 Vul, B0.5 IVe) showed one of the weakest wind modulations; at the other extreme, our coolest star ψ Per (B IIIe-shell) displayed the greatest wind variation.

R. Ignace: Can you comment on what the V/R ratio (e.g., in Hα) does or is expected to do for your three classes of Be stars?
G. Peters: In the NRP case one sometimes observes V/R variations in the He I lines (also paralleled in Hα) that have the same period, and are in phase with the photospheric NRP. This is the so-called "searchlight effect", where the enhanced FUV flux from the hot crest produces greater ionization in the portion of the disk just above the crest. V/R tends to be ~1.0 at NRP phases of 0.25 and 0.75, when the hot/cool patches are aligned along our line-of-sight. A good example is 28 Cyg. In the other two cases one might see either short-term or long-term V/R changes. Several of the Be-shell stars (e.g., 48 Lib, ζ Tau) display long-term V/R variations in Hα that are likely due to precessing spiral density waves. I would think a fast-moving stream of material from an active region on a star (case 2) would make conditions favorable for generating such spiral waves.

P. Koubský: ψ Per is known to be probably the strongest Be-radio star. Is it possible that there is a connection between the radio emission and the deviation of ψ Per on the diagram you have just shown?
G. Peters: The radio emission would come from a large region of space; but if, as the UV data imply, there is an active region of mass loss in the photosphere (perhaps due to a magnetic field) a great deal of material could be built up at large distances from the star.

Short and Medium Term Variability of Emission Lines in Selected Southern Be Stars

Thomas Rivinius

Landessternwarte Königstuhl, D-69117 Heidelberg, Germany

Abstract. We observed a sample of several southern Be stars from 1995 to 1997 typically for several months in each season using our spectrograph HEROS. One of these stars, μ Cen, was found to be in the process of continued gradual recovery of the Hydrogen emitting disk which had been lost from 1977-1989. During the monitoring period numerous line emission outbursts were observed. A generalized pattern of an outburst cycle is derived from observations of different circumstellar lines at times of various levels of emission from the disk. Relative quiescence in which mostly periodic varaibiltiy is seen, rapid decreases of emission (precursor), outburst, and subsequent relaxation can be distinguished as the main constituing phases, even though there are distinct differences between different groups of spectral lines. Based on this empirical phenomenology, a schematic picture of the associated ejection of matter into a near-stellar orbit is sketched and similarities between μ Cen and other stars will be outlined.

1 Observed variability

As a part for a search for multiperiodic line profile variability (Rivinius et al., 1998a, 1998b, Štefl, this Volume, e.g.), we observed several southern stars Be stars with our fiber linked spectrograph HEROS. HEROS covers more than the entire optical range (3450 Å to 8600 Å) with a resolving power of 20 000. A detailed description of the instrument is given by Kaufer (1998). The spectra were taken typically at least daily. The observations are summarized in the upper part of Table. 1.

Photospheric periods: For all four stars we found coherent photospheric periods. The multiperiodicity of μ Cen is disussed by Rivinius et al. 1998b. For η Cen Janot-Pacheco et al. presented a poster during this meeting in which they announced several periods. At least the previously known period 0.64 day and a period of 0.57 day is also present in our data. Since the data of Janot-Pacheco et al. was taken with a entirely different sampling and equipment, we regard this as another case of a multiperiodic Be star. In the case of ω CMa only the well known 1.36 day periodicity could be confirmed. For FW CMa we were able to detect a previously unknown period of 0.83 day, behaving spectroscopically similar to ω CMa.

With the exception of FW CMa, these periods can also be detected in the emission. In the case of η Cen this is clearly seen in Hα, while for μ Cen this is revealed by the power distribution across the line profile, that reaches

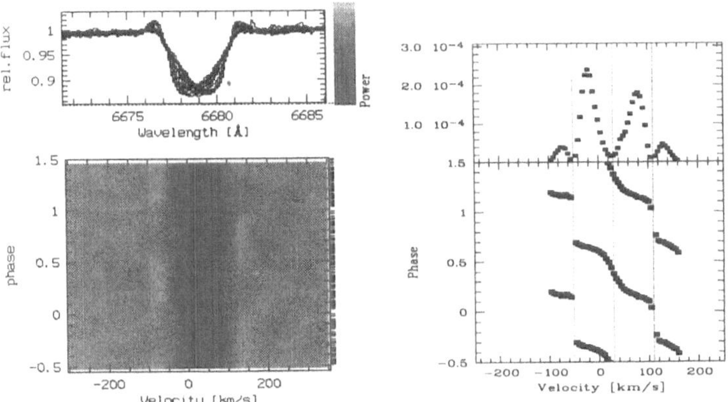

Fig. 1. The phased spectra of HeIλ6678 of ω CMa in 1997 and the results of a Fourier analysis

a secondary maximum at the positions of the emission peaks. Most clearly this behaviour however is seen in ω CMa (cf. Fig. 1).

Transient periods: For a description of this phenomenon please refer to Štefl's paper in this Volume.

Rapid decreases of emission: Just at the very beginning of an outburst, the emission peak heights of the Balmer lines in μ Cen decrease. This decrease can be as strong as 30 % in Hα but less in the higher Balmer lines. Simultaneous to the peak decay however, the broad photospheric wings are filled up, so that the overall variation in equivalent width (EW) is not necessarily large. In fact, these wings increase in comparable strength in all Balmer lines, so that from Hβ or Hγ on the EW actually *decreases* (counting emission as negative EW). A similar event was seen in ω CMa. Although the overall emission level is low in η Cen, a decrease might be seen in our data prior to an outburst. As fourth case found by inspection of our database we mention κ CMa.

Slow line width variations: During an outburst, the peak separation of optical thin lines like FeIIλ5169 may increase within a few days by more than 100km/s in the case of μ Cen. Afterwards, the peaks return to their initial positions gradually over the course of months. This behaviour has been observed also for η Cen. There is also some indication for such an event in ω CMa, but due to the rather narrow lines the variations are harder to detect than for the more equator-on cases.

As summarized in Table 1, only FW CMa seems to exhibit a non-variable circumstellar environment. If this is however a true physical stability or if the variations just can't be observed under such a polar orientation because of averaging effects, or because the strong emission (the highest in our sample), or even just because it is the least intense one observed by us cannot be judged.

Table 1. The obtained number of spectra and observed variability types

Object		η Cen	μ Cen	ω CMa	FW CMa
Sp. type		B1.5Vne	B2IV-Ve	B2IV-Ve	B2.5Ve
Inclination		equator on	quite low	low	nearly polar
1992		–	27	–	–
1995		46	96	–	–
1996		294	202	99	8
1997		80	81	128	48
Photosph.	in absorption	multi	multi	mono	mono
periods	in emission	yes	yes	yes	no
Circumst.	Transient periods	maybe	yes	yes	no
emission	rapid decreases	maybe	yes	yes	no
variability	base width	yes	yes	maybe	no
	slow decay	yes	yes	maybe	no

2 The Proposed Disk Scenario

Considering the observations above, at least for the cases of μ Cen and η Cen we propose a, yet speculative, disk formation "fueling" scenario.

Since the base width of an optical thin emission line is given by the material closest to the star, the variability of this width indicates that the disk might be usually detached from the stellar surface. Only during outburst events, in which the disk is replenished, a considerable amount of matter is ejected to the close circumstellar environment. (Outbursts in μ Cen seem triggered by the photospheric multiperiodicity cf. e.g. Baade, this Volume). During the outbursts, one ore more clouds are ejected to close orbits, causing the transient periods. The emission height decrease/wing increase is produced by shielding of ionizing radiation towards the disk by some less dense pseudophotosphere (e.g. blue supergiant photospheres also exhibit broad emission wings in all Balmer lines, cf. e.g. Kudritzki, this Volume). After a few days, axisymmetry is attained and the ejected material (partly?) moves outwards and merges with the detached disk. During this process, the profiles return to their initial state. A more detailed description of this scenario based on μ Cen data only is given by Rivinius et al. 1998a.

References

Kaufer A., 1998, In: *Reviews in Modern Astronomy* 11, Schielicke R. (ed.), p. 177

Rivinius Th., Baade D., Štefl S., et al., 1998a, A&A **333**, 125

Rivinius Th., Baade D., Štefl S., et al., 1998b, A&A **336**, 177

Discussion

G. Koenigsberger: What is the time scale over which the periods you mention are coherent? Why is an explanation in terms of the presence of a binary companion excluded? How can the mechanism you describe for the variability lead to such coherent periodicities?
T. Rivinius: At least ten years. The observed multi-periodicity is hardly explainable by a singly periodic physical process of orbital motion. I am rather optimistic for the cases of μ Cen and η Cen that it is non-radial pulsation. For ω CMa and FW CMa there is less evidence, but NRP is still a mechanism to be considered.

S. Shore: It seems that you are asking for a lot of work to be done by a little cloud: i.e., to shield a local region and then maintain shielding after spreading out. Have you calculated the optical depths of the clouds? How much mass is involved?
T. Rivinius: We have not yet modelled this hypothesis, but the broad wings observed during the initial phases of an outburst of μ Cen (cf. Rivinius et al. 1998a) might indicate a dense layer that starts to get away from the central star while in this layer temperature and density conditions as in late B supergiants prevail.

Stanislav Štefl and Steve Shore

Disk Winds of B[e] Supergiants

Franz-Josef Zickgraf

Observatoire Astronomique de Strasbourg, 11 rue de l'Université,
F-67000 Strasbourg, France

Abstract. The class of B[e] supegiants is characterized by a two-component stellar wind consisting of a normal hot star wind in the polar zone and a slow and dense disk-like wind in the equatorial region. The properties of the disk wind are discussed using satellite UV spectra of stars seen edge-on, i.e. through the equatorial disk. These observations show that the disk winds are extremely slow, $v_\infty \simeq 50 - 90\,\mathrm{km\,s^{-1}}$, i.e. a factor of ~ 10 slower than expected from the spectral types. Optical emission lines provide a further means to study the disk wind. This is discussed for line profiles of forbidden lines formed in the disk.

1 Introduction

Radiation pressure is accepted as the dominant driving mechanism in the mass loss phenomenon of hot stars, especially in the upper part of the Hertzsprung-Russell (H-R) diagram. Likewise, the existence of an upper boundary of stellar luminosities in the H-R diagram is a well established observational fact (Humphreys & Davidson 1979) which in the hot stars is believed to be related to a stability limit also caused by radiation pressure, i.e. the Eddington limit. There is now, however, increasing evidence that in addition to radiation pressure also rotation plays an important role in the mass loss process in this part of the H-R diagram.

If rotation plays a major role in the mass loss of massive stars, then effects on the circumstellar environment should be observable. It would modify the stability limit by reducing the effective gravity and thereby influence the mass loss process. The mass-loss rate should then vary with stellar latitude and therefore lead to some kind of (observable) non-sphericity.

In recent years observational evidence is indeed mounting that many hot supergiants in the upper part of the H-R diagram exhibit axial symmetry in their circumstellar environments. Likewise, indication of non-spherity has been found in their descendants, the Wolf-Rayet stars. Hence, rotation is certainly an important parameter in these stars.

Among the luminous stars the probably most spectacular object is the Luminous Blue Variable (LBV) η Car for which recent *HST* images clearly showed a bipolar structure consisting of two polar lobes and an equatorial "disk". Other LBVs like R 127 and AG Car also show signs of non-sphericity (e.g. Clampin et al. 1993, Schulte-Ladbeck et al. 1994). A particularly interesting group of stars are the *B[e] supergiants*, which most probably have non-spherical stellar winds caused by rotation. The B[e] supergiants represent a post-main sequence evolutionary stage of massive ($M \gtrsim 8M_\odot$) stars. At this time 15 of these stars are known in the Magellanic Clouds (MCs), 4 in the SMC and 11 in the LMC (cf. Zickgraf 1998, and references therein). The observations strongly suggest that the B[e] supergiants are characterized by a two-component stellar wind comprising in particular a disk-like, slow and dense equatorial wind which is basically distinguished from the winds observed usually in hot supergiants in the same part of the H-R diagram. An empirical model suggested by Zickgraf et al. (1985) for this group of stars is described in Sect. 2. Spectroscopic observations of the disk winds in the satellite UV are discussed in Sect. 3. Optical emission line profiles of forbidden lines originating in the disk wind are discussed in Sect. 4.

2 Empirical model for B[e] supergiants

Spectroscopically and photometrically B[e] supergiants are characterized by strong Balmer emission lines, narrow permitted and forbidden low-excitation emission lines of e.g. Fe II, [Fe II] and [O I], and by a strong mid-IR excess which is attributed to hot circumstellar dust with a typical temperature of 1000 K. Most B[e] supergiants have early-B spectral types. An important result of extensive spectroscopic observations in the optical and satellite UV region was that a subgroup of B[e] supergiants comprising the larger fraction ($\approx 70-80\%$) of these stars shows *hybrid* spectra (Zickgraf et al. 1985, 1986). This term means the simultaneous presence of both, narrow low-excitation lines and broad high-excitation absorption features of C IV, Si IV, and N V in the satellite UV and/or of He I in the optical region. The high ionization lines show wind expansion velocities typical for early B supergiants on the order of $\sim 1000-2000\,\mathrm{km\,s^{-1}}$, in contrast to emission-line widths of not more than several $10\,\mathrm{km\,s^{-1}}$. An example of this class of objects is R 126 in the LMC which was investigated in detail by Zickgraf et al. (1985).

The smaller fraction of B[e] supergiants does not show the signatures of a high velocity wind but only exhibits the narrow low-excitation emission lines. In most of these cases narrow and nearly unshifted absorption features of singly ionized metals similar to shell-type absorptions observed frequently in classical Be stars were found at high spectral resolution in the visual wavelength region. A typical instance of this class is R 50 in the SMC.

These properties were explained by Zickgraf et al. (1985, 1986) in terms of a two-component stellar wind model consisting of a radiation-driven CAK-type wind as observed in all hot high-luminosity stars (Castor et al. 1975) from the poles and an additional slow disk-forming wind from the equatorial region of the star. The model assumes that the hot and fast polar wind gives rise to the broad high-excitation absorption features whereas the low-excitation lines and the dust are formed in a cool, dense, and slowly expanding equatorial disk wind (Zickgraf et al. 1985).

The marked differences in the spectral appearance between individual B[e] supergiants are interpreted in this "unified" model by assuming different aspect angles between the stars' equatorial plane and the line of sight. In this picture stars showing the broad high-excitation absorption features are viewed more or less pole-on. Those stars only showing narrow emission lines or shell-type absorption lines but no signatures of a hot high velocity wind are supposed to be seen edge-on. Support for this model comes from polarimetric observations by Magalhaes (1992). He found significant intrinsic polarisation for MC B[e] supergiants viewed edge-on according to their spectroscopic characteristics, *viz.* R 50 in the SMC, and the two LMC stars R 82 and Hen S22. These stars are the ideal targets to study the exotic disk winds.

3 UV observations of the disk wind

In order to study the properties of the outflowing disk winds we observed these three B[e] supergiants in the satellite UV (Zickgraf et al. 1996). The observations were carried out in 1991 with the *International Ultraviolet Explorer* (IUE) in the LWP range (λ = 1800 - 3200 Å) using the high resolution mode of the spectrograph. Exposure times of up to 13 hours were required in order to obtain sufficient S/N ratio. Sections of the spectra are displayed in Figs. 1 and 2.

Inspection of the spectra shows that they exhibit common properties, like e.g. narrow absorption features of singly ionized metals, in particular of Fe II. However, there are also individual characteristics which differ from one object to the other. Whereas in the spectrum of R 82 the emission components are absent or only weak with the exception of Mg II, R 50, and also Hen S22, exhibit much stronger emission components of the P Cygni profiles. The absorption components of R 50 on the other hand are considerable weaker than in the two other stars, even for the Fe II of multiplet 1. This may partly be due to a combination of a low wind velocity and the resolution of the IUE spectra. The difference between R 50 and R 82 is surprising because in the visual wavelength region both stars exhibit very similar spectra (Muratorio 1981, Zickgraf et al. 1986). A comparison of the spectra of R 82 and Hen S22 with those of LBVs reveals certain similarities. This is especially evident in the case of R 82. The IUE UV spectrum of this star strongly resembles the spectrum of S Dor observed during outburst phase (cf. Leitherer et al. 1985).

The wind velocities were measured from the blue edges of the absorption components. In the following all velocities are given relative to the systemic velocity taken from Zickgraf et al. (1986). The edge velocity measured for Hen S22 from the blue edges of the P Cygni absorption components of strong Fe II multiplets 1, 62, and 63 is $v_{\rm edge} \approx -120$ km s^{-1}. The centers of the P Cygni absorptions components of the strongest Fe II lines are blueshifted

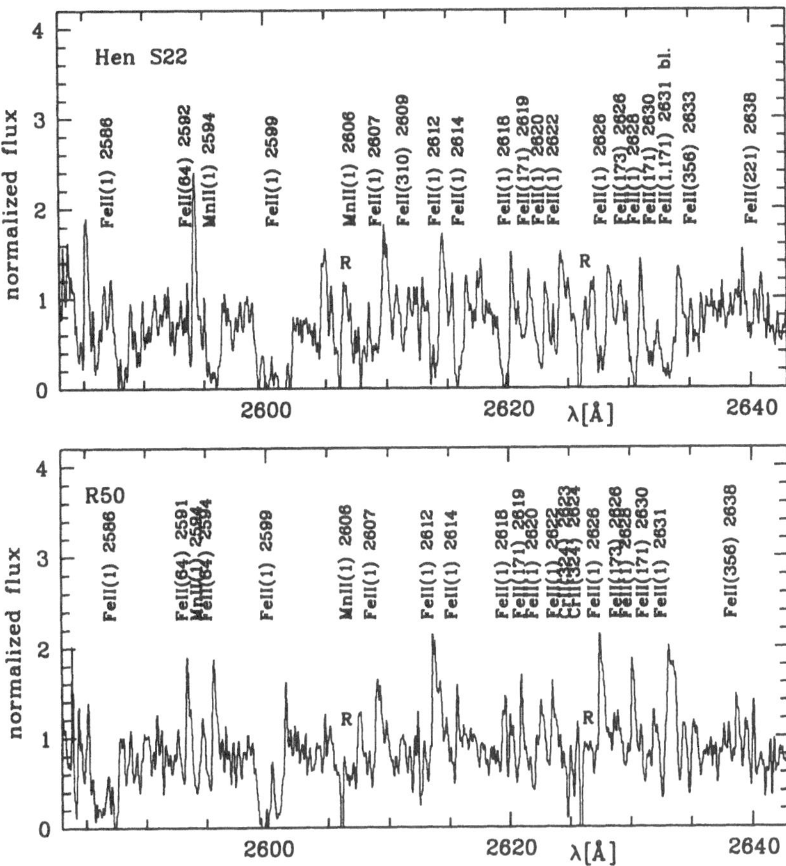

Fig. 1. Section of the IUE-LWP spectra of Hen S22 and R 50. The spectra are dominated by numerous lines of Fe II. Some lines exhibit P Cygni profiles. "R" denotes reseau marks.

with respect to the systemic velocity by $v_{\rm exp} = -60\ {\rm km\,s^{-1}}$. The absorption lines in the spectrum of R 82 are also shifted to the blue with respect to the systemic velocity, however, slightly less than for Hen S22. The edge velocity is $v_{\rm edge} \approx -100\ {\rm km\,s^{-1}}$ for the strongest lines. The expansion velocity at the centers of these lines is $v_{\rm exp} = -40\ {\rm km\,s^{-1}}$ only. Although the absorption components of the P Cygni profiles of Fe II in the spectrum of R 50 are weaker than in the previous two stars we could nevertheless measure the expansion and edge velocities for several Fe II lines. An edge velocity of $v_{\rm edge} = -75\ {\rm km\,s^{-1}}$ was determined from the blue edges of the strongest lines of Fe II. The expansion velocity measured at the line centers of Fe II is $-27\ {\rm km\,s^{-1}}$.

The edge velocities overestimate the terminal wind velocity due to turbulent motions in the winds. Improved values for the terminal velocities of the

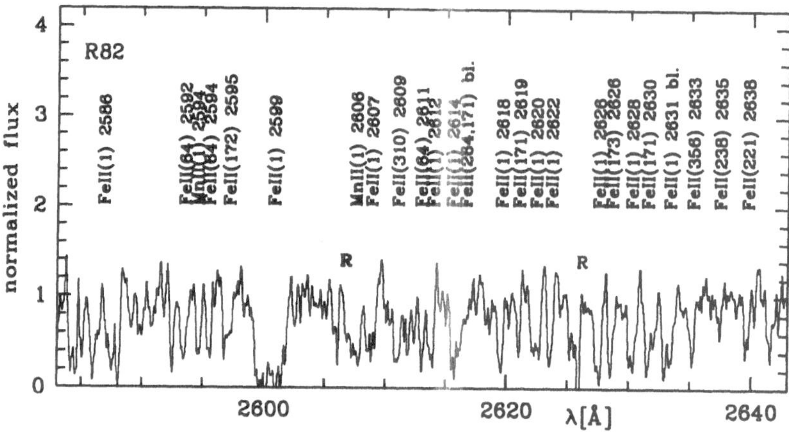

Fig. 2. Section of the IUE-LWP spectra of R 82. The spectrum is dominated by numerous narrow absorption lines of Fe II, very similar to S Dor.

winds were derived from line fits for the Fe II lines. We used the SEI method for the calculation of the line profiles (Lamers et al. 1987) which takes this effect into account. A disk geometry was assumed with a disk opening angle of 30°. Turbulence velocities v_{D} are on the order of 0.2 to $0.3 v_\infty$. The terminal velocity is then $v_\infty \simeq v_{\mathrm{edge}} - 2\,v_{\mathrm{D}}$. This leads to terminal velocities smaller than measured from the blue absorption edges, i.e for Hen S22 $v_\infty = 85\,\mathrm{km\,s^{-1}}$, for R 82 $v_\infty = 70\,\mathrm{km\,s^{-1}}$, and for R 50 $v_\infty = 50\,\mathrm{km\,s^{-1}}$.

The results are summarized in Tab. 1. They show that the disk winds are rather extreme compared to other object classes like LBVs, and B-A super- and hypergiants. The velocities derived for the B[e] supergiants are about a factor of 10 smaller than expected from their spectral types. They have even slower winds than the LBVs and the A-type hypergiants.

Adopting a ratio $v_\infty / v_{\mathrm{esc}} \simeq 1.3$ (Lamers et al. 1995) allows to estimate $\log g_{\mathrm{eff}}$ and from this quantity with mass $M \simeq 2/3\,M_{\mathrm{ZAMS}}$, $\Gamma = \Gamma_{\mathrm{rad}} + \Gamma_{\mathrm{rot}}$, where $g_{\mathrm{eff}} = (1 - \Gamma)\,g_{\mathrm{grav}}$. Γ_{rad} and Γ_{rot} are due to radiation pressure and rotation, respectively. This leads to $\log g_{\mathrm{eff}}$ value of 0.2 – 0.7. Rotation at a speed of $\lesssim 220\,\mathrm{km\,s^{-1}}$, i.e. 70-80% of break-up velocity, would be sufficient to cause these low effective gravities.

4 Emission lines from the disk wind

The properties of the disk wind can also be studied using the emission line profiles of the low-excitation lines which originate in the disk. Of particular interest are forbidden transitions because they are optically thin and therefore radiation transfer does not complicate the interpretation of the line profiles. For this purpose selected lines, mainly forbidden lines of [OI], [Fe II], [N II],

Table 1. Wind velocities of the B[e] supergiants as derived from the blue absorption edges, $v_{\rm edge}$, and v_∞ derived from the SEI line fits together with velocities obtained for LBVs and A-type hypergiants. Included are also average wind velocities of normal B- to A-type supergiants.

Star	Sp. type	$v_{\rm edge}$ [$\rm km\,s^{-1}$]	v_∞ [$\rm km\,s^{-1}$]	reference
Hen S22	B[e] 0 - 0.5	120	85	Zickgraf et al. (1996)
R 82	B[e] 2 - 3	100	70	"
R 50	B[e] 2 - 3	75	50	"
R 71	LBV	127		Wolf et al. (1981)
S Dor	LBV	140		Leitherer et al. (1985)
R 127	LBV	150		Wolf et al. (1988)
R 45	A3Ia-O	200		Stahl et al. (1991)
R 76	A0Ia-O	200		"
	B0	1970		Panagia & Macchetto (1982)
	B1	830		"
	B5	500		"
	A0	185		"

but also permitted lines of Fe II, were observed in a sample of MC B[e] supergiants with the coudé echelle spectrometer (CES) at the 1.4m CAT at ESO, La Silla. The spectra have a spectral resolution of $R = 50\,000$, i.e. a velocity resolution of $\Delta v = 6\,\rm km\,s^{-1}$.

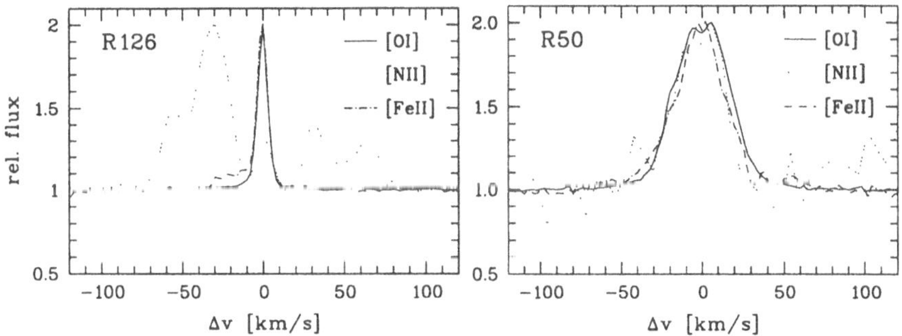

Fig. 3. Profiles of forbidden lines of R 126 (pole-on) and R 50 (edge-on), observed with a velocity resolution of $6\,\rm km\,s^{-1}$. The profiles have been normalized to the maximum line flux. Note the complex structure of the profile of [N II] of R 126. R 50 shows a split [O I] profile ($\Delta v = 15\,\rm km\,s^{-1}$).

The high-resolution spectra showed that the structure of the line forming zone may be rather complex, probably more complex than assumed in the empirical model discussed in Sect. 2. As examples line profiles of the pole-on star R 126 and the edge-on star R 50 are displayed in Fig. 3. The [N II] profile of R 126 is rather complicated. It is split in five components, one unshifted at the systemic velocity, and two on the red and the blue side, respectively. The [O I] and [Fe II] lines show only one component at the systemic velocity. The profile of [N II] can at least qualitatively be understood if the splitting of the blue and red component is neglected. R 126 is very likely seen pole-on. Assuming a radially expanding disk wind with a velocity on the order of 100 km s^{-1}, a disk opening angle of 30°, and an inclination angle of 0°, a velocity component perpendicular to the plane of the disk of about 50 km s^{-1} would occur. The blue and the red emission components could then be formed in layers at the edges of the disk. Disk expansion perpendicular to the disk could also explain the observed absorption dip in Hα (Zickgraf et al. 1985). The unshifted component of [N II], and the [O I] and [Fe II] lines should originate close to the plane of the disk. Rotation could also play a role in the line forming region. R 50 shows a split line profile of [O I]. The line splitting is 15 km s^{-1}. It was interpretated by Zickgraf (1988) as being due to a rotating disk seen edge-on. Although the general appearance of the profiles can be qualitatively understood with the disk model the complex line profile of [N II] of R 126 indicates a more complicated structure than that of a simple disk.

References

Castor J.P., Abbott D.C., Klein R.I., 1975, ApJ 195, 157

Clampin M., Nota A., Golimowski D.A., Leitherer C., Durrance S.T., 1993, ApJ Letters 410, L35

Humphreys R.M., Davidson K., 1979, ApJ 232, 409

Lamers H.J.G.L.M., Cerruti-Sola M. Perinotto M., 1987, ApJ 314, 726

Lamers H.J.G.L.M., Snow T.P, Lindholm D.M., 1995, ApJ 455, 269

Leitherer C., Appenzeller I., Klare G., Lamers H.J.G.L.M. Stahl O., Waters, L.B.F.M., Wolf B., 1985, A&A 153, 168

Magalhaes A.M., 1992, ApJ 398, 286

Muratorio G. 1981, A&AS 43, 111

Panagia N., Macchetto F., 1982, A&A 106, 266

Schulte-Ladbeck R.E., Clayton C.G., Hillier D.J., Harries T.J., Howarth I.D., 1994, ApJ 429, 846

Stahl O., Aab O., Smolinski J., Wolf B., 1991, A&A 252, 693

Wolf B., Appenzeller I., Stahl O., 1981, A&AS 74, 239

Wolf B., Stahl O., Smolinski J., Cassatella A., 1988, A&A 103, 94

Zickgraf F.-J., 1988, in: L. Bianchi and R. Gilmozzi (eds.) *Mass Outflows from Stars and Galactic Nuclei*, Kluwer Academic Publishers, Dordrecht, p. 211

Zickgraf, F.-J., 1998. In: *B[e] stars*, eds. A.M. Hubert and C. Jaschek, Kluwer Academic Publishers, in press

Zickgraf, F.-J., Wolf, B., Stahl, O., Leitherer, C., Klare, G., 1985, A&A 143, 421

Zickgraf, F.-J., Wolf, B., Stahl, O., Leitherer, C., Appenzeller, I., 1986, A&A 163, 119

Zickgraf F.-J., Humphreys R.M., Lamers H.J.G.L.M., Smolinski J., Wolf B., Stahl O., 1996, A&A, 315, 520

Discussion

M. Friedjung: Can rotation affect the line profiles? You might have a combination of rotational plus wind line broadening and the relation between the emission widths and the wind velocity could be coincidental.
F.-J. Zickgraf: Yes, rotation can affect the line profiles. I have in fact modelled emission-line profiles assuming a rotating disk in order to explain double-peaked profiles (Zickgraf 1988, in: Mass Outflows from Stars and Galactic Nuclei, L. Bianchi and R. Gilmozzi, Kluwer, p. 211). However, for the fitting of the UV P Cygni profiles we have assumed purely radial expansion.

A. Maeder: First, I would like to emphasise that your two-component stellar wind is quite consistent with the prediction of the so-called von Zeipel theorem for rotating stars. Now, can you give estimates for the ratio of the polar to equatorial mass loss rates? This is quite important for stellar evolution, because polar mass loss does not remove angular momentum while equatorial mass loss does.
F.-J. Zickgraf: The mass flux ratio between the equatorial and polar wind is likely on the order of a factor of 10, maybe even higher.

S. Shore: You know that there is another way to get displaced narrow lines: take $\beta \ll 1$ (say, 0.2 or so) and you will get narrow lines (e.g., slab calculations for MWC 560), so your assumption of $\beta \sim 4$ may be too restrictive. Also, why are you throwing away the absolute flux calibration of the IUE data? Normalisation can be very misleading.
F.-J. Zickgraf: The value $\beta \sim 4$ is suggested by the similarity of the B[e] spectra with those of A hypergiants for which $\beta \sim 4$ was derived. The flux calibration was not thrown away. However, for the line fitting procedure we were primarily interested in the ratio of line flux to continuum flux (F_λ / F_c) rather than calculating absolute fluxes.

A. Moffat: In your cartoon you showed that dust forms in the disk. Although no one knows (in detail) how to form dust in hot environments, all would agree that you need high compression (plus shielding). So would this not suggest that the disk is quite thin?
F.-J. Zickgraf: The assumption that the density in the disk is constant perpendicular to the plane of the disk is very likely too simple. It seems more realistic to assume a density enhancement toward the mid-plane of the disk, i.e., a decreasing density perpendicular to the disk plane. Therefore, close to the mid-plane of the disk much higher densities could be found than implied by the simplified picture presented in the cartoon.

J. Bjorkman: Your Fe II mass-loss rate places a lower limit on the electron scattering optical depth in the disk. If $\tau_{es} \sim 1$ and the opening angle is about 30° then the polarisation will be around 6 % or so. Is this consistent with polarisation observations of B[e] supergiants?
F.-J. Zickgraf: The polarisation is as much as 2 – 3 % (cf. also Magalhães & Rodrigues, these proceedings).

P. Stee: How do you determine the continuum level in order to obtain UV edges? Do you take into account the underlying photospheric lines in your simulations? We have developed a "two component" radiative wind model (these proceedings) where the wind is pushed by optically thin lines at the equator and optically thick ones in the polar regions leading to a $v_\infty^{pole}/v_\infty^{eq} \sim 10$ with $v_\infty^{eq} \sim 100 - 200$ km/s and a mass flux ratio of $\Phi_{eq}/\Phi_{pole} \sim 30$, in agreement with your observations (cf. Stee et al. 1995, A&A, 300, 219; Stee 1998, A&A, in press).
F.-J. Zickgraf: The normalisation was done in the two-dimensional order-wavelength space of the high-resolution IUE echelle spectra. Photospheric lines were not taken into account for the profile calculations. Photospheric Fe II is not expected to contribute significantly in early B supergiants.

Franz-Josef Zickgraf and Immo Appenzeller

Polarimetric Evidence of Non-spherical Winds

Antônio Mário Magalhães[1] and Cláudia V. Rodrigues[2]

[1] IAG, Univ. São Paulo, Caixa Postal 3386 - São Paulo SP 01060-970 - Brazil
[2] INPE, Av. dos Astronautas, 1758 - São José dos Campos SP 12227-900 - Brazil

Abstract. Polarization observations yield otherwise unobtainable information about the geometrical structure of unresolved objects. In this talk we review the evidences for non-spherically symmetric structures around Luminous Hot Stars from polarimetry and what we can learn with this technique. Polarimetry has added a new dimension to the study of the envelopes of Luminous Blue Variables, Wolf-Rayet stars and B[e] stars, all of which are discussed in some detail.

1 Introduction

In the past few years there has been mounting evidence that the mass loss in Luminous Hot Stars (LHS) is non-spherically symmetric and this meeting is in fact a testimony to that. In addition, the abundance of free electrons in the winds of such objects makes Thomson scattering an important opacity source. This combination of asymmetry and scattered (hence polarized) light may result in an observed degree of polarization in the radiation we detect from LHS. Polarization observations carry then great potential to explore the environment of LHS.

In this talk we review the evidences for non-spherically symmetric structures around LHS from polarimetry and what we can learn from such data about the physics of such structures. Recent related reviews include those of Bjorkman (1994) and Schulte-Ladbeck (1997). Several talks in this conference also have direct bearing on the topic (K. Bjorkman, Brown and Ignace, Eversberg et al., Rodrigues and Magalhães, Schulte-Ladbeck et al.).

2 Some Polarimetry Basics

One great asset of polarization observations is that they yield diagnostics related to the geometrical structure of unresolvable objects. Generally, it can be said that the polarization is the ratio between the scattered flux and the total flux from the object. The polarization from a stellar envelope will depend in detail on the density and geometrical distribution of matter around the star (e.g., Wood et al. 1996). Techniques for measuring polarization in the UV-optical-IR have greatly advanced in recent years (Roberge and Whittet 1996; Magalhães et al. 1996).

The polarimetric wavelength dependence may be modified by any competing opacity and any unpolarized, diluting light from the star and/or wind.

Examples include hydrogen bound-free and free-free opacities as well as line opacity such as from iron. Hydrogen recombination line emission tends to decrease the polarization across corresponding features, such as Balmer lines. All this provides valuable wind diagnostics. Dust scattering can also play a role in the outskirts of evolved objects. The wavelength dependence of single dust scattering depends on the nature of the grains and their size.

While this review will be concerned mostly with linear polarization, circular polarization may also in principle arise from processes such as multiple dust scattering in an envelope or magneto-emission from stellar spots. Electron scattering produces no circular polarization by itself.

Intrinsic polarization may be detected from the time variability of the observed polarization. In addition, the scatter of the data points in the Q-U diagram (Q = P.cos(2θ) and U = P.sin(2θ), where P=percent polarization, θ=position angle) will tell whether there is a preferred plane of symmetry or not. Binary stars where the scattering envelope surrounds one of them will show up as loops in the Q-U diagram (Brown et al. 1978). Intrinsic polarization may also show up through spectropolarimetry. If the observed polarization varies across a line, such as Hα, the vector difference in the Q-U plane of the continuum and line polarizations will provide the position angle (PA) of the intrinsic polarization (e.g., Schulte-Ladbeck et al. 1992).

3 Observations of Luminous Hot Stars

3.1 Luminous Blue Variables

Luminous Blue Variables (LBV) represent an intermediate stage between OB and WR stars (Maeder 1996). Direct evidence for asymmetric outflows comes from imaging (cf. Nota, these proceedings). In this case, spectropolarimetry has been used to probe mass loss on small spatial scales.

The **P Cyg** nebula has been resolved by direct imaging by Leitherer and Zickgraf (1987). P Cyg shows stochastic changes in its optical linear polarization (Hayes 1985), with night to night changes of 0.2% and 6° in the polarization degree and PA, respectively.

Taylor et al. (1991a) have obtained spectropolarimetry of P Cyg for 20 nights during the 1989-1990 season. The observed polarization showed no preferred plane, consistent with random ejections of matter from the star. No correlation between increased line emission and polarization was observed. This was interpreted as a result from the time lag between these events, since about 40^d are required for a mass ejection to travel out to a distance of about 3 R_*, within which the polarization is thought to be produced.

Further constraints on P Cygni's envelope came from UV spectropolarimetry with WUPPE (Taylor et al. 1991b). A broad dip in the polarization around 2600-3000Å suggested the existence of an absorptive opacity by FeII lines in the envelope. High resolution imaging of P Cyg (Nota et al. 1995;

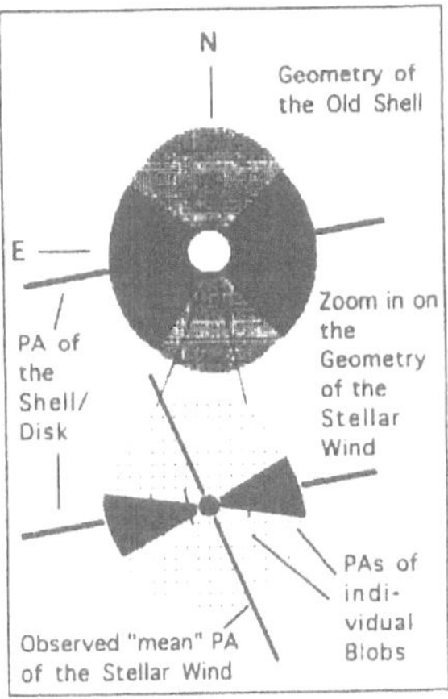

Fig. 1. Geometry of circumstellar wind of R127 (Schulte-Ladbeck et al. 1993).

Nota, these proceedings) shows that the structure of the envelope is indeed clumpy, nicely consistent with the structures seen much closer to the star in the polarimetry data.

The Large Magellanic Cloud LBV **R 127** has been observed for spectropolarimetry by Schulte-Ladbeck et al. (1993). The intrinsic polarization, indicated by the line effect at Hα, showed a level around 1.5% and was suggestive of electron scattering with possible FeII depression from within the envelope. The polarization was variable but with PA values restricted within a 'cone', with the interstellar value as apex, in the Q-U diagram.

The observed nebula (Clampin et al. 1993) is about 2 pc in size and $\approx 10^4$ yr old. There are symmetric enhancements in the (coronographic) image along a direction $\approx 90°$ from the polarization PA value. The suggested geometry for R127 (Schulte-Ladbeck et al. 1993; Fig. 1) is then that of a mass ejection in a preferred plane. The present geometry (from imaging) is defined by events taking place very close to the star (from polarimetry).

A few other LBV have been observed polarimetrically. In AG Car (Leitherer et al. 1994; Schulte-Ladbeck et al. 1994b), the geometry of the nebula shows an alignment with the PA derived from spectropolarimetry, with broad, polarized wings across Hα suggesting electron scattering. In HR Car (Clampin et al. 1995) the PA from imaging and that from polarimetry are actually the same, about 30°. However, we note that, according to Weis et al. (1997), the bipolar nebula has actually its axis at PA$\approx$125° and the imaging and polarimetry data are again consistent. Further monitoring of these objects to confirm the ejections in a preferred plane would be highly desirable.

In summary, polarimetry indicates that LBV may show either stochastic ejections (P Cyg) or, more commonly, a preferred plane for mass loss. In any case, the geometry present in the observed nebulae is already present in (and presumably imposed by) the wind very close to the star. Possible sources for this density contrast have been conjectured by Nota et al. (1995) but it is not possible yet to discern among them.

3.2 Wolf-Rayet Stars

Wolf-Rayet (WR) stars are the polarimetrically best studied class among the LHS (e.g., Robert et al. 1989; Moffat and Robert 1991; Drissen et al. 1992). For (presumed) **single** WR stars, there is a range in the observed variations of optical linear polarization: (a) WN stars vary more than WC ones in a given subclass; (b) Cooler sub-types (i.e., slower winds) vary more, although a few (≈20%) WR show no variability; (c) Polarization variations have time scale of days and are wavelength independent; (d) Most WR show no preferred plane, but there are a few exceptions. Intraday variability is still poorly known.

For **binary** WR stars, cyclic variations of polarization with binary phase are often seen. This is due to the O-star light scattered off the dense WR wind. Mass loss rates can be derived (St.-Louis et al. 1988) as well as the inclination of the systems (Brown et al. 1978), providing important information about WR masses.

Circular polarimetry has been looked for in EZ Cma (Robert et al. 1992) with negative results, suggesting that the star does not show activity related to strong magnetic fields.

Harries et al. (1998) performed a spectropolarimetric survey of 16 WR. Their data are consistent with a distribution of intrinsic polarizations biased towards small values, with only ≈ 20% of stars with $P \geq 0.3\%$. Radiative transfer models suggest equator-to-pole density contrast of 2-3. Combining their results with literature data, for a total of 29 stars, the 5 known objects with 'line effects' cluster around the high mass loss & luminosity part of the $\dot{M}$-L diagram (Fig. 2). Also, the $\dot{M}$ values from radio and optical are in good agreement, suggesting that the wind structures have density contrast independent of radius.

The results of Harries et al. (1998) seem to suggest that the global wind asymmetries in WR winds arise only in the fastest rotators (Ignace et al. 1996). Specially in view of the distribution of 'line effect' stars in the $\dot{M}$-L diagram, we feel that this is also supported by the fact that rotating stars evolve towards higher luminosity (Fliegner et al. 1996).

As the O component screens the WR envelope in an eclipse, the observed polarization may change dramatically and it can be used to model the WR wind (e.g., St-Louis et al. 1993; Rodrigues and Magalhães 1995). Spectropolarimetry across eclipses holds also great potential for probing the ionization structure of the wind.

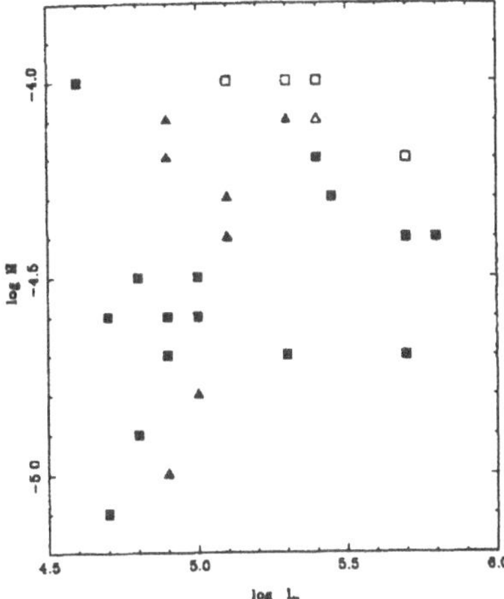

Fig. 2. Spherical (filled symbols) and non-spherical (open symbols) WR stars in the $\dot{M}$-L diagram (Harries et al. 1998).

3.3 B[e] Supergiants

These objects show evidence of a two-component wind: a hot, fast polar wind and a denser, slow equatorial wind (see Zickgraf, these proceedings; de Araújo et al. 1994). Magalhães (1992) showed that the Magellanic B[e] supergiants do present intrinsic polarization, lending further support to the model put forward by Zickgraf et al. In addition, the higher intrinsic polarization values were all associated with objects spectroscopically found to be edge-on. The polarization of these systems, $P_{edge-on}$, correlated some with the average electron density N_e of the envelopes but it correlated somewhat better with the IR [K-L] dust excesses.

Magalhães (1992) pointed out however that in the IR we tend to detect the larger grains, which are poor scatterers in the optical and might not polarize; instead, electrons closer to the star might be operative. Interestingly, the $P_{edge-on} - N_e$ correlation of Magalhães (1992) with the AV16/R4 (a binary, Zickgraf et al. 1996b) point removed becomes actually the tightest one. Spectropolarimetry of the most highly polarized object (S22, Schulte-Ladbeck and Clayton 1993) showed that electron scattering is indeed present, at least for that object. Monte Carlo scattering models (Melgarejo et al. 1999) suggest that *homogeneous* disks fit the polarization data for B[e] stars well. This is consistent with the very slow winds observed and modeled by Zickgraf et al. (1996a) from spectroscopic data, providing another interesting link between the different types of observations.

Three Magellanic B[e] stars have shown variability in polarization (S22, Schulte-Ladbeck and Clayton 1993; S18, Schulte-Ladbeck et al. 1994a) and photometry and spectroscopy (R4, Zickgraf et al. 1996b) similarly to LBV stars. While further scrutiny may show others to be variable too, Gayley & Owocki 1995 have shown that the B[e] class actually extends to luminosities much lower ($\log L/L_{\odot} \approx 4$) than their supergiant counterparts ($\log L/L_{\odot} \approx 5.5 - 6.0$).

3.4 Other Objects

Lupie and Nordsieck (1987) showed that OB stars have intrinsic polarization. An on-going spectropolarimetric survey of OB supergiants is being conducted by Karen Bjorkman (Bjorkman 1994). The observed random PA values suggest that instabilities in an otherwise spherical wind (rather than in a disk) are the cause of the variations. The less luminous Be stars, which show disks, are discussed by K. Bjorkman elsewhere in these proceedings.

Another class of LHS is the Ofpe/WN9 stars, of which ten or so are known in the Magellanic Clouds. They may be O stars in transition to the WR Stage that experience an LBV stage with Ofpe/WN9 characteristics in quiescence (Crowther and Smith 1997). R127 (section 3.1) has actually become an LBV from an Ofpe/WN9 object (Stahl et al. 1983). Pasquali et al. (1997) showed that HDE 269445 has a two component wind. Undoubtedly this class as a whole would be a prime target for polarimetric studies.

4 Conclusions

Imaging and spectropolarimetry data indicate that non-spherically symmetric winds about LHS are the norm. In addition to the suggested systematic observations, other new polarimetric techniques, such as using the Hanle effect in the UV for sensitive detection of magnetic fields (Nordsieck and Harris 1996) look promising. In addition, the new generation of large aperture telescopes such as Gemini and VLT will offer polarimetric capabilities that will be important particularly for the study of objects in the Magellanic Clouds. At the same time, detailed envelope modeling is just becoming possible especially due to Monte Carlo techniques, providing an important feedback on theoretical models. The next few years are bound to witness the coming of age of polarimetry of Luminous Hot Stars and the tapping of its full potential.

AMM thanks the SOC for the invitation and acknowledges support from Fapesp (grants 97/11299-2 and 98/04267-0) and CNPq. CVR has received financial support from Fapesp (grants 98/1443-1 and 92/1812-0).

References

de Araújo, F.X., de Freitas- Pacheco, J.A., Petrini, D. (1994): MNRAS **267**, 501

Bjorkman, K. S. (1994): Ap&SS **221**,335
Brown, J. C., McLean, I. S., Emslie, A. G. (1978): A&A **68**, 415
Clampin, M. et al. (1993): ApJ **410**, L35
Clampin, M. et al. (1995): AJ **110**, 251
Crowther, P.A., Smith, L.J. (1997): A&A **320**, 500
Drissen, L., Robert, C., Moffat, A. F. J. (1992): ApJ **386**, 288
Fliegner, J., Langer, N., Venn, K.A. (1996): A&A **308**, L13
Gummersbach, C.A., Zickgraf, F.-J., Wolf, B. (1995): A&A **302**, 409
Harries, T. J., Hillier, D. J., Howarth, I. D. (1998): MNRAS **296**, 1072
Hayes, D. P. (1985): ApJ **289**, 726
Ignace, R., Cassinelli, J. P., Bjorkman, J. E. (1996): ApJ **459**, 671
Leitherer, C. Zickgraf, F.-J. (1987): A&A **174**, 103
Leitherer, C. et al. (1994): ApJ **428**, 292
Lupie, O. L., Nordsieck, K. H. (1987): AJ **93**, 214
Maeder, A. (1996): In: Leitherer, C., Fritze-von-Alvensleben, U., Huchra, J. (eds.) From Stars to Galaxies. ASP Conf. Ser. **98**, San Francisco, p. 141
Magalhães, A. M. (1992): ApJ **398**, 286
Magalhães A. M. et al. (1996): In: Roberge W. G., Whittet D. C. B. (eds.) Polarimetry of the Interstellar Medium. ASP Conf. Ser. **97**, San Francisco, p. 118
Melgarejo, R., Magalhães, A.M., Rodrigues, C.V. (1999): in preparation.
Moffat, A.F.J., Robert, C. (1991) In: van der Hucht, K. A., Williams, P. M. (eds.) Wolf-Rayet: Binaries, colliding winds and evolution. Dordrecht, Kluwer, p. 109
Nordsieck, K. H., Harris, W. M. (1996): In: Roberge W. G., Whittet D. C. B. (eds.) Polarimetry of the Interstellar Medium. ASP Conf. Ser. **97**, San Francisco, p. 100
Nota, A. et al. (1995): ApJ **448**, 788
Pasquali, A. et al. (1997): A&A **327**, 265
Roberge, W.G, Whittet, D.C.B. (1996) (eds.) Polarimetry of the Interstellar Medium. ASP Conf. Series **97**, San Francisco.
Robert, C. et al. (1989): ApJ **347**, 1034
Robert, C. et al. (1992): ApJ **397**, 277
Rodrigues, C. V., Magalhães, A. M. (1995): In: van der Hucht, K. A., Williams, P. M. (eds.) Wolf-Rayet: Binaries, colliding winds and evolution. Dordrecht, Kluwer, p. 260
St.-Louis, N. et al. (1988): ApJ **330**, 286
St.-Louis, N. et al. (1993): ApJ **410**, 342
Schulte-Ladbeck, R. E. (1997): Rev. Mod. Astron. **10**, 135
Schulte-Ladbeck, R. E., Clayton, G. C. (1993): AJ **106**, 790
Schulte-Ladbeck, R. E. et al. (1992): ApJ **387**, 347
Schulte-Ladbeck, R. E. et al. (1993): ApJ **407**, 723
Schulte-Ladbeck, R. E. et al. (1994a): Space Sci. Rev. **66**, 193
Schulte-Ladbeck, R. E. et al. (1994b): ApJ **429**, 846
Stahl, O. et al. (1983): A&A **127**, 49
Taylor, M. et al. (1991a): AJ **102**, 1197
Taylor, M. et al. (1991b): ApJ **382**, L85
Weis, K. et al. (1997): A&A **320**, 568
Wood, K. et al. (1996): ApJ **461**, 828
Zickgraf, F.-J. et al. (1996a): A&A **315**, 510
Zickgraf, F.-J. et al. (1996b): A&A **309**, 505

Discussion

H. Henrichs: What about O stars?

M. Magalhães: The work of Lupie and Nordsieck (1987) showed that 6 OB stars (out of 8, mostly supergiants) had intrinsic polarisation. Their polarimetric variability correlated directly with the average Hα equivalent widths and inversely with $v \sin i$. From the detected variability they concluded that the wind structures in the envelopes were short lived and formed at different distances from the stars.
In the on-going spectropolarimetric survey of hot stars at the University of Wisconsin conducted by Karen Bjorkman, the OB supergiants show significant polarisation variability, with 9 out of 18 objects showing the effect in more than one observation. The observed random position angles again basically agree with the above conclusions.

R. Schulte-Ladbeck: I was thinking about how the B[e] polarisation might help us learn about disk properties. Currently, observations of the linear polarisation have two problems: firstly, the only spectropolarimetric data set is that of Schulte-Ladbeck & Clayton. In this case, the H_2 line effect allows us to estimate the interstellar polarisation so that we may determine the amount of intrinsic polarisation. All other stars have only broad-band polarimetry and the interstellar polarisation correction is less secure.
Secondly, the only public-user polarimeter on the southern hemisphere is on the AAT. The throughput of the system, unlike HPOL, does not remain high shortward of 4000 Å. Therefore no data exists for the polarisation variation across the Balmer discontinuity. One might try in the future to obtain measurements across the Paschen jump.

Cláudia Rodrigues and Mário Magalães

Wolf-Rayet Wind Models: Photometric and Polarimetric Variability

Cláudia V. Rodrigues[1,2] and Antônio Mário Magalhães[2]

[1] INPE, Av. dos Astronautas, 1758 - 12227-900 São José dos Campos SP - Brazil
[2] IAG-USP, Av. Miguel Stefano, 4200 - 04301-904 São Paulo SP - Brazil

Abstract. We have modelled the observed random variation in broad band intensity and polarization of some isolated Wolf-Rayet stars assuming that their winds have localized, enhanced density regions (blobs). Our model is based on a Monte Carlo code that treats all Stokes parameters of the radiation bundle. This study indicates that the blobs must have sizes comparable to the stellar dimension and be near the base of the envelope. These blobs can be interpreted as a variable structure of large geometric cross section causing the observed polarimetric and photometric variability.

1 Introduction

Wolf-Rayet (WR) stars can present random fluctuations in broad band flux and polarization and also in spectral line profiles. The broad band variations can reach 10% in flux (Antokhin et al. 1995; Marchenko et al. 1998) and 0.5% in polarization (St.-Louis et al. 1987; Drissen et al. 1987; Robert et al. 1989). The changes in the spectral line profiles can be divided in two types. One of them is the small moving bumps which appear in some optical emission lines (Robert 1994). They may be associated to small-scale instabilities intrinsic to a radiative wind (Owocki 1994; Gayley & Owocki 1995). The discrete absorption components (DACs) are also present in WR stars (e.g., Prinja & Smith 1992). They comprise a larger portion of the profile and may be associated with a large amount of mass (Massa, Prinja & Fullerton 1995 and references therein). These structures may have an external origin (relative to the wind) as, for instance, rotation, binarity and/or photospheric processes (Owocki 1994).

2 The model

To study the random variability in WR stars, we have assumed that the envelope has regions of enhanced density which we call blobs. Our goal is to constrain the physical characteristics the blobs may have in order to explain the observed broad band variability. We have solved the radiative transfer in an electron scattering envelope using the Monte Carlo code described in Rodrigues (1997).

The blobs have been assumed spherical and immersed in a spherical envelope. The density law of the envelope can be chosen among many analytical expressions. The density inside the blob follows the same law as the envelope, but it is multiplied by an arbitrary factor which introduces the density enhancement. We are able to treat an arbitrary number of blobs as well. The only source of radiation is a spherical central source emitting isotropically. The use of a Monte Carlo code has allowed us to consider optically thick envelopes characteristic of WR winds.

The photopolarimetric variability may arise in two situations: (1) if the wind changes from a homogeneous configuration to an inhomogeneous one; (2) if the wind is always inhomogeneous, but with a moving blob whose relative position to the source and/or to the line of sight is variable.

3 Results

The code provides us with values of the flux, linear polarization and its position angle as a function of the line of sight under which the system is observed. In order to simplify the analysis, each model has been characterized by only two values: the minimum flux normalized to that of a homogeneous envelope, ΔI; and the maximum polarization, ΔP. In doing that, we have assumed that the flux variation is caused by extinction so that a decrease in flux can only happen if the blob is in the line connecting the source and the observer. In general, the blob also scatters light to any direction and this can produce an increase in the flux (relative to the homogeneous case). However, this increase in our model barely reaches 1%.

We find that the variation in flux does not constrain the physical properties of the blob. An extinction of 10% is achieved for practically any model by simply adjusting the optical depth of the blob. On the other hand, most models tend to produce a polarization smaller than that observed. A value of $\approx 0.5\%$ is only obtained for very specific conditions: blobs of dimensions similar to the star and which are near the base of the envelope. An example of a model which fits the observed values is presented in Tab. 1. The blob in this model covers an solid angle of 0.32 steradians, which is equivalent to 2.5% of total solid angle of the envelope (4π).

The interpretation of these results is that the blobs should have a large geometric cross section (blob radius $\approx$ stellar radius) in order to produce the observed values of polarization. A structure having a smaller cross section could not produce 0.5% of polarization. This result does not depend on the blob density.

4 Conclusions

This work has shown that the structure causing polarization must be relatively large, with its size similar to that of the star. This does not necessarily

Table 1. An example of a model which reproduces the random broad band variability of Wolf-Rayet stars

Parameter	Value
Radius of the envelope	10 $R_\star$
Optical depth of the envelope	2.0
Optical depth of the blob	5.0
Blob position	3 $R_\star$
Radius of the blob	1 $R_\star$
ΔI	11%
ΔP	0.55%
Blob mass	1% of the mass of the envelope

mean that there must be a single huge blob, but that the average enhancement of the density is spread over an large area of the wind. In a forthcoming paper, we will also study how these results correlate with the spectral variations observed in WR stars.

5 Acknowledgments

CVR has received financial support from FAPESP (grants 98/1443-1 and 92/1812-0). AMM acknowledges support from FAPESP (grants 97/11299-2 and 98/04267-0) and CNPq (grant 30.1558/79-5/FA).

References

Antokhin, I., Bertrand, J. F., Lamontagne, R., Moffat, A. F. J. & Matthews, J. 1995: AJ **109**, 817

Drissen, L., St.-Louis, N., Moffat, A. F. J. & Bastien, P. 1987. ApJ **322**, 888

Gayley, K. G. & Owocki, S. P. 1995: ApJ **446**, 801

Marchenko, S. V., Moffat, A. F. J., Eversberg, T., Hill, G. M., Tovmassian, G. H., Morel, T. & Seggewiss, W. 1997: MNRAS **294**, 642

Massa, D., Prinja, R. K. & Fullerton, A. W. 1995: ApJ **452**, 842

Owocki, S. P. 1994: Ap&SS **221**, 3

Prinja, R. K. & Smith, L. J. 1992: A&A **266**, 377

Robert, C. 1994: Ap&SS **221**, 137

Robert, C., Moffat, A. F. J., Bastien, P., Drissen, L. & St.-Louis, N. 1989: ApJ **347**, 1034

Rodrigues, C.V. (1997): *Radiative Transfer in Hot Star Envelopes using the Monte Carlo Method*, PhD Thesis, IAG-USP

St.-Louis, N., Drissen, L., Moffat, A. F. J., Bastien, P. & Tapia, S. 1987: ApJ **322**, 870

Discussion

M.-M. MacLow: How would your results change if your blobs were sheets instead of spheres; that is, partial shells with, e.g., much less than one steradian solid angle coverage?
C. Rodrigues: For optically thin structures, the geometric depth may be important; for instance, spherical blobs have more scattering material than partial shells of the same cross section. However, for optically thick structures, only the cross section is important (see below).

J. Brown: I would like to comment that Richardson, Brown, & Simmons treated the same problem analytically but for small blobs and small optical depth. We found that, in this case, the only way to make the polarisation variations small enough compared to the photometric variations was for the blobs to be very dense and to produce significant emission (which you neglect). It would be interesting to check whether your high τ blobs really do not contribute much to the total emission.
C. Rodrigues: You are right in the sense that the emission of the blobs (and also from the envelope) must be considered. But in your work you have considered that the photometric variations were caused by scattering out of the line of sight of the blobs; i.e., they represent an increase in flux relative to a "homogneous envelope". In this case $\Delta \mathrm{I}/\Delta \mathrm{P}$ must be around 1 based on the scattering properties of electrons. In our work, we have considered $\Delta \mathrm{I}$ to be caused by extinction. In that case, $\Delta \mathrm{I}$ is always greater that $\Delta \mathrm{P}$.

J. Cassinelli: You say that the change in polarisation depends on the angular size of the blob. However, isn't there also a strong dependence on the optical depth $\Delta\tau$ of the blob (or shock fragment)?
C. Rodrigues: In the optically thin regime, the polarisation grows with angular size and optical depth. However, for optically thick blobs, only the cross section is important because in this case multiple scattering occurs. In other words, for optically thick blobs only the region facing the WR photosphere "produces photons" which have been scattered only once. These are the photons producing polarisation.

P. Veen: The 10 % flux variation of WR stars is wavelength dependent, being higher in the violet. What extinction law is expected in your model?
C. Rodrigues: Our model does not predict any wavelength dependence on extinction since we do not consider emission and/or absorption in the envelope. However, these processes may be included and a wavelength dependence can thus arise.

J. Bjorkman: You point out that the problem you have producing polarisation variations as large as 0.5 % is caused by not having enough surface area in the blob. Could you solve this problem by having three or four blobs?

C. Rodrigues: Partially. If you have two blobs in opposite directions you can enhance the polarisation. But if you think about randomly oriented blobs the polarisation may decrease with a higher number of condensations because we are approaching a more "symmetric" configuration.

A. Moffat: I was somewhat surprised to hear you say that photometric/polarimetric broad-band observations are not seeing the same thing as the spectroscopic observations. After all, both have similarities: e.g., time scales, randomness, etc. On the other hand, our recent spectroscopic analysis of clumps (cf. my talk) shows that they are not optically thick, contrary to your broad-band clumps.

C. Rodrigues: We suggest that the relatively large sizes of the blobs producing polarisation seems to indicate that they are not the small instabilities expected to arise in radiatively driven winds (which may be related to the sub-peaks present in the optical emission lines). More probably, the polarimetric variation may be related to spectroscopic features caused by large-scale strucures. Maybe the small and large-scale structures are not uncorrelated, but have the same physical origin.

About the optical thickness of the blobs: there must be a relatively high density contrast for a measurable polarisation to exist. If the density contrast is decreased, the blob size must be increased in order to get the same amount of polarisation. Anyway, we think that the structures causing the sub-peaks in the emission lines are not the carriers of the polarisation. They are very small and even with a considerable density contrast they can be optically thin.

S. Owocki: It may be true that while continuum polarisation is most sensitive to large-scale blobs, the line-profile bumps are easier to detect for more localised, smaller-scale blobs. Overall, the wind may have a continuous distribution of scales.

A. Moffat: I agree. In fact, we do not have serious enough constraints on clump parameters to be able to give meaningful interpretations of polarisation variability. We might have to wait for a direct resolution of the wind by interferometry.

Anisotropic Outflows from LBVs and Ofpe/WN9 Stars

Antonella Nota[1,2]

[1] Space Telescope Science Institute, 3700 San Martin Drive, Baltimore MD 21218, USA
[2] Astrophysics Division, Space Science Department of the European Space Agency

Abstract. In the past decade, the increasing sophistication of the high resolution techniques from the ground, and the advent of HST have allowed a systematic study of the outflows around evolved massive stars. It has been established (Nota & Clampin 1997) that most LBVs and Ofpe/WN9 stars are surrounded by associated circumstellar nebulae which have been ejected in some previous phase of their evolution. These nebulae are the fossil record of the interactions of previous winds and of the violent ejections in which the stars most likely have shed their outer layers. The study of the morphology, kinematics and chemical composition of the ejected material has allowed us to gain deep insight in the ejection and shaping mechanism.

1 Are outflows isotropic?

In this paper, I will review the observational evidence for the isotropy, or lack thereof, of outflows from luminous, evolved stars. I will be mainly concentrating on the large scale structures of nebulae around LBVs and Ofpe/WN9 stars, in the Milky Way and in the LMC. In order to establish whether the distribution of ejected material is isotropic or not, we study a) the morphology of the ionized and neutral material, from direct images in the optical and IR, b) the kinematics of the gas, which provides information on the nebular expansion motions.

In terms of morphology, LBV and Ofpe/WN9 associated nebulae always show the presence of a preferential symmetry axis. In Figure 1, we show three examples of the most *"bipolar"* nebulae, where such terminology is drawn from the PN classification. In such a structure, two expanding bubbles can be distinguished, aligned with a preferential axis, associated with a narrow waist. η Carinae (Davidson et al. 1997; Schulte-Ladbeck et al. 1988b) is the textbook case of such a morphological type (Figure 1, left panel). The nebulae around HR Carinae (Hutsemékers & Van Drom 1991a; Clampin et al. 1995; Weis et el. 1997; Nota et al. 1997; Hulbert et al. 1998) and the LBV candidate HD168625 (Hutsemékers et al. 1994; Nota et al. 1997; Robberto & Herbst 1998) are also bipolar, although in these two cases the morphology is less visually striking. In the case of HR Carinae, the nebula is much older, and only the southern bubble is visible, and fragmented. As we will discuss later, the kinematic data confirms presence of a truly bipolar outflow. In the case of HD168625, the compact elliptical nebula in the center is likely to be the "*waist*" of the bipolar distribution. Such a waist is resolved into a horn shaped structure in mid-IR images obtained by Robberto & Herbst (1998).

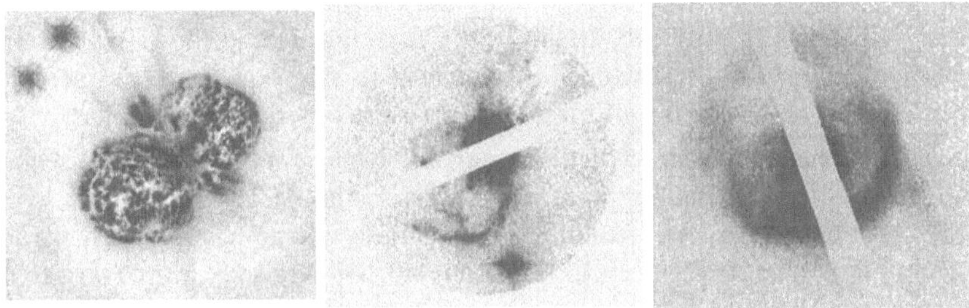

Fig. 1. Three examples of galactic extreme bipolar nebulae, in the light of $H\alpha$: η Car (courtesy of R. White), HR Car, HD168625. In all images, North is up and East to the left.

Most of the remaining known nebulae display much milder asymmetries. AG Carinae (Smith et al. 1997; Trams et al. 1996; Schulte-Ladbeck et al. 1994; Leitherer et al. 1994), R127 (Walborn 1982; Stahl 1987; Clampin 1993; Smith et al. 1998) and Wra 751 (Hu et al. 1990; Hutsemékers & van Drom 1991b; de Winter et al. 1992) are three representatives of the more conspicuos class of nebulae where a preferred axis can still be noted but the overall symmetry is elliptical (Figure 2).

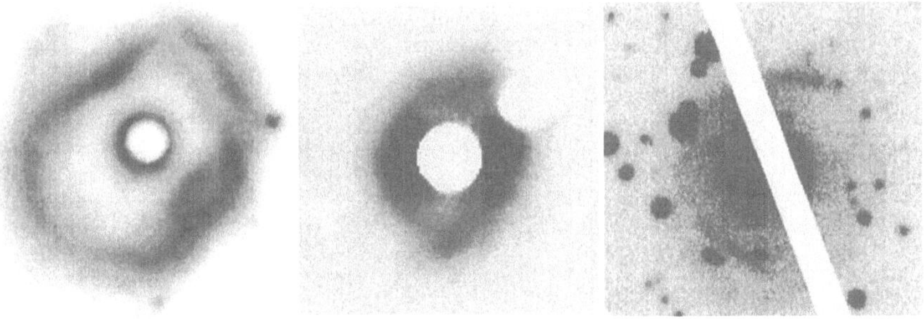

Fig. 2. Examples of elliptical nebulae in the light of $H\alpha$: AG Carinae (galactic), R127 (LMC), Wra 751 (galactic). In all images North is up and East to the left. The images are not equally scaled.

AG Carinae and R127 are remarkably similar: AG Carinae is galactic, $\simeq 36'' \times 36''$ in size, while R127 is in LMC, and its apparent size is much smaller ($\simeq 8'' \times 9''$) (Figure 2: left and middle). Both nebulae are quite old (several tens of thousand years). In both cases it is possible to note the presence of a preferred axis (the major axis of the ellipse) and of two symmetrical regions of enhanced brightness, aligned with the minor axis.

In the currently accepted scenario that ejected nebulae are shaped by interacting winds (Nota et al. 1995; Frank, Balick & Davidson 1995; García-Segura et al. 1996), the ratio of the density in the equatorial waist with respect to the polar region ultimately defines the observed morphology.

The fast blue supergiant wind is more effective at sweeping a region where the density is low, producing a more elliptical nebula when the density contrast is lower, and a more exteme bipolar nebula when the density contrast is higher. It is also interesting to note that all nebulae in Figure 1 are relatively young (100-5000 yrs) while the nebulae in Figure 2 are ten times older and more, indicating that evolution also plays a role in the formation of the currently observed morphologies.

Kinematical studies of these nebulae have subsequently confirmed what had been previously proposed on the basis of morphological considerations alone. For all known LBVs and Ofpe/WN9 associated nebulae, detailed kinematical studies can be found in the literature. In Figure 3, we show the kinematical maps for the two extreme cases: a bipolar outflow on the left panel (HR Carinae, Nota et al. 1997) and an expanding shell, on the right panel (S61, Pasquali et al. 1998). In both maps the radial velocity (reduced to the heliocentric system, in km s^{-1}) is plotted as a function of position with respect to the star (in arcseconds, east to the left, west to the right). In the case of S61, the data (filled dots) are modelled with a spherically symmetrical expansion law (empty circles). For most nebulae around LBVs and Ofpe/WN9 the kinematic maps fall in between these two extreme cases.

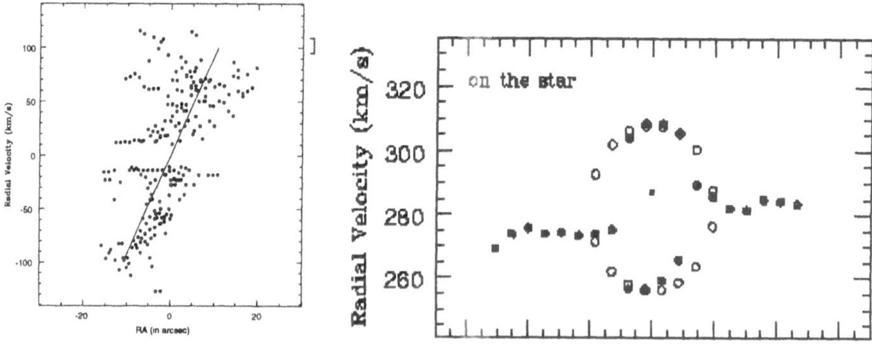

Fig. 3. Kinematic map for a) a bipolar nebula (HR Carinae) b) an expanding shell (S61)

It is worth mentioning at this point the case of the WNL star Wra 124 which is surrounded by a large clumpy nebula, studied at high resolution both from the ground (Sirianni et al. 1998) and with HST (Grosdidier et al. 1998). On the basis of morphology alone it is difficult to establish the nebular structure: while the inner nebula seems to be dominated by a bipolar morphology (Sirianni et al. 1998), the large scale HST images reveal a miriad of fragmented clumps, arches and filament with no apparent overall structure (Grosdidier et al. 1998). However, the study of the kinematic properties help to discriminate at least two motions: an expanding shell and a superimposed bipolar outflow (Figure 4). Most likely, even more complex motions are present in the circumstellar environment of Wra 124.

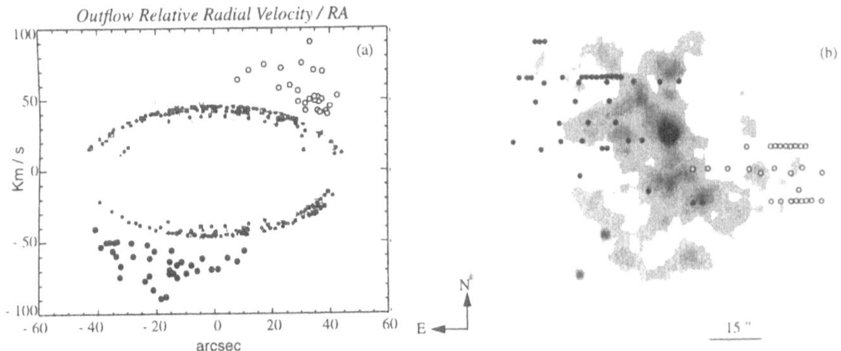

Fig. 4. Radial velocity map of the M1-67 nebula (left panel). The bipolar outflow is seen as the residual to the elliptical model of the expanding shell. The regions defining the bipolar outflow are marked on the Hα image of the nebula (right panel).

1.1 The exception: P Cygni

Although P Cygni is the prototypical Luminous Blue Variable, its surrounding nebula is certainly not the protypical LBV nebula: it is the only one to display a true spherical symmetry, and a highly nonhomogeneous distribution of clumps within a spherical envelope (Figure 5: left panel). The kinematic information (Nota et al. 1998) confirms the suggested scenario: in Figure 5 (right panel) we show the radial velocity map obtained with the long slit positioned at 8″ N, intercepting the four brightest clumps in the northern hemisphere (B, D, F, H - Nota et al. 1998). While the overall profile is reminiscent of a spherically expanding shell, a more complex structure is dicernible in the line profile of the line Fe[II] λ 4287. The clumps appear to be both at the surface and inside the expanding shell. This finding is in agreement with Skinner et al. (1998) who, on the basis of radio data, concluded that 1) the circumstellar environment of P Cygni is very clumpy, 2) the clumps appear randomly distributed, and 3) in close proximity to the central star they are variable with timescale of days.

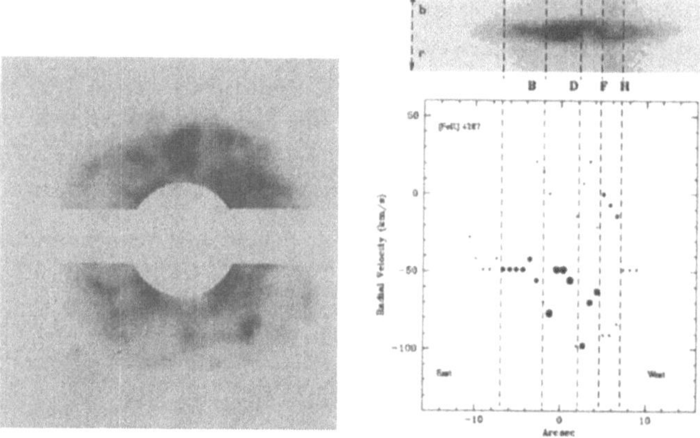

Fig. 5. Image of the P Cygni nebula in the light of Hα (left panel); a subset of the long slit image centered on the Fe[II] λ4287 line (right panel). Note the complex structure of the nebular line, projected onto its corresponding kinematic profile.

2 The connection between large scale structure and the stellar wind

For a number of LBVs, a connection has been found between the large scale structure of the nebulae, up to a few parsecs from the star, and the structure of the stellar wind. In the case of AG Carinae, for example, Leitherer et al. (1994) noted that the stellar spectrum in the optical and the UV wavelengths is dominated by the effects of a massive stellar wind. They detected two wind components: a slow dense wind, where the bulk of the recombination radiation is emitted, and a faster, less dense wind, visible in the absorption components of the UV P Cygni profiles (with velocities up to -1000 km s^{-1}). At the same time, Schulte-Ladbeck et al. (1994) found very large variations of the polarization with time, along a preferred direction which is aligned with the major axis of the optical nebula (Figure 2, left panel). These two independent observations seem to indicate that the asymmetries noted in the ejected nebula are also present in the stellar wind, very close to the star.

Spectropolarimetry of the LMC LBV R127 in UBVRI, Hα, [NII] has also shown large intrinsic continuum polarization (1-1.5 %) indicating both the presence of a substantial quantity of free electrons in the wind and a very large deviation from spherical symmetry very close to the star. Again, the direction of the polarization vector matches the nebular symmetry axis (Schulte-Ladbeck et al. 1993) from the optical images (Figure 2: middle panel).

Spectropolarimetry has proven to be a fundamental diagnostic tool also in cases where no asymmetries are present. Well before any images of the nebula around P Cygni had been obtained, Taylor et al. (1991) gathered optical spectropolarimetric measurements of P Cygni and found intrinsic polarization variable with time, on a short time scale, indicating that the scattering material lied close to the star. Moreover, they found a variation in the position angle of the polarization vector, indicating no preferential direction, and therefore they ruled out a spherically symmetric wind or an axisymmetric shell. When the first optical images of the nebula were obtained (Barlow et al. 1994; Nota et al. 1995), it was clear that spectropolarimetry had proven right: the nebula was composed of clumps, distributed within a spherically symmetrical envelope, withot any preferential axisymmetry. A few years later, Skinner et al. (1998) found that the radio emission close to the star is also highly inhomogeneous, with timescales of a few days. The connection between the stellar wind structure and the outer nebula holds for P Cygni, although in the context of a completely different morphological structure.

3 A new result: the *true* R143 nebula

R143 was identified by Parker (1993) as a LBV. R143 was believed to be surrounded by a circumstellar nebula, whose morphology was significantly different from all other observed LBV nebulae: three filaments (the *"fingers*) emerged from the star, ending in unresolved clumps (Figure 6, left panel): a very different morphology from all other known LBV nebulae.

From the discovery paper, it was not clear whether the nebula and the central star were physically associated: inspection of the surrounding region showed that R143 was located in an extended HII region, rich in gas and dust. Smith et al. (1998) therefore took long slit spectra to assess the nature of the nebula and found the *three fingers* not to be associated with the central star, both in terms of their kinematics and chemical composition, but to belong to the underlying HII region. However, they did find a compact nitrogen enriched region (Figure 6, right panel), in close proximity to the star, with a spatial extension of 2-4″, which was subsequently resolved in HST/WFPC2 images as a small compact nebula most likely recently ejected by the central LBV (Schulte-Ladbeck et al. 1998a).

4 Conclusions

In this paper, I have reviewed the current observational evidence related to anisotropic outflows in LBVs and Ofpe/WN9 stars. Most luminous stars in

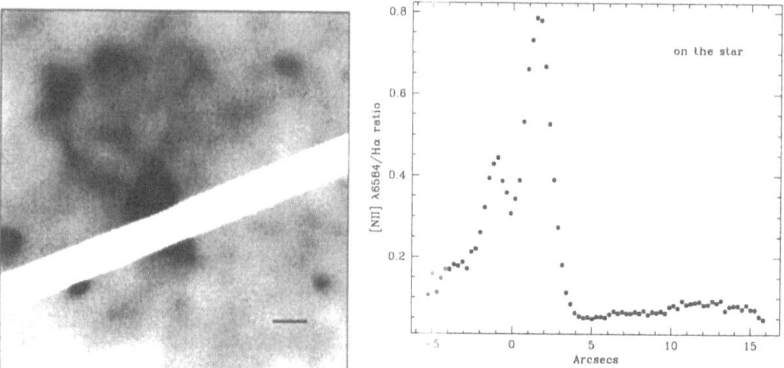

Fig. 6. Hα image of the R143 circumstellar region (left panel). The N/H ratio is shown in the right panel as a function of position with respect to the star, indicating the presence of a nitrogen enriched region in close proximity to the central star.

these two categories are surrounded by spectacular circumstellar nebulae of ejected material. With the notable exception of one object, P Cygni, all nebulae display some level of axisymmetry, ranging from the true bipolar morphology of η Carinae, to the more frequent elliptical nebulae. The kinematic information available confirms the morphological evidence that the outflows are not isotropic. In a few cases, using spectropolarimetric techniques, it has been possible to associate the structure of the stellar wind with the large scale structure of the ejected nebulae: in all these cases, a close analogy has been shown. In this context, ejected nebulae have shown to be truly the fossil record of the history of their previous winds. Hydrodynamical models have been able so far to reproduce qualitatively, quite successfully, the morphology of these nebulae. The time is now ripe to proceed to a quantitative application of the models, with the objective of reconstructing the nebular origin, and therefore, the evolutionary history of the central stars.

5 Acknowledgements

AN is grateful to friends and colleagues who have been integral part of the LBV nebulae project: Mark Clampin, Anna Pasquali, Regina Schulte-Ladbeck, Linda Smith.

References

Barlow, M.J., Drew,J.E., Meaburn,J. & Massey, R.M. 1994, MNRAS 268, L29

Clampin, M., Nota, A., Golimowski, D.A., Leitherer, C. & Durrance, S.T. 1993, ApJ 410, L35

Clampin, M., Schulte-Ladbeck, R.E., Nota, A., Robberto, M., Paresce, F., & Clayton, G. 1995 AJ 110, 251

Davidson, K., Ebbets, D., Johansson, S., Morse, J.A., Hamann, F.W., Balick, B., Humphreys, R., Weigelt, G. & Frank, A. 1997, AJ 113, 335

de Winter, D., Peréz, M.R., Hu, J.H. & Thé, P.S. 1992, A&A 257, 632

Frank, A., Balick, B. & Davidson, K. 1995, AJ 109, 178

García-Segura, G., Langer, N. & MacLow, M.,-M. 1996, A&A 316, 133

Grosdidier, Y. Moffat, A.F.J., Joncas, G. & Acker, A. 1998, ApJ 506, L127

Hu, J.H., de Winter, D., Thé, P.S. & Peréz, M.R. 1990, A&A 227, L17

Hulbert, S., Nota, A., Clampin, M., Leitherer, C., Pasquali, A., Langer, N. & Schulte-Ladbeck, R.E. 1998, this conference

Hutsemékers, D. & Van Drom, E. 1991a A&A 248, 141

Hutsemékers, D. & Van Drom, E. 1991b, A&A 251, 620

Hutsemékers, D., Van Drom, E., Gosset, E. & Melnick, J. 1994 A&A 290, 906

Leitherer, C., Allen, R., Altner, B., Damineli, A., Drissen, L., Idiart, T., Lupie, O., Nota, A., Robert, C., Schmutz, W. & Shore, S.N. 1994, ApJ 428, 292

Nota, A., Livio, M., Clampin, M. & Schulte-Ladbeck, R.E. 1995, ApJ 448, 788

Nota, A., Pasquali, A., Clampin, M., Pollacco, D., Scuderi, S. & Livio, M. 1996 ApJ 473, 946

Nota, A. & Clampin, M. 1997: *Luminous Blue Variables: Massive Stars in Transition* eds. A.Nota, H.J.L.M. Lamers (PASP), p.303

Nota, A., Smith, L.J., Pasquali, A., Clampin, M., & Stroud, M. 1997 ApJ 486, 338

Nota, A., Clampin, M., Pasquali, A., Robberto, M., Pollacco, D., Ligori, S., Paresce, F. & Staude, J. 1998 in preparation.

Parker, J.Wm., Clayton, G.C., Winge, C. & Conti, P. 1993, ApJ 409, 770

Robberto, M., & Herbst, T.M. 1998, ApJ 498, 400

Schulte-Ladbeck, R.E., Leitherer, C., Clayton, G.C., Robert, C., Meade, M.R., Drissen, L. Nota, A. & Schmutz, W. 1993, ApJ 407, 723

Schulte-Ladbeck, R.E., Clayton, G.C., Hillier, D.J., Harries, T.J. & Howarth, I.D. 1994, ApJ 429, 846

Schulte-Ladbeck, R.E., Pasquali, A., Clampin, M., Nota, A., Hillier, D.J.& Lupie, O. 1998, this conference.

Sirianni, M., Nota, A., Pasquali, A. & Clampin, M. 1998, A&A 335, 1029

Skinner, C., Becker, R.H., White, R.L., Exter, K.M., Barlow, M.J. & Davis, R.J. 1998, MNRAS 296, 669

Smith, L.J., Stroud, M.P., Esteban, C., Vilchez, J.M. 1997, MNRAS 290, 265

Smith, L.J., Nota, A., Pasquali, A., Leitherer, C., Clampin, M. & Crowther, P.A. 1998 ApJ 503, 278

Stahl, O. 1987 A&A 182, 229

Taylor, M., Nordsieck, K.H., Schulte-Ladbeck, R.E. & Bjorkman, K.S. 1991 AJ 102, 1197

Trams, N.R., Waters, L.B.F.M. & Voors, R.H.M. 1996, A&A 315, L213

Walborn, N.R. 1982, ApJ 256, 452

Weis, K., Duschl, W.L., Bomans, D.J., Chu, Y.-H., & Joner, M.D. 1997 A&A 320, 568

Discussion

K. Weis: For the sake of completeness I would like to comment that Weis et al. (1997, A&A 320, 568; April 1997) already showed the bipolar structure of the nebula around HR Car before Nota et al. (1997, ApJ 486, 338; September 1997).

S. Shore: It is worth noting that for several LBVs the bolometric luminosity remains constant. This implies that the outflows are, if not spherical, at least optically thick right down near the star.
MWC 560 has not been mentioned at this meeting, but is well known to show a detached absorption line. Yes, it is a symbiotic and $L \sim 10^3 L_{\odot}$, but don't hold that against it!

Antonella Nota and Anna Pasquali

Non-isotropic Outflows in the Infrared: ISO Imaging of LBVs

Norman R. Trams[1], C.I. van Tuyll[2], Robert H.M. Voors[3], Alex de Koter[2], Laurens B.F.M. Waters[2], and Patrick W. Morris[4]

[1] Astrophysics Division, ESA Space Science Department, ESTEC, PO Box 299, 2200 AG Noordwijk, The Netherlands
[2] Astronomical Institute, University of Amsterdam, Kruislaan 403, NL-1098 SJ Amsterdam, The Netherlands
[3] Astronomical Institute Utrecht, Princetonplein 5, Utrecht, The Netherlands
[4] SRON Laboratory for Space Research, Sorbonnelaan 2, NL-3584 CA, Utrecht, The Netherlands

Abstract. In this review we present the ISO[1] imaging of nebulae around Luminous Blue Variables. Three LBVs have been imaged with ISO: HR Car, AG Car and the LBV candidate G79.29+0.46. The ISOCAM instrument did not resolve the nebula around HR Car. However some nebular emission lines are seen in the spectral energy distribution of this source. A proper deconvolution of the images may resolve the nebula. For AG Car and G79.29+0.46 the nebula is clearly resolved. The structure and intensity distribution of the nebular emission is dependent on the wavelength, indicating a separation between the ionised matter and the dust in the nebulae. Some model calculations are presented for the G79.29+0.46 nebula.

1 Introduction

1.1 LBV nebulae

Most Luminous Blue Variables (LBVs) are surrounded by nebulae (see Smith, 1994 and Nota and Clampin, 1997 for reviews). These nebulae are thought to be the result of the interaction between the strong LBV wind and a surrounding medium that is due to an older mass loss phase of the star. Most of the galactic LBV nebulae have been discovered using optical imaging. For the more distant LBVs (e.g. in the LMC) the presence of the nebulae has been inferred from optical spectroscopy. In some cases these nebulae have been confirmed by Hubble Space Telescope images.

Using optical imaging in emission lines (Hα) the mass of the ionized matter in the nebula can be determined. The ionized masses are typically a few solar masses (Nota and Clampin, 1997). Most nebulae display some degree

[1] Based on observations with the Infrared Space Observatory (ISO). ISO is an ESA project with instruments funded by the ESA member states (especially the PI countries: France, Germany, The Netherlands and the United Kingdom) and with the participation of ISAS and NASA

of bipolarity, which strengthens the interacting wind model for the formation of the nebulae. In this model the correlation between nebular mass and age points to the formation in subsequent mass loss episodes. Abundance studies of the nebula (Smith et al., 1997) and studies of the dust content of the nebulae (Waters et al., 1998) indicate that they are expelled in a Red Supergiant (RSG) phase. In that case the interacting wind model would point to interactions between the current LBV wind and the remnant of the RSG wind.

The dust in LBV nebulae has been discussed by Hutsemékers (1997) and Waters et al. (1998). The dust masses seen in these nebulae range from 0.0003 to about 0.04 solar masses. Hutsemékers (1997) finds a correlation between stellar luminosity and nebular dust mass. Large grains seem to dominate in the nebulae, and give rise to the observed dust scattering in AG Car and HD168625 (Viotti et al., 1988 and Hutsemékers et al., 1994). Waters et al. (1998) find evidence for crystalline silicates in a number of LBV nebulae from ISO SWS spectra (see also Waters et al., this volume).

1.2 IR imaging of LBV nebulae

Not many imaging observations in the infrared spectral region have been published. Most IR imaging has been done at near-infrared wavelengths (2.2 μm, e.g. McGregor et al., 1988; Hutsemékers et al., 1994). At these wavelengths the nebular emission is mostly due to scattering of stellar radiation by the dust grains in the nebula. Only at longer wavelengths beyond 10 micron direct imaging of the nebular dust content is possible, since here the thermal continuum of the dust starts to dominate (e.g. Humphreys and Davidson, 1994).

Most LBVs in the galaxy and in the Magellanic Clouds have been detected by IRAS, however the spatial resolution of that instrument was too low to resolve the nebula. For AG Car McGregor et al. (1988) find extended emission from the nebula at 50 and 100 μm using the KAO. For the LBV candidate G79.29+0.46 (Higgs et al., 1994) IRAS maps at 12, 25 and 60 micron resolving the nebula were published by Waters et al. (1996). Recently some LBVs have been imaged using ground based instrumentation. Voors et al. (1997) discuss TIMMI images of HR Car at 10 and 12.8 μm. Their images show a small (15") clumpy nebula. The most prominent features seen in the 10 micron image are an arclike structure about 3" SE of the star and a bright blob about 1.5" NW of the star. Interestingly enough the ionized matter seen in this image and the optical images (Clampin et al., 1995) show that the dust and ionized gas close to the star is more clumpy than the gas distribution further away. Voors et al. estimate the age of the inner nebula to be around 850 years.

Robberto and Herbst (1998) in a recent paper show IR imaging of the LBV HD168625 at 2.16, 4.7, 10, 11.6 and 20 μm using a new instrument on UKIRT (MAX). These high resolution images (pixel size of 0.265 by 0.265

arcsec) clearly resolve the nebula around this star at both wavelengths. They conlude that the IR emission of this nebula comes from a geometrically thin layer of dust on the outer edges of the lobes seen in the nebula of this star. They also conclude that the dust in the nebula around this star consists of large grains.

Unfortunately all ground based imaging observations of these stars at longer wavelengths are hampered by the presence of the earths atmosphere making the observations of these sometimes faint nebulae very time consuming.

1.3 ISO imaging observations of LBVs

The Infrared Space Observatory (ISO) was an ESA satellite launched in November 1995. It consisted of a 60'cm superfluid helium cooled telescope and a complement of 4 spectroscopic and photometric instruments for the wavelength range from 2 to 200 μm (see Kessler et al., 1996). The satellite worked for more than 2 years, and was finally switched off after the helium ran out in May 1998. Two ISO instruments were used for imaging: The ISOCAM instrument for the wavelength range 2 to 17 μm with a pixel size of 1.5 to 6 arcsec (see Cesarski et al., 1996) and the ISOPHOT instrument for the longer wavelengths (20 to 200 μm) with a resolution of between 10 and 60 arcsec, using a raster scanning mode (see Lemke et al., 1996).

Only a few LBVs have been imaged using ISO. The main reason for this is that the resolution of the instruments was not very high (the diffraction pattern of the telescope was about 8 arcsec wide at 10 μm), so only the largest nebula could be resolved, and also the sensitivity of especially the ISOCAM instrument was so high that the central star of many of these nebulae would completely saturate the detectors, since no coronographic mode was available. Only HR Car, AG Car and G79.29+0.46 have been imaged using ISOCAM. For those observations the Circular Variable Filters (CVFs) of ISOCAM were used as narrow band filters to minimise the risk of saturation by the central star. For G79.29+0.46 mapping using the ISOPHOT instrument was performed as well at longer wavelengths by Wendker and Molthagen (1998).

2 Discussion on individual objects

2.1 HR Car

The observations of HR Car were performed using ISOCAM, with the CVF set at 8 wavelenghts (8.689 μm, 8.993 μm (Ar III), 11.480 μm, 12.410 μm, 12.820 μm (Ne II), 13.530 μm, 15.580 μm (Ne III) and 15.960 μm). Staring mode was used with a 6 arcsec pixel field of view, giving a total field of 180 by 180 arcsec. The observations were performed on February 19, 1996.

The data were processed using the CAM Interactive Analysis (CIA[2]). The nebula is not clearly resolved, although there is a hint for non sphericity towards the SE of the star (corresponding to the arc observed by Voors et al. in the TIMMI images). A spectral energy distibution of the central source and the surrounding area however clearly shows the nebular [Ne II] emission line at 12.82 μm, indicating that the nebula is seen in the images. Also the continuum slope of the surrounding area is different compared to the continuum of the central source (see Trams et al., 1996). Unfortunately at this time no point spread function for the CVF of ISOCAM is available, therefore no deconvolution of the images can be performed to increase the resolution. A proper deconvolution may be able to resolve the nebula in this case.

2.2 AG Car

The observations of AG Car were performed using the same settings as the observations for HR Car. These observations were performed on February 17, 1996. Contour plots of the resulting images are given in Figure 1.

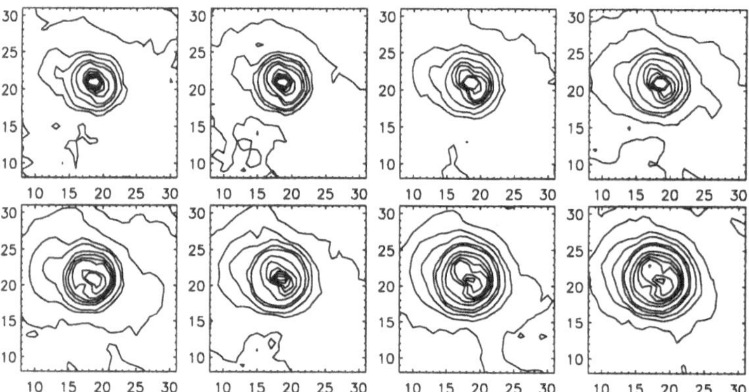

Fig. 1. Contour plots of the ISOCAM images of AG Car. North is to the top and East is on the left. Contours are for 0, 5, 10, 20, 30, 40, 50, 75, 100, 200, 300, 400, 500, 750 and 1000 mJy. The wavelengths of the images are from top left to bottom right: 8.689 μm, 8.993 μm, 11.480 μm, 12.410 μm, 12.820 μm, 13.530 μm, 15.580 μm and 15.960 μm.

The ring nebula is clearly seen in the CAM images. Also the bipolar structure that is clear in the optical images is found in the IR. The nebula

[2] CIA is a joint developement by the ESA Astrophysics Division and the ISOCAM consortium led by the ISOCAM PI, C. Cesarsky, Direction des Sciences de la Materie, C.E.A., France.

is especially strong in the [Ne II] line at 12.82 μm. Interestingly the bipolar structure is strong in the continuum, i.e. in the thermal dust emission. Images of the [Ne II] line emission (with the continuum subtracted) and the PAH emission band (the image at 11.48 μm includes part of this band) show a clear difference in geometry. The ionized gas emission is more ring like, whereas the dust particles are concentrated in the bipolar structure. This is similar to the images shown by Nota et al. (1995), who show a clumpy bipolar nebula in the visual continuum and a smooth ring like structure in the Hα image.

2.3 G79.29+0.46

The nebula around the LBV candidate G79.29+0.46 is about 3 arcmin in diameter (Higgs et al., 1994). This is too large to fit in a single ISOCAM frame. We therefore observed this star using a 3 by 3 map with half a beam step size and 6 arcsec pixels. This gives a total field of view of 360 by 360 arcsec. The images were obtained at ten wavelengths. Apart from the ones mentioned above we also obtained images at 10 and at 17 μm. The images are shown as contour plots in Figure 2.

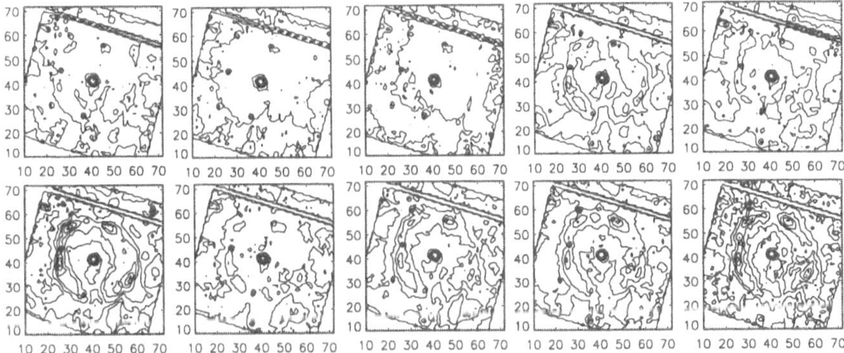

Fig. 2. Contour plots of the ISOCAM images of G79.29+0.46. North is to the top and East is to the left. Contours are for 0, 10, 20, 30, 40, 50, 75, 100, 200 and 500 mJy. The wavelengths of the images are from top left to bottom right: 8.689 μm, 8.993 μm, 10 μm, 11.480 μm, 12.410 μm, 12.820 μm, 13.530 μm, 15.580 μm, 15.960 μm and 17 μm.

The ring nebula is clearly seen in our ISOCAM images. Also some structure in the nebula is seen. The nebula is clumpy with very striking holes in the ring nebula on the north and southern side at all wavelengths. Since these holes are seen at all wavelenghts, they cannot be the result of an ionisation difference, but must be due to a difference in column density. Furthermore comparing the ISO images with larger IRAS maps of this area of the sky, the holes are seen towards the highest density in the Interstellar medium (the object is close to a molecular cloud that shows up as a clear rise in 60 μm flux). This points to the fact that the nebula is probably the result of an interaction between (asymmetric) winds instead of an interaction between a wind and the local ISM.

The [Ne II] line emission image and that of the PAH emission were compared with the radio map presented by Higgs et al. (1994) and the 25 μm map presented by Waters et al. (1996) and Wendker and Molthagen (1998). The [Ne II] emission coincides with the radio (free-free) emission and that the PAH emission coincides with the 25 μm emission (dust). Where the [Ne II] image and the radio image show bright areas, the PAH and 25 μm image show weaker emission. This is consistent with an ionisation of the nebula by the central star, since in that case in the dusty areas the UV radiation is shielded and therefore the ionisation of the nebula would be lower.

In the spectral energy distributions we calculated for this star, we can see that apart from the presence of the [Ne II] line in the nebula, there is also [Ne II] emission on the position of the central star. Furthermore the continuum at longer wavelengths is not stellar, but more nebular in shape (similar to HR Car). We therefore conclude that a small and unresolved nebula is present in G79.29+0.46 very close to the star (probably within 6-10 arcsec radius).

3 A model for G79.29+0.46

We have simulaneously modeled the central star G79.29+0.46 and its dusty ring nebula. The low surface gravities and dense winds of LBV stars are expected to yield a significant increase of the stellar radius – starting at near IR wavelengths – due to free-free processes. We have modeled these processes in a consistent manner, using the most recent version of the non-LTE model atmosphere code ISA-Wind of de Koter et al. (1993,1997). This code treats the photosphere and stellar wind in a unified manner, i.e. it makes no artificial separation as is done in core-halo approaches. The adopted stellar parameters are $L = 3 \times 10^5\ L_\odot$, placing the star at 1800 pc, and a core radius of 45 $R_\odot$. To fit the slope of the near IR continuum up to ± 10 μm a mass loss of $1.3 \times 10^5\ M_\odot$/yr was required. The terminal velocity was set at 100 km/sec, consistent with the width of Hα (see Voors et al., 1999, in preparation). This model yields a peak strength in Hα about 30 times continuum, in fair agreement with observations. The slope of the velocity law was set at $\beta = 1$. Indeed, these parameters imply a wind that is so dense that at a thermalization optical depth $1/\sqrt{3}$ at the center of the V band (see de Koter et al., 1996) the flow velocity is already 6 km/sec. In other words, the stellar photosphere is formed in the wind – G79 shows a so-called 'pseudo photosphere'. The resulting stellar radius is 65 $R_\odot$, yielding a $T_{\rm eff} = 16.6$ kK.

The observed and predicted energy distributions are presented in Figure 3. For the fits of the energy distribution we also used the ISO SWS spectrum from Morris et al. (1998, in preparation). The model fits both the optical and IR continuum very well, when assuming an extinction E(B-V) = 3.9. Also the absorption at 10 μm, due to interstellar amorphous silicates, is in good agreement. Note the absorption at 3 μm and the broad absorption between $\pm$6 and 8 μm are not reproduced by the model. These are likely due to water ices. At 4 μm there is also absorption due to CO_2 ices. These ices are probably located in the foreground radio source DR15.

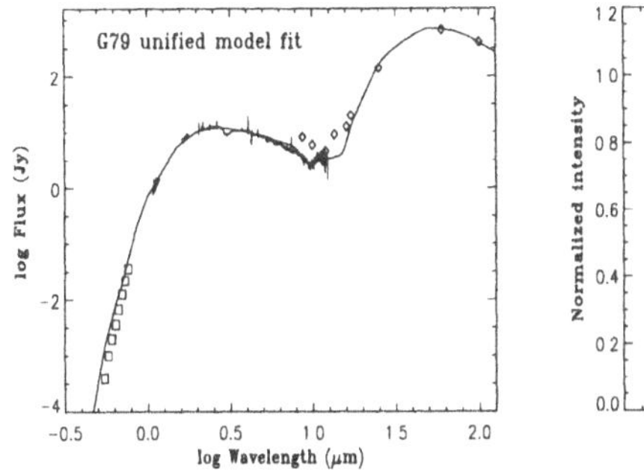

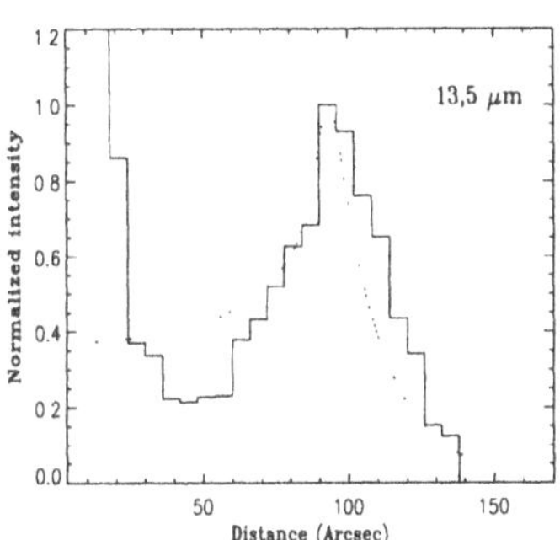

Fig. 3. Left: observed and predicted energy distribution from our model of the stellar atmosphere and the nebular dust shell. Right: Observed and fitted azimuthal integrated intensity profile for the G79.29+0.46 nebula.

The circumstellar material has been modeled assuming an optically thin dust shell, irradiated by the unified model. We have assumed the material to be composed of amorphous olivine grains. The grains are distributed in size from 0.01 to 0.1 micron following a power law distribution with index m = -3.5. We require this model to fit *both* the energy distribution as well as the azimuthally integrated intensity profile as obtained from ISOCAM at continuum wavelengths 8.689, 10.0, 11.480, 12.410, 13.530, 15.960 and 17.0 μm. This last constraint yields a firm grip on the geometrical extension of the dust shell. We find an inner and outer radius of the dust of 2.48×10^{18} and 1.10×10^{19} cm respectively. The density distribution in the shell is assumed to be proportional to r^{-2}. The dust temperatures range between T=61 (large grains) and 64 K (small grains) at the inner boundary to T=38 to 40 K at the maximum radius.

The model that fits the ISOCAM & IRAS points best contains a dust mass of 0.15 $M_\odot$. Assuming a dust to gas ratio of 0.01, the nebula contains 15 $M_\odot$ of material. The adopted stellar luminosity implies that G79 would have started out the main sequence as a 40 $M_\odot$ star. Evolutionairy tracks of Meynet et al. (1994) show that when such a star leaves the main sequence and starts its secular redward motion it first reaches $T_{eff} = 20$ kK when it has lost ± 7 $M_\odot$. After a brief RSG phase, the star again increases in temperature as progressively deeper layers are exposed due to mass loss. As it now passes $T_{eff} = 20$ kK on its blueward motion, the star has already lost ± 21 $M_\odot$. Near and during the RSG phase the star therefore looses some 14 $M_\odot$. Taking the derived shell mass at face value, this would imply G79 is most likely a post-RSG star. The star will likely evolve into a WNL star.

The azimuthally integrated ISOCAM images also allow one to derive information on the geometry of the dust region. ISOCAM images show the dust region is fairly circular in shape. So, the dust may be in a spherical shell or in a more flattened disk-shape geometry, which we view pole on. Figure 3 shows the radial intensity profile at 13.5 μm. The observations are represented by the solid line. The dotted line gives the radial intensity of our spherical symmetric model. The ISOCAM resolution of 6" at 15 μm convolves the intensity of the central star out to several tens of arcsec. In the model this part of the intensity profile is excluded. The observations peak at 90", which corresponds to the distance of the inner radius of the shell, where the density, temperature and line of sight through the dust region are largest. At smaller distances from the core, the model intensity drops of to twice the emission of a radial column. If the observed emission drops below this limit - as is the case in G79 - this means that the dust distribution can not be spherical. The dust must be concentrated towards the equatorial plane, i.e. the geometry should be disk-shaped. We are currently investigating the precise geometrical distribution of the dust. Beyond 90", the intensity drops as the line of sight through the shell/disk decreases. The reasonable fit of our model to this part of the intensity profile suggests that in the inner radial segment of the dust shell (from 90" to 140") the density distribution is reasonably well represented by the adopted r^{-2} dependence.

References

Cesarsky, C.J., Abergel, A., Agnese, P., et al. (1996): A&A **315**, L32

Clampin, M., Schulte-Ladbeck, R.E., Nota, A., Robberto, M., Paresce, F., Clayton, G.C. (1995): AJ **110**, 251

de Koter, A., Schmutz, W., Lamers, H.J.G.L.M. (1993): A&A **277**, 561

de Koter, A., Lamers, H.J.G.L.M., Schmutz, W. (1996): A&A **306**, 501

de Koter, A., Heap, S.R., Hubeny, I. (1997): ApJ **477**, 792

Higgs, L.A., Wendker, H.J., Landecker, T.L. (1994): A&A **291**, 295

Humphreys, R.M., Davidson, K. (1994): PASP **106**, 1025

Hutsemékers, D. (1997): in "Luminous Blue Variables: Massive stars in Transition", eds. Nota and Lamers, ASP Conf. Ser. 120, p.316

Hutsemékers, D., Van Drom, E., Gosset, E., Melnick, J. (1994): A&A **290**, 906

Kessler, M.F., Steinz, J.A., Anderegg, M.E. et al. (1996): A&A **315**, L27

Lemke, D., Klaas, U., Abolins, J. et al. (1996): A&A **315**, L64

McGregor, P.J., Finlayson, K., Hyland, A.R., Joy, M., Harvey, P.M., Lester, D.F. (1988): ApJ **329**, 874

Meynet, G., Maeder, A., Schaller, D., Schaerer, D., Charbonnel, C. (1994): A&AS **103**, 97
Nota, A., Livio, M., Clampin, M. (1995): ApJ **448**, 788
Nota, A., Clampin, M. (1997): in "Luminous Blue Variables: Massive stars in Transition", eds. Nota and Lamers, ASP Conf. Ser. 120, p.303
Robberto, M., Herbst, T.M (1998): ApJ **498**, 400
Smith, L.J. (1994): in "Circumstellar media in late stages of stellar evolution", eds. Clegg et al., Cambridge University Press, p.64
Smith, L.J., Stroud, M.P., Esteban, C., Vilchez, J.M. (1997): MNRAS **290**, 265
Trams, N.R., Waters, L.B.F.M., Voors, R.H.M. (1996): A&A **315**, L213
Viotti, R., Cassatella, A., Ponz, D., Thé, P.S. (1988): A&A **190**, 333
Voors, R.H.M., Waters, L.B.F.M., Trams, N.R., Käufl, H.U. (1997): A&A **321**, L21
Waters, L.B.F.M., Izumiura, H., Zaal, P.A., Geballe, T.R., Kester, D.J.M., Bontekoe, Tj.R. (1996): A&A **313**, 866
Waters, L.B.F.M., Morris, P.W., Voors, R.H.M., Lamers, H.J.G.L.M., Trams, N.R. (1998): Ap&SS **255**, 179
Wendker, H.J., Molthagen, K. (1998): Ap&SS **255**, 187

Discussion

G. Koenigsberger: In those cases where the nebula has a "horseshoe" shape, do you have an estimate for the density contrast required so that there is greatly reduced emission in certain portions of the shell?
N. Trams: The density contrast required would be similar to the contrast required to explain the empty central part of the nebula. This is rather high (1/10 to 1/100). The interesting thing in G 79 is that the highest density ISM is on the side of the gap in the nebula, therefore excluding the possibility of a nebula due to swept-up ISM.

M.-M. MacLow: Couldn't a thin shell seen in projection (e.g., with $\Delta R/R \sim 12$, appropriate for an adiabatic shock) produce an observed circular shell with an apparently empty interior?
N. Trams: We have just started doing model calculations for the shell seen in G 79. The first calculations show that a thin spherical shell with a thickness of about 1/10 of its radius still produces significant emission in the central area.

Radio Evidence for Non-isotropic Outflows from Hot Stars

Stephen White

Dept. of Astronomy, University of Maryland,
College Park, MD 20742

Abstract. Radio diagnostics for outflows from hot stars are reviewed and three well–studied examples (WR 140, η Car and MWC 349) are discussed in some detail. The three systems are very different and their radio properties are diverse, but in each case, the radio data provide strong evidence for the presence of equatorially–enhanced outflows.

1 Introduction

Radio observations provide a powerful tool for the study of hot stars. Radio emission can penetrate the clouds of dust which tend to obscure the locations of massive stars, and are capable of high spatial resolution. On the other hand, there are some notable limitations to what radio observations can do. We first discuss the general properties of radio emission from hot stars and then discuss three examples in more detail.

Two very different emission mechanisms are responsible for radio emission from hot stars, and both arise in their outflows. The most common is *thermal free–free* or *bremsstrahlung*, which is responsible for the classic stellar–wind radio emission. The opacity for this mechanism varies as $n_e^2\,T^{-1.5}\,\nu^{-2}$, where n_e is the electron density, T the electron temperature and ν the radio frequency. The fact that the opacity decreases as frequency increases, while density in an outflow decreases with radius, leads to a fundamental property of free–free radio emission from stellar winds: higher frequencies probe deeper into the stellar wind. For a constant–velocity wind ($n_e \propto r^{-2}$) the radius of the optically thick surface, which limits how deeply we can see, scales with frequency as $\nu^{-0.7}$. The combination of this scaling of optically–thick source dimension with frequency and the ν^{+2}–dependence of the black–body emission law produces the classic $\nu^{0.6}$ spectrum of a constant–velocity free–free–emitting stellar wind (Olnon 1975, Panagia & Felli 1975, Wright & Barlow 1975). This law does not require a spherically–homogeneous outflow: the outflow can be asymmetric and maintain the $\nu^{0.6}$ spectrum as long as wind motion is radial and the temperature and ionization state remain unchanged with radius (Schmid-Burgk 1982).

The second mechanism is *nonthermal synchrotron emission*, more familiar from supernova remnants and quasars. It is important in hot stars because of the facility with which shocks can form in the powerful winds: the characteristic speed of a hot star wind is generally in excess of the ambient sound speed in the wind, so that any significant velocity fluctuation, such as a fast knot overtaking a slower one or two winds colliding, has the potential to result in a shock. Once a shock forms, electron acceleration apparently takes place (by an as yet not completely understood mechanism). A high Mach–number shock produces a characteristic power–law energy distribution of spectral index -2, which results in a $\nu^{-0.5}$ radio flux spectrum (in the optically–thin limit believed appropriate to these sources). A detailed model for nonthermal emission from a single–star wind carrying random shocks was worked out by Rick White (1985). Note that the acceleration must take place at some distance from the star: the powerful radiation field of a hot star can quench shock acceleration close to the star where the inverse–Compton mechanism depletes energy from a high–energy electron faster than a shock can supply it (Chen & White 1994). This is a mild problem for models in which the magnetic field in the synchrotron source is a stellar field carried out by the wind, since it will diminish with distance from the star. An alternative is for the magnetic field also to be generated in the shocks as a byproduct of the plasma turbulence there.

2 MWC 349

MWC 349 is one of the most extreme galactic members of the class of luminous B stars showing forbidden emission lines, known as B[e] stars. These stars are characterized by very strong Balmer lines, forbidden FeII lines and very strong emission from hot dust at infrared wavelengths. These features are assumed to derive from the combination of a very dense equatorial disk, which provides the Balmer emission and the dust, together with a powerful ionized polar outflow which is responsible for the forbidden lines (e.g., see the review by Conti 1997).

MWC 349's radio spectral index of +0.6 was the prime evidence for the correctness of the stellar–wind outflow interpretation of the radio emission. However, the original model of a spherically–symmetric outflow had to be abandoned when VLA observations showed remarkable structure in the radio emission (Cohen et al. 1985, White & Becker 1985). A high–spatial–resolution image of the 8 GHz radio continuum is shown in Fig. 1. It is seen to break up into two sets of curved "horns" on either side of the star with much weaker emission between. The peak brightness temperature in the horns is 10^4 K, while it drops to 4400 K at the inferred position of the star. The interpretation of this image is that the outflow from MWC 349 has low velocity but high density in the equatorial plane (the low–brightness region between the horns), and higher velocity but lower density in the polar directions (the directions of the horns). The density is so high in the equatorial plane that the gas there has completely recombined and is neutral, while in the polar directions

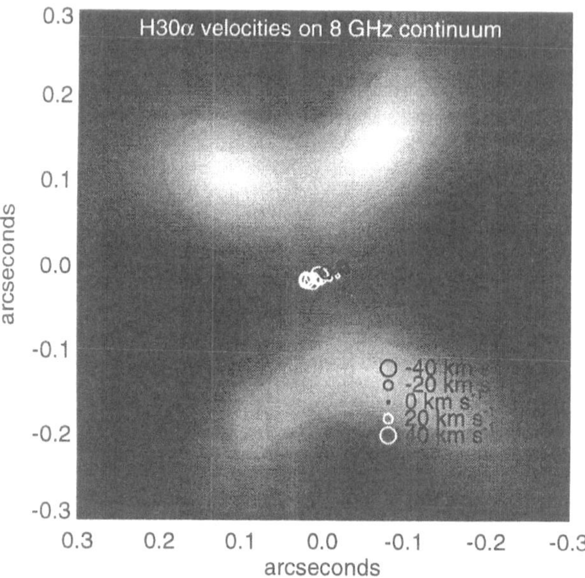

Fig. 1. The positions of different velocities in the masing 232 GHz H30α recombination line (White et al. 1998) plotted over a greyscale display of the 8 GHz VLA continuum image of MWC 349 (data from Rodríguez & Bastian 1994), showing the relationship of the line emission in the disk to the continuum emission from the polar outflow. Note that the bulk of the material in the equatorial plane is believed to be neutral and hence produces little microwave emission, whereas the polar outflow is believed to be ionized.

the gas is still ionized and thus can be optically thick at the temperature characteristic of a photo–ionized wind.

MWC 349 is also famous among radio astronomers as the first source to show masing (and recently, lasing) recombination lines (Martín–Pintado et al. 1989, Strelnitski et al. 1996). It has long been recognized that population inversion occurs in Hydrogen level populations in photoionized sources, due to the fact that a recombining electron generally re–enters the energy–level ladder at the top, and the rate of transitions increases as it drops down the ladder to lower energy levels. However, population inversion sufficient to produce a significant amplification was first seen in MWC 349, and seems to occur there for all $\Delta n = 1$ transitions with $n < 36$ ($f > 135$ GHz). At 232 GHz the H30α line shows two very strong features at +32 km s^{-1} and -16 km s^{-1}, and Fig. 1 also shows the positions of different velocities within this recombination line, from a recent observation with the BIMA array (White et al. 1998). These data confirm that the masing recom line arises in the equatorial plane, not in the bipolar horns where the continuum emission is strongest. The position–velocity diagram across the disk is consistent with the masing emission arising at a constant–radius surface, presumably where the continuum becomes optically thick. Such masing lines promise to be a valuable diagnostic for hot star winds in the future: they are strongest at millimeter and far infra–red wavelengths where dust obscuration is low and high spatial resolution can be achieved, and the line data can be used to derive kinematic information for stellar outflows.

3 WR 140 = HD 193793

This remarkable system demonstrates that a wealth of information on the outflows from hot stars can be gained even from radio data in from a system which is spatially unresolved. WR 140 consists of a (secondary) WC7 Wolf–Rayet (WR) star and an O4-5 star (primary) in a highly elliptical orbit with a period of 7.9 years. Early radio surveys of WR stars revealed that WR 140 was a strong and interesting source with a nonthermal spectrum, and subsequent observations have only reinforced this view (e.g., Becker & White 1985, Williams et al. 1990, White & Becker 1995). However, at a distance of 1.3 kpc the major axis of the orbit, 29 AU, subtends just 2 millarcseconds, which is barely feasible with VLBI techniques, and to date only radio light curves are available.

White & Becker (1995) monitored the radio emission from WR 140 at three frequencies (1.4, 5 and 15 GHz) with the VLA every month for 8 years. The light curves show a wealth of structure. The basic model is that the nonthermal emission arises due to synchrotron emission from energetic electrons accelerated in the shock where the powerful winds from the two stars collide. There is also some free–free emission from the winds, and the nonthermal radio emission is modulated by free–free absorption as the location of the shock changes with respect to the winds. The shock strengthens near periastron and weakens near apastron. The orientation of the orbit is such that the O star is closest to us during most of the orbit; the WC7 star passes between us and the O star just after periastron, at phase 0.01. The O star wind has a terminal velocity of 3100 km s^{-1} with $\dot{M}$ = 1.8 $\times$ 10^{-6} M$_\odot$ yr^{-1}, while the WR wind has a velocity of 2600 km s^{-1} with $\dot{M}$ = 6 $\times$ 10^{-5} M$_\odot$ yr^{-1}. In principle, to model this system one needs to know all the physical propoerties of the binary orbit and the winds, and understand the shock physics and its variation with stellar separation, electron acceleration at the shock, the source of magnetic fields and absorption in the wind; conversely, all these phenomena affect the radio light curves.

The light curves are plotted in Figure 2. Free–free absorption should be largest at the lowest frequency and the 1.5 GHz light curve is consistent with this expectation: it shows no significant flux except for a narrow peak near phase 0.8. At 5 GHz the peak is much broader, with only weak emission just after phase zero when the strong wind of the WR star is between us and the wind–interaction region. The 15 GHz emission, which should suffer the least

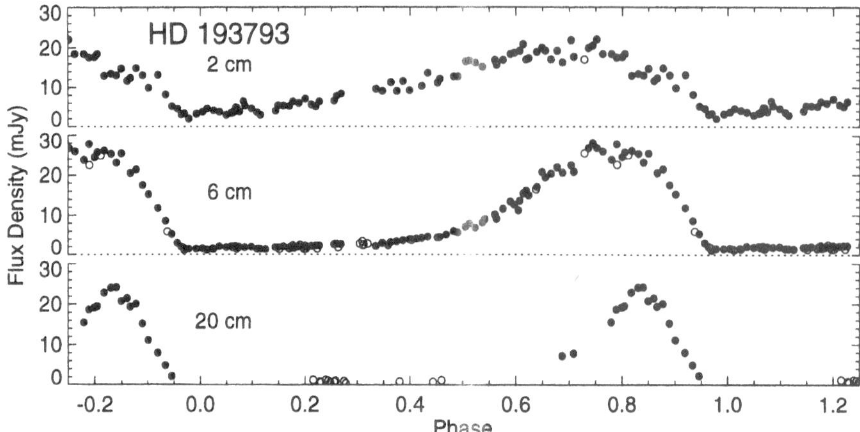

Fig. 2. The radio light curve for WR 140 plotted against orbital phase over the 7.9–year orbit at each of three wavelengths. Filled circles are VLA data taken every month by White & Becker (1995); open circles represent other data compiled by Williams et al. (1990) covering more than one cycle, indicating that the light curve repeats almost perfectly from one cycle to the next. (Figure provided by Rick White.)

absorption and therefore best represent the intrinsic emission profile, peaks at phase 0.70, somewhat earlier than the 1.5 GHz peak: between phases 0.70 and 0.85 the 15 GHz flux drops while the 1.5 GHz emission rises. From phase 0 to 0.5, the spectral index is positive, consistent with thermal free–free emission.

The fact that there is any 1.5 GHz emission detected at all is difficult to explain, since the parameters for the WR wind imply that it should be optically thick at 1.5 GHz well outside the binary orbit. Williams et al. (1990) showed that a model with spherically–symmetric winds could not explain the light curves and argued that the shadow of the O star creates a cavity in the WR wind where the opacity is greatly reduced. White & Becker (1995) argued that even the O star wind will be optically thick and absorb any 20 cm emission behind it; they suggest instead that the WR wind is confined to an equatorial plane inclined with respect to the orbital plane. In this geometry the WR wind sweeps across the position of the O star twice per orbit; one of these occasions is presumably at phase 0.70, where the radio flux is maximum. Thus in this system an equatorially–enhanced wind for the WR star is not resolved directly but is inferred from the light curves.

4 η Carinae

η Carinae as a radio source shows an interesting amalgam of the properties of the two previous examples. Like MWC 349, the radio emission from η Car is entirely thermal free–free emission (ignoring some nonthermal emission from

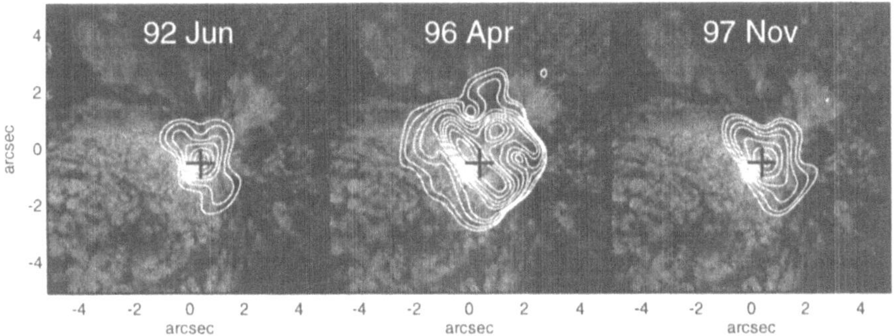

Fig. 3. Contours of 8 GHz radio emission overlaid on the HST image of the core of the η Carina nebula, at three different epochs. The resolution of the radio data is of order 0.3″. The bipolar lobes of the Homunculus nebula are seen in the lower–left and upper–right corners; the "equatorial skirt" lies just above the diagonal running from the upper left corner to the bottom right corner.

features in the outer nebula) but like WR 140 the emission is apparently strongly influenced by the binary nature of the system.

Figure 3 shows how the spatial distribution of radio emission from η Car evolves over the 5.5 year cycle discovered by Damineli (1996). The period is believed to represent the orbital period of a binary system consisting of an LBV star and an O star in a highly eccentric orbit, with our line of sight, when projected onto the orbital plane, coinciding almost exactly with the major axis of the orbit (Damineli, Conti & Lopes 1997, Davidson 1997). The radio light curve shows a large variation over the 5.5–year cycle, from 0.7 Jy at 8.8 GHz in 1992 June (corresponding to periastron for the binary, i.e., the time of minimum excitation in the optical spectrum noted by Damineli) to 3 Jy in 1995 (near apastron) and back down to 0.9 Jy in January 1998 (periastron again).

The images show that the changes in the radio flux are due to the appearance of extended emission which is not present at periastron (White et al. 1994, Duncan et al. 1995, Duncan, White & Lim 1997). The radio spectrum of all of this emission is consistent with thermal free–free emission: the brightest sources have brightness temperatures of 10^4 K, and the weaker extended emission has the flat spectrum expected of optically–thin free–free emission. In addition, we observe recombination–line emission from the extended emission, implying photo–ionized gas. The implication is that all of this additional emission comes from gas which is neutral when the binary is at periastron, and ionized when the system is near apastron.

The location and kinematics of the gas indicate its nature: it lies almost entirely to the north–west of the star, and the recombination line data indicate that all the gas is blue–shifted with respect to the systemic velocity. The lobe of the Homunculus nebula which is projected to the north–west of the star is known to be receding from us, so the velocity of the radio-emitting gas is inconsistent with a location there: instead, it must be in the equatorial skirt. Material in the skirt which appears projected to the north–west of the star should indeed be moving towards us. We argue that the gas in the outflow from η Car (estimated to be 10^{-3} $M_\odot$ yr^{-1} at 500 km s^{-1}) is so dense that it recombines quite close to the star(s), and hence most of the gas surrounding the system is neutral. The ionizing photons from the LBV star never penetrate very far through its massive outflow. Near periastron the hot companion star is also enveloped in the outflow of the LBV star, but away from periastron the LBV is on the far side of the O star and ionizing photons from the O star can reach gas in the equatorial skirt in our direction, ionizing it and causing it to produce radio emission. When periastron approaches the LBV wind again cuts off the ionizing flux from the companion and the neutral gas recombines and vanishes from the radio images.

References

Becker, R. H., & White, R. L. 1985, ApJ, **297**, 649.
Chen, W., & White, R. L. 1994, Ap&SS, **221**, 259.
Cohen, M., Bieging, J. H., Dreher, J. W., & Welch, W. J. 1985, ApJ, **292**, 249.
Conti, P. S. 1997, in Luminous Blue Variables: Massive Stars in Transition, ed. A. Nota & H. J. G. L. M. Lamers (San Francisco: Astron. Soc. Pac.), p. 161.
Damineli, A. 1996, ApJL, **460**, L49.
Damineli, A., Conti, P. S., & Lopes, D. F. 1997, New Astron., **2**, 107.
Davidson, K. 1997, New Astron., **2**, 387.
Duncan, R. A., White, S. M., & Lim, J. 1997, MNRAS, **290**, 680.
Duncan, R. A., White, S. M., Lim, J., Nelson, G. J., Drake, S. A., & Kundu, M. R. 1995, ApJL, **441**, L73.
Martín-Pintado, J., Bachiller, R., Thum, C., & Walmsley, C. M. 1989, A&A, **215**, L13.
Olnon, F. M. 1975, A&A, **39**, 217.
Panagia, N., & Felli, M. 1975, A&A, **39**, 1.
Rodríguez, L. F., & Bastian, T. S. 1994, ApJ, **428**, 324.
Schmid-Burgk, J. 1982, A&A, **108**, 169.
Strelnitski, V. S., Haas, M. R., Smith, H. A., Erickson, E. F., Colgan, S. W. J., & Hollenbach, D. J. 1996, Science, **272**, 1459.
White, R. L. 1985, ApJ, **289**, 698.
White, R. L., & Becker, R. H. 1985, ApJ, **297**, 677.
White, R. L., & Becker, R. H. 1995, ApJ, **451**, 352.
White, S. M., Duncan, R. A., Lim, J., Nelson, G. J., Drake, S. A., & Kundu, M. R. 1994, ApJ, **429**, 380.
White, S. M., Welch, W. J., Vogel, S. N., & Lim, J. 1998, ApJ, in preparation.
Williams, P. M., van der Hucht, K. A., Pollock, A. M. T., Florkowski, D. R., van der Woerd, H., & Wamsteker, W. M. 1990, MNRAS, **243**, 662.
Wright, A. E., & Barlow, M. J. 1975, MNRAS, **170**, 41.

Optical Interferometry of Non-spherical Winds

Farrokh Vakili, Denis Mourard, Philippe Stee, and Daniel Bonneau

Observatoire de la Côte d'Azur, Fresnel-GI2T, F-06460 Caussols, France

Abstract. Optical Long Baseline Interferometry (OLBI) combined with spectral resolution is capable to sounding the stratification of luminous star winds. In this paper we review some few, yet unique results obtained on P Cyg and Be stars from OLBI observations. We also examine the prospects for studying hot stars winds offered by optical synthesis arrays presently under construction.

1 Introduction

High angular resolution has always opened up new vistas in astronomy and this is particularly true for studying the physics of hot luminous stars and their winds. While the HST has sharpened our view of the nebulosities around LBV's, WR and other exotic objects (see the superb results presented during this colloquium), zooming into the immediate surroundings of their central engines will need sub-mas spatial resolution only achievable by means of OLBI. Hopefully, the planned operation of more than 10 interferometers within 10 years, working at both the visible and IR wavelengths (for more details see SPIE proceedings 1998) will offer the tools to improving our understanding of the mechanisms producing massive stars photometric, spectrometric and polarimetric variability.

2 Spectrally resolved interferometry: how it works?

Michelson interferometers recombine the stellar light from widely separated telecopes in a central focal plane where the estimate of the so-called fringe visibility informs about the angular extent of the source in the direction of the interferometric baseline. The fringes are classically analysed through a chromatic filter which has the property to average the angular extent of the source over the filter bandpass. In this process and in the case of a 2 telescope interferometer any information about the position and asymmetry of the celestial source is lost unless one can record simultaneously the fringe signal from a pointlike reference source within the interferometric field of view. It was recognized in early days of modern OLBI (Labeyrie 1980) that measuring the fringes of an emission line star through a spectrograph is similar to having such a reference provided the star is unresolved at the continuum wavelengths. In this way it becomes possible to derive both the size and position of the emitting regions around a star as a function of Doppler-shift accross its spectral lines. We call this technique Interferometric Doppler Imaging (IDI) by analogy to classical spectroscopic Doppler-imaging where the temporal changes of line profiles are replaced by relative spatial positions obtained from the interferometer. For technical aspects of IDI we invite the reader to see Vakili et al. 1997 and recall that IDI can achieve sub-mas super-resolution on the kinematics of hot star winds with optical interferometers using decametric baselines.

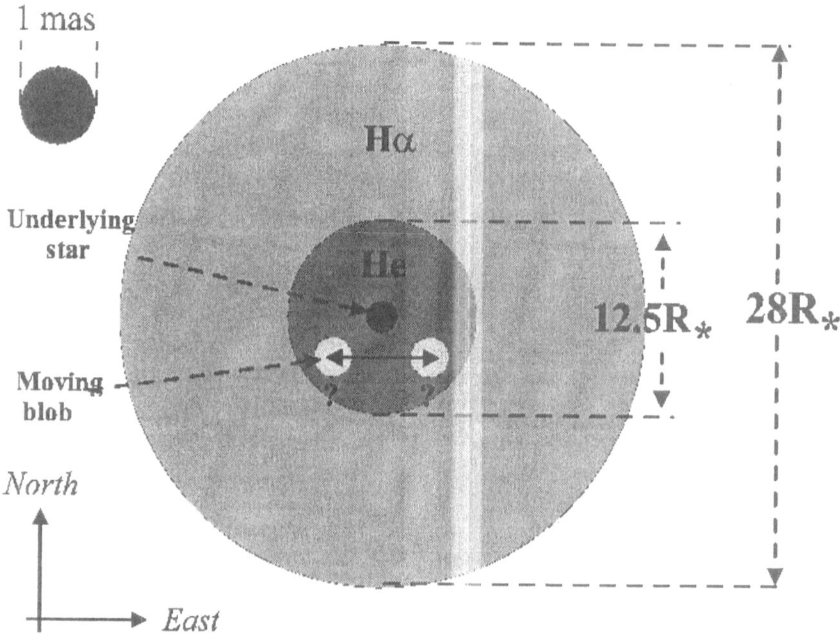

Fig. 1. Schematic morphology of P Cyg according to GI2T spectrally resolved interferometry in '93. The Hα and HeI6678 emitting envelopes given in units of stellar radius ($76 \pm 15 R_\odot$ corresponding to 0.4 mas for a distance of 1.8 kpc). The projected position of the Hα emitting blob detected by the GI2T is at 4 stellar radii to the south of the central star BUT its E.-W. position (noted by ?) is unknown.

3 Asymmetry in the wind of P Cyg

The prototype LBV star P Cyg constitutes a favorite target for spectrally resolved OLBI. Following earlier attempts, the GI2T-team was able to successfully observe P Cyg in August,1994 in both Hα and HeI6678 lines with an effective spectral resolution of 1.7 A. Based upon robust calibrations on a number of reference stars, Vakili et al. 1997, estimated the equivalent uniform disk diameter of P Cyg in Hα as 5.52 $\pm$ 0.47 mas corresponding to 28 stellar radii at a distance of 1.8 kpc. By sampling the interferometric signal across Hα using spectral windows of 3.4 A wide, starting from the blue and finishing in the red wings of the line, an asymetrical structure was discovered at RV $= -208$ km s^{-1}. The north-south projected position of this structure, presumably a bright blob in P Cyg's wind, was found at 4 stellar radii to the south of the central star (Fig. 1). Vakili et al. gave also a rough estimate of the HeI6678 emitting envelope as 12.5 stellar radii on the order of the distance of the detected blob and argued about possible relation of asymetries at the basis of P Cyg wind and the fine structures at larger distance from the central star detected by HST coronography (Nota et al. 1995).

4 Interferometry of Be star disks

OLBI observations of Be star disks have been made so far by 3 different interferometers (see Vakili et al. 1994 and references therein). Following earlier observations from the I2T and GI2T interferometers in the southern France (Thom et al. 1986, Mourard et al. 1989) Quirrenbach and colleagues carried a systematic survey of the brightest northern hemisphere Be stars up to the 4th visual magnitude (see Stefl's review in these proceedings). The survey was made using earth-rotation synthesis possibilities offered by the Mark III interferometer at Mount Wilson (Quirrenbach et al. 1997). These remarkable observations proved that the extent of Be star envelopes exceed tens of stellar radii and are indeed flattened disks with an excellent correlation of their elongation positionnal angle and spectropolarimetric measurements.
The Mark III group gave the extent of Be star disks through a 1nm filter which was narrower than their Hα emission for a number of them, thus masking their envelope highest RV regions. Also no information was given on the relative position and extent of emitting iso-radial velocity regions inside the envelope. Thus, despite its excellent sensitivity, Mark III could not determine any departure from azimuthal symetry of Be star equatorial disks. This shortcoming can be particularly misguiding in the case of binary Be stars (ϕ Per) or those objects which exhibit indirect evidence for azimuthal asymetry (ζ Tau, see below) or co-rotating structures inferred from X-ray and UV spectroscopy (Smith et al. 1998). Despite this, Mark III observations provide a solid ground upon which future interferometric work aiming to map Be star disks by means of aperture synthesis imaging can be built.

Based on a radiative transfer model, more recent studies of γ Cas in Hα (Stee et al. 1995), Hβ and HeI6678 (Stee et al. '98) give the most complete picture of γ Cas envelope constrained from GI2T spectrally resolved observations (Stee et al. these proceedings).

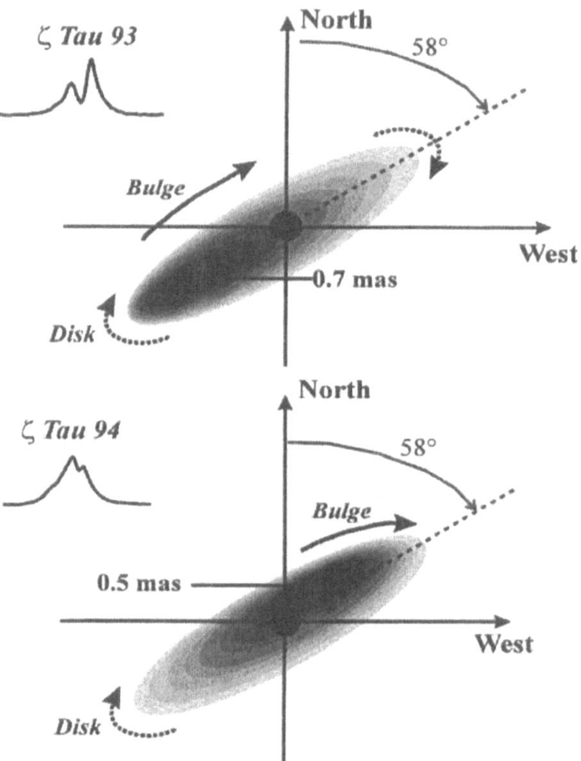

Fig. 2. ζ Tau's long term variability representation according to GI2T. In November '93 Hα has a VR=.57 profile and the extra R emission in the disk originates in a region whose N-S projected position is at 0.7 mas South of the central star. In October '94 VR=1.26 and this region has a N- projected position at 0.5 mas North of the central star. The elongation angle of the envelope projected on the sky is at 58°West from the North. The curved arrows marked *Disk* and *Bulge* depict the absolute direction of rotation of ζ Tau's disk and the density wave inside this disk (prograde motion).

4.1 Polarimetry AND interferometry of γ Cas

In principle there is no obstacle to equip the focal instrument of an interferometer by polarimetric optics. Added to spectral resolution, a polarimeter

enables OLBI to study the extent and position of emitting regions of the wind as a function of wavelength AND of polarization, either linear or circular. Pilote experiments of this kind on the GI2T (Rousselet-Perraut et al. '97) have been carried on γ Cas in '94 using quite simple polarizing filters. Although these observations failed to detect any noticeable effect between linear polarizations of this star, they determined its angular size in natural light at λ=660nm and $\Delta\lambda$=3.4nm) as 2.9 ± 0.5mas. This extent is somehow larger than the value usually admitted in the litterature (Stee et al. 1995). If confirmed by future interferometric observations they it would mean that the continuum emission of the envelope (free-free, free-bound, scattering) would be larger than classically thought (Waters et al. '92).
The difficulty of spectro-polarimetric interferometry is due to their optical complexity which demand numerous and stringent calibrations of internal effects in order to attain the desired and unbiased accuracies to determine any polarization effect from stellar origines. This difficulty is somehow counterbalanced by the interferometric angular resolution of local polarized structures of the star's wind otherwise averaged over its extent when using classical spectropolarimetry at low spatial resolution. To better understand this property, one can think of the special case of a dipolar magnetic star having identical rotation and magnetic axes perpendicular to observer's line of sight. No Zeeman signature could be observed by spectropolarimetry whilst an interferometric signal could in principle be detected from the different spatial locations of σ_+ and σ_- components of a the star across its spectral lines.

4.2 The case of the Be eclipsing binary β Lyr

The Be star β Lyr, also known for its strong radio emission, is a puzzling eclipsing binary (B0V+B6IIp) whose resolution constitutes a challinging problem to optical interferometers. In a coordinated multi-site/technique campaign organized on '94, the GI2T-team (Harmanec et al. 1996) could find the existence of jet-like structure perpendicular to the accretion disk around the B0V component. This conclusion came from the fact that, although the β Lyr complex remained unresolved as a binary system by the GI2T, its Hα flow was partially resolved in the north-south direction without any net dependence on the orbital phase.
In a recent paper the University of Winconsin team (Hoffman et al. 1998) reported the analysis of UV and visual spectropolarimetric data on β Lyr between '92 and '94. They found the continuum polarization to be at 90° of the Balmer lines polarization which remained constant whatever the orbital phase thus confirming the bipolar nature of the Hα emission discovered by the GI2T-team. The complex morphological picture of β Lyr derived from these observations predicts significantly different spatial locations of the emitting regions in the continuum and spectral lines AND as a function of polarization and orbital phase. Indeed these locations could be determined from IDI

observations on the GI2T which has a Wollaston-based polarimeter coupled to its spectrograph (Mourard et al. 1998).

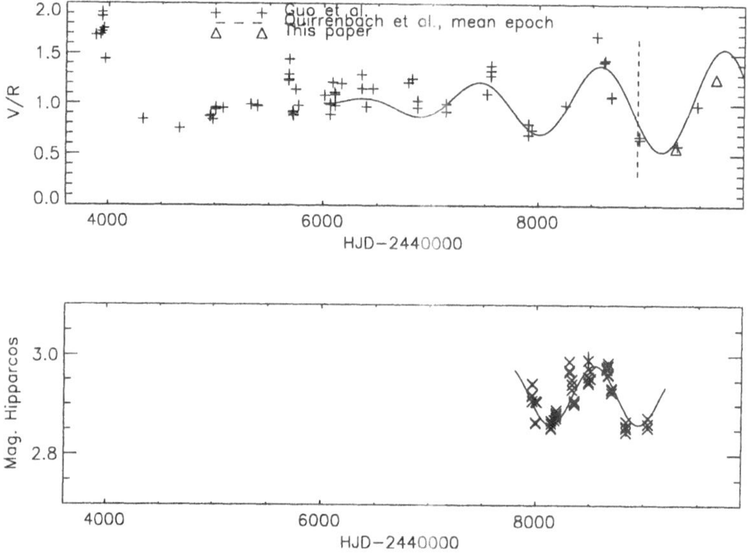

Fig. 3. Top: long term Hα V/R variability of ζ Tau. The continuous line corresponds to the fit of an amplifying sine wave after 1988. Bottom: photometric data from the Hipparcos mission. Note the relatively good correlation between the V/R and photometric cyclicities.

4.3 Interferometry and long term Be variability

One of the long standing problems of Be stars (Telting et al. 1994) has been their observed cyclic V/R variability (V/R: Violet to Red intensity ratio of spectral emission profile wings). The V/R cycles are on the order of a few years, rarely exceed ten years and can be followed/preceeded by active periods where the Be star presents LBV-like eruptions (Marlborough 1997). Plausible scenarios accounting for V/R cyclicity have only recently emerged from theoretical studies of global oscillations of Be star equatorial disks (Papaloizou 1992, Okazaki 1997). Following our earlier suggestion (Vakili et al. 1994) to check such theories against OLBI observations and using GI2T interferometry of ζ Tau on November '93 and October '94, we determined the apparent motion of a prograde density wave in the equatorial disk of this Be shell star (Fig. 2). For our epochs of observation ζ Tau presented Hα V/R values of 0.57 and 1.26 (derived from the GI2T's spectrograph itself) which

nicely agreed with those of Guo and co-workers' spectroscopy until '93 (Guo et al. 1995). Using these data, we derived a 3.1 years cycle for ζ Tau's V/R variability in the range of the predicted values by the one-armed oscillation theory. We used the 58° elongation angle of ζ Tau given by the Mark III interferometer (Quirrenbach et al. 1997) and phasing our interferometric data of '93 and '94 with the V/R cyclicity, determined the absolute positions of the enhanced density region across the equatorial disk of this Be shell star. The scaled geometry of ζ Tau projected on the sky plane is depicted in Fig. 2. Our GI2T findings agree surprisingly well with Hipparcos photometric data obtained during '92 and '93. Indeed, the partial eclipse expected to occur when the high density equatorial region passed behind the central star of ζ Tau, thus increasing its apparent magnitude (Mennickent et al. 1997) is visible on Hipparcos wide band photometry (Fig. 3).
This isolated result clearly shows that patrol OLBI observations must be applied to other objects, specially to pole-on Be stars like η Tau, for which V/R variability cannot, or very hardly, be detected by spectroscopy. However, in order to interpret OLBI observations in terms of physical parameters of these stars, adequate models including hydrodynamics and radiative transfer coupling should produce interferometric observables from monochromatic maps across emission lines (Stee et al. 1995).

5 Discussion

The few results outlined above demonstrate that a number of important discoveries can be foreseen from the operation of optical synthesis arrays like the VLTI, CHARA and the Hawaii Keck interferometer in the coming decade. Meanwhile, the use of a full-fledged GI2T and its spectro-polarimetric focal instrument should bring new insights to the understanding of hot luminous stars and the physics of their winds. Hereafter is a non exhaustive list of problems that OLBI could adress:
a)Determination of the size of Be underlying (flattened?) stars: this is possible by spectrally resolved interferometry of photospheric absorption lines using earth-rotation synthesis (a 3 telescope interferometer is more efficient).
b)Detailed imaging of Be star equatorial disks and their kinematics: this is better done using the dignosis value of optically thin FeII emission lines (Hanuschick '96) and also carrying simultaneous visible and infrared OLBI.
c)The role played by NRP and magnetic activity in the variability of massive star winds: this is possible with mas resolution to sound the inner regions of the wind in Balmer series and spectral lines of He, Si, O, C, etc..
d)Multiplicity of massive stars: interferometry can detect unseen companions and "weight" the individual masses. For close binaries the detailed geometry of the system can be mapped as a function of orbital phase to better understand mass exchange mechanisms for instance.

e)Direct detection of inhomogenities at the basis of massive star winds (like for P Cyg): this probably needs OLBI observations to be alerted by spectro/photometry to follow the spatial change and location of the inhomogeneity until its effect, if any, becomes observable at lower spatial resolutions achievable by HST or adaptive optics on 4-8m class ground telescopes.
f)Last but not least, for a number of luminous hot stars OLBI can directly determine parameters like R and L of the Wind-Momentum Relation $\dot{M}v_\infty R^{0.5} \propto L^x$ as defined by R. Kudritzki during this colloquium.

References

Guo Y., et al., 1995, A&A. Suppl., 112

Harmanec P., et al., 1996, A&A, 312, 879

Hanuschik R.W., 1996, A&A, 308, 170

Hipparcos and Tycho Catalogues, ESA SP-1200, June 1997, Vol. 12, page C38

Hoffman, J.L., et al., 1998, ApJ., 115, 1576

Labeyrie A., Proceedings of "Optical and IR Telescopes for 1990's", Vol. II, Jan. 1980, Tucson, USA, Ed. A. Hewitt, 786

Marlborough, M., 1997, A&A, 317, L17

Mennickent R.E., Sterken C. and Vogt, 1997, A&A, 326, 1167

Mourard, D., et al., 1989, Nature, 342, 520

Mourard, D., et al., 1994, A&A, 283, 705

Mourard, D., et al., 1998, Proceedings of SPIE conference, Kona, Hawaii, Vol. 3350, p517

Nota, A., et al. 1995, ApJ, 448, 788

Okazaki A.T., 1997, A&A, 318, 548

Papaloizou, J.C., et al., 1992, A&A, 265, L45

Quirrenbach A., et al., 1997, ApJ., 479, 477

Rousselet-Perraut, et al., 1997, A&A. Suppl., 123, 173

Smith, M., et al. ApJ, under press

SPIE conference proceedings on "Astronomical Interferometry", Kona, Hawaii, 1998, Vol. 3350

Stee, Ph., et al., 1995, A&A, 300, 219

Stee, Ph. 1998, A&A, 336, 980

Telting J.H., Heemsherk, et al. 1994, A&A, 288, 558

Thom C., et al., 1986, A&A, 165, L13

Vakili, F., et al., 1994, "Pulsation, Rotation and Mass-Loss in Early-type Stars", Proceedings of IAU Symposium 162, Eds. L. Balona et al., Kluwer Acad. Publ.

Vakili F., et al. 1997, A&A, 323, 183

Vakili F., et al. 1998, A&A, 335, 261

Waters, L.B.F.M., 1992, A&A, 253, L25

Direct Observational Evidence for Magnetic Fields in Hot Stars

Gautier Mathys

European Southern Observatory, Casilla 19001, Santiago 19, Chile

Abstract. Attempts at achieving direct detections of magnetic fields in hot stars are reviewed. The techniques used in these observations and their analysis are described, with emphasis on the physical situations to which they are relevant. Results of projects carried out in the last couple of years are reported, and prospects for future investigations are briefly considered.

1 Introduction

As emphasized in many contributions at this meeting, an increasing number of observations of hot stars provide *indirect* evidence that magnetic fields must be present in those stars. Such observations include, for instance, rotationally modulated winds in O stars, X-ray emission in Be stars, and transient features in the profiles of absorption lines in the visible spectrum of Be stars. This has triggered, in the last few years, an intensification of the efforts towards *direct* detections and measurements of magnetic fields in those stars. In this contribution, I review the current status of our knowledge in this area.

2 Diagnosis Methods and Brief History

Stellar magnetic fields are accessible to direct observation through the Zeeman effect that they induce in spectral lines. The most straightforward manifestation of this effect is the splitting of the lines into several components. The wavelength separation of adjacent split components depends on the atomic properties of the levels involved in the considered transition, but typically is of the order of $4.67\,10^{-13}\,\lambda^2\,\langle H\rangle$, where λ is the nominal wavelength of the transition (expressed in Å) and $\langle H\rangle$ is the mean magnetic field modulus (in G), that is, the line-intensity weighted average over the visible stellar hemisphere of the modulus of the magnetic vector. Hence for a line at, say, 5000 Å, the separation of magnetically split components per kG of field strength is of the order of 10^{-2} Å or, in velocity units, 0.6 km s^{-1}. Not only can this effect be observed only in spectra recorded at sufficiently high resolution of stars with fairly considerable and uniform[1] magnetic fields, but also it

[1] If there is a wide dispersion of field strengths over the stellar surface, the contributions to the observed line coming from different parts of the stellar surface

may easily be smeared out through broadening of the magnetically split line components by other effects such as, primarily, Doppler effect due to stellar rotation and/or local atmospheric motions. In the kind of stars considered in this meeting, such motions are often quite significant and ruin virtually every chance of direct observation of magnetic line splitting.

In principle, even when magnetic line splitting cannot be directly observed, it remains possible to detect stellar magnetic fields from observations of the differential broadening effect that they induce in lines of different magnetic sensitivity. This approach has been successfully applied to a number of stars, mostly either active late-type stars or magnetic Ap and Bp stars. But again, its suitability for the study of the stars of interest at this colloquium is limited. Indeed, although it should in principle be possible to diagnose the magnetic field from only two lines of very different magnetic sensitivity, in practice, our ability to do this is limited by the uncertainties introduced by possibly unrecognized effects that may alter the shapes of either of the two lines and may be misinterpreted as due to a magnetic field. This possibility appears especially critical for hot stars with extended atmospheres, where different lines are often formed at different depths, possibly in different velocity fields, with different departures from LTE, etc. The usual workaround is to use a larger number of lines (ideally, a statistically significant sample) for the differential magnetic broadening study. This workaround is bound to be much less successful for hot stars than for solar-type or Ap stars, the spectra of which contain an incomparably larger number of lines suitable to magnetic field diagnosis.

Therefore, the best chances for a successful diagnosis of the magnetic fields of hot stars rest on the exploitation of the polarization induced in the spectral lines by the Zeeman effect. To first order, the result of the latter is a shift in the wavelength of the centre of gravity of a line between observations performed in right and left circular polarizations (RCP and LCP, resp.). The order of magnitude of this shift is $9.34\,10^{-13}\,\lambda^2\,\langle H_z\rangle$, which at first glance is similar to the separation of adjacent components of resolved lines. However, there are two differences. First, rather than splitting *within* a line, the effect is, in the present case, the *global* shift of a *whole* line. The latter is much easier to detect than the former, because in particular other effects affecting the line shape are essentially similar in both circular polarizations, hence do not mask the magnetically induced differences between the latter. Yet, while the magnetic splitting is proportional to the mean field modulus $\langle H\rangle$, the observed wavelength shift between RCP and LCP is proportional to the mean longitudinal field $\langle H_z\rangle$. The latter is the line-intensity weighted average over the visible stellar disk of the component of the magnetic vector along the line of sight. Due to its vectorial nature, this average will differ from zero

---

will be characterized by different line splittings, which once integrated over the visible stellar disk may broaden the split line components to such an extent that they are no longer resolved.

only provided that the star's magnetic field has a suitable large-scale organization: for complex field structures (e.g., such as in the sun), where the star is covered by a large number of small regions where fields of opposite polarities but similar strengths prevail, the contributions of these small regions to the disk-integrated spectral lines average out to a large extent, so that the lines show no net wavelength shift between RCP and LCP. Admittedly, in some circumstances (thanks, e.g., to different rotational Doppler shifts of regions of different field polarities), higher-order effects may still detectable as differences between the line profiles recorded in RCP and LCP: this possibility, which anyway involves significant additional complexity, will be briefly considered later in this contribution.

The observation of global wavelength shifts between spectral lines recorded in RCP and LCP (and the related applications, such as Balmer line photopolarimetry: see Landstreet 1982) has been so far the main source of stellar magnetic field measurements. The vast majority of the stars in which definitive magnetic field detections have been achieved through this technique are Ap and Bp stars. These stars have magnetic fields which, in first approximation, are nearly dipolar; the dipole axis is generally inclined at some non-zero angle with respect to the stellar rotation axis. This simple dipolar field structure is a very favourable one for the observation of a global wavelength shift of disk-integrated spectral lines between RCP and LCP.

The lower temperature limit of the magnetic Ap star phenomenon lies around spectral type F0 and is understood to result from the onset in cooler stars of significant convection in outer layers. There is no such discontinuity in the stellar physical properties towards the high temperatures: therefore it appears rather natural, in a first step, to try to detect magnetic fields in hot stars through the techniques that have been successfully applied for Ap and Bp stars. In particular, the hottest magnetic Bp stars currently known are of spectral type B1, and there is no reason to believe a priori that large-scale organized magnetic fields and/or chemical photospheric peculiarities should not exist in higher-temperature stars. It has, more plausibly, been argued that there must be less detections of such phenomena in the hottest B stars and in O stars than in lower-temperature stars as a result of the combination of the smaller total number of OB stars and of the increased difficulty to detect fields and spectral anomalies in them, due to the specific characteristics of their spectra (low density of spectral lines, broadened and/or distorted by various dynamical effects). Accordingly, all the early attempts at detecting magnetic fields in hot stars have aimed at determining their mean longitudinal magnetic field from the wavelength shifts of their spectral lines between opposite circular polarizations. H.W. Babcock had investigated approximately a dozen Oe/Be stars in his seminal survey of stars in search of magnetic fields (Babcock 1958): two of them only showed fairly doubtful hints of the presence of a magnetic field. Among the various attempts made since Babcock's work until the last couple of years, some of the best ones (with measurement uncertainties of the order of 100 G or less) include the work of Barker et al. (1981) on the O4ef star ζ Pup, Landstreet's (1982) investigation of a sample of apparently normal upper main-sequence stars, and the study of a sample of Be stars by Barker et al. (1985). Neither of these (nor any of the contemporary works not explicitly listed here) yielded any definite field detection.

3 Recent Searches for Magnetic Fields in Hot Stars

Not only do the attempts at magnetic field detections in hot stars described in the previous section all aim at exploiting the same manifestation of the Zeeman effect, but also they all were made using basically only two types of instruments. One is the original Babcock-type photographic Zeeman analyzer, through the use of which two high-dispersion stellar spectra corresponding to incoming light of opposite circular polarizations are simultaneously recorded next to each other on a photographic plate. The other is the Balmer line photopolarimeter, which transmits alternatively to a photomultiplier tube RCP and LCP light issued from a narrow region of the wing of a Balmer line of hydrogen, isolated by use of an interference filter. The difference between the signals corresponding to incoming RCP and LCP is computed on the fly by the electronics associated with the photomultiplier tube, which is equivalent to measuring the wavelength shift of the line between the two polarizations. While the Babcock-type analyzer suffers from the inherent limitations of the photographic plate in terms of noise, the achievable ratio between polarimetric signal and noise in Balmer line photopolarimetry is limited by the smallness of the former, as a result of the fact that Balmer lines are very broad in early-type stars.

In recent years, however, a number of new instruments have entered into operation, through which high resolution spectra are recorded simultaneously in two mutually orthogonal (circular or linear) polarizations on a CCD, hence combining the strengths of the Babcock-type and Balmer line analyzers and overcoming their limitations to a large extent. Several of these new instruments have been used by various groups for studies of magnetic fields of hot stars: the Zeeman analyzer of the ESO CASPEC spectrograph (described by Mathys and Hubrig 1997), the polarimeter built in Meudon for use on various spectrographs (Donati et al. 1997), the MuSiCoS spectropolarimeter at the Pic-du-Midi Observatory (Donati et al. 1998), and the William-Wehlau spectropolarimeter (Eversberg et al. 1998).

In parallel, new methods have been developed for the analysis of the data, which allow one not only to interpret the global wavelength shift of lines between RCP and LCP, but also to exploit the additional information that may exist in differences between the *shapes* of the lines in the two polarizations. These techniques have already proved successful in works on other types of stars. The *Zeeman-Doppler imaging* (Semel 1989) has allowed Donati et al. (1990) to achieve the first spectropolarimetric determinations of magnetic fields in active late-type stars. The *moment technique* (Mathys 1988) has proved quite effective in providing new constraints on the geometrical structure of the magnetic fields of Ap and Bp stars (Mathys 1993). A first glimpse of the large potential of the *least-square deconvolution method* has been given by Donati et al. (1997). On the basis of these positive results, these new techniques appear very promising for studies of hot star magnetic fields. They only start to be applied in such studies, but this is undoubtedly one of the ways to go for the future. Yet, until now, most of the results obtained with the new generation instruments have followed from the classical approach of measuring the wavelength shifts of lines between RCP and LCP. These recent results are briefly reviewed below.

The hot star for which the largest number of attempts at detecting a magnetic field have been made in the last few years is definitely the O7V star θ^1 Ori C. The interest for this star has been triggered by the reports by the Heidelberg group that it shows a phenomenology very similar to that of magnetic oblique rotators of the Ap and Bp group, such as, in particular, periodic variations of line intensities. This raised high hopes that θ^1 Ori C might be the first representative of the long sought after hot extension of the magnetic chemically peculiar A and B stars into the O spectral type class. Should this be the case, the star would logically be expected to harbour a strong, roughly dipolar magnetic field. However, the attempts made by Donati and Wade (1998) with MuSiCoS, Eversberg et al. (1998) with the William-Wehlau polarimeter, and myself with various collaborators using CASPEC (Mathys et al., in preparation), all yielded null measurements of the mean longitudinal field, with typical uncertainties of the order of 250 G. Donati and Wade compute that, for a dipolar field geometry, this corresponds to an upper limit of the polar field strength of 1.6 kG. Incidentally, these authors also detected a strong circular polarization of the continuum, which definitely does not arise from a magnetic field but whose origin remains for the time being an enigma. Another promising candidate, the O7.5 giant ξ Per, whose wind undergoes cyclical variability, has been observed at the CFHT through Balmer line photopolarimetry by Henrichs et al. (1998), who failed to detect a longitudinal magnetic field with 1σ error bars of the order of 70 G.

Lex Kaper and myself conducted a search for magnetic fields in a sample of 10 fairly slowly rotating O stars of various types, using CASPEC spectra. We determined the longitudinal fields with, in the best cases (half of the sample) 1σ uncertainties of the order of 250–300 G, but we failed to achieve any definite detection. Some marginal measurements (between 2 and 3σ) suggest that weak fields might nevertheless possibly be present, which could become detectable through a fairly modest improvement (say, a factor of 2) of the measurement accuracies. A more detailed account of this work will be given in a forthcoming paper.

The above-described investigations of O stars were aimed at detecting global fields with sufficient organization at the scale of the whole star: in that respect, they were quite similar to the older studies such as those mentioned in Sect. 2. In such a study based on Balmer line photopolarimetry, Barker et al. (1985) had derived an upper limit of 250 G for the longitudinal field of the Be star λ Eri. However, these observations hardly constrain the possible existence of *localized* magnetic fields, which are often assumed to be responsible for the occurrence of transient phenomena in the spectrum of this star, such as ephemeral "dimples" or weak emissions in the line He I λ 6678 and in other He I lines. If this interpretation is correct, detection of such localized fields should in principle be achievable by observations of the line transients that they trigger, since the latter should indeed arise from the location of the confined magnetic spot, without dilution from the rest of the stellar disk. An attempt to apply this approach, made by Myron Smith and myself with CASPEC, was severely hampered by bad weather conditions, and the small number of observations that could actually be performed did not allow us to follow any transient from its appearance to its disappearance. However, some transients were observed for part of their existence. The analysis of these data is still in a rather preliminary stage. So far no clear evidence of detection of a magnetic field has been obtained, but final, more quantitative conclusions require further analysis work.

A definite detection in a class of stars not previously known to harbour magnetic fields has been achieved by Donati (1998), who applied the least-square deconvolution technique to measure a longitudinal field of the order of 50 G in the Herbig Ae star HD 104237, thereby confirming the reality of a marginal detection achieved earlier (Donati et al. 1997). However, Herbig Ae stars are admittedly somewhat cooler than the vast majority of the stars of interest at this meeting.

4 Prospects and Suggestions for Future Work

The almost entirely negative outcome of the attempts made so far to detect magnetic fields in hot stars with non-axisymmetric and/or variable winds may seem discouraging. Should one jump to the conclusion that these stars do not have magnetic fields and give up all attempts to detect the latter? In my opinion, this would be premature: there are a number of reasons to believe that magnetic fields may indeed be present in hot stars although they have so far escaped detection.

On the one hand, magnetic fields have definitely been observed in a number of helium peculiar stars between spectral types B1 and B6. These stars do have trapped circumstellar plasmas and non-asymmetric winds governed by their magnetic fields (Shore 1998 and references therein). Their mass-loss rate may be lower than in the most extreme objects discussed in this colloquium, but they certainly have much in common with the latter in terms of underlying physics. Further interest arises from the fact that some at least of these stars may not be just oblique magnetic rotators, as widely accepted in the classical view that their group constitutes a hot extension of that of the magnetic Ap stars, but that instead they may be the first examples of an entirely new class of magnetic stars, in which the observed magnetic variations are due to stellar pulsation and not the result of the changing aspect of the visible stellar hemisphere as a result of stellar rotation. Arguments supporting this view come from the fact that the magnetic variations of the B2p He strong star HD 96446 are inconsistent with the oblique rotator model (Mathys 1994).

On the other hand, the upper limits derived so far for the magnetic fields of hot stars are not yet low enough to be inconsistent with the magnetic field intensities advocated to explain various observational manifestations taken as indirect evidence of the presence of fields. Achievement of better sensitivity in searches for magnetic fields may well allow detections to be finally made. While for cooler stars, such improvement has often been achieved by the simultaneous consideration of more spectral lines, in the present case, it should rather come from higher signal-to-noise single line observations. Potential progress may also come from detailed studies of the polarized line profiles, in the eventuality that the fields of hot stars have structures significantly more complex than those of the magnetic Ap and Bp stars. In summary, there remain a wealth of directions to be explored, and whether they eventually lead to magnetic field detections or not, they will without doubt yield new and deeper insight into the physics of the considered stars.

References

Babcock H.W., 1958, ApJS 3, 141

Barker P.K., Landstreet J.D., Marlborough J.M., Thompson I., Maza J., 1981, ApJ 250, 300

Barker P.K., Landstreet J.D., Marlborough J.M., Thompson I.B., 1985, ApJ 288, 741

Donati J.-F., 1998, private communication

Donati J.-F., Catala C., Wade G.A., et al., 1998, A&A (in press)

Donati J.-F., Semel M., Carter B.D., Rees D.E., Cameron A.C., 1997, MNRAS 291, 658

Donati J.-F., Semel M., Rees D.E., Taylor K., Robinson R.D., 1990, A&A 232, L1

Donati J.-F., Wade G.A., 1998, A&A (in press)

Eversberg T., Moffat A.F.J., Debruyne M., et al., 1998, these proceedings

Henrichs H.F., de Jong J.A., Nichols J.S., et al., 1998, preprint

Landstreet J.D., 1982, ApJ 258, 639

Mathys G., 1988, A&A 189, 179

Mathys G., 1993, in IAU Coll. No. 138, ASP Conf. Series vol. 44, p. 232

Mathys G., 1994, in IAU Symp. No. 162, Kluwer, Dordrecht, p. 169

Mathys G., Hubrig S., 997, A&AS 124, 475

Semel M., 1989, A&A 225, 456

Shore S.N., 1998, these proceedings

Discussion

T. Eversberg: The Landstreet equation is used for absorption-line stars. Is it reasonable to apply this formalism to emission lines coming from expanding winds?
G. Mathys: Yes, polarisation effects due to the magnetic field are similar in emission or in absorption. However, emission likely originates from higher layers than absorption. Due to magnetic flux conservation, magnetic fields in regions where emission lines are formed should be weaker than in the regions of formation of absorption lines: accordingly, it is probably even more difficult to detect a magnetic field in emission lines than in absorption lines.

C. Rodrigues: Why couldn't cyclotron emission explain the high circular polarisation of θ^1 Ori C?
G. Mathys: To achieve the observed level of continuum circular polarisation by a cyclotron effect, the magnetic field would be so strong that the lines would be badly distorted: this is not observed.

P. Stee: Maybe the proper technique will be to observe with the VLTI in polarised light, because the global polarisation can be zero whereas locally you can have a polarisation of 100 %.
G. Mathys: I agree.

A. Moffat: What is the typical S/N you have in your spectra to search for circular polarisation? Why not go for several 1000?
G. Mathys: The spectra that I discussed have S/N ratios of ~400. It is limited by the instrumental configuration used, in particular the inter-order scattered light of the non-dispersed echelle format. For hot stars, the broad wavelength coverage provided by this format is not a strong asset, since only very few diagnostic lines are available in any case. Accordingly, it might be preferable to observe a single line or a very limited number of lines in a restricted wavelength range with a standard spectrograph. The instrumental signature in the latter can probably be handled better, which would allow one to achieve higher (very high) S/N, which of course is desirable.

HST WFPCII Observations of the Inner HR Car Nebula

Stephen Hulbert[1], Antonella Nota[1], Mark Clampin[1], Claus Leitherer[1], Anna Pasquali[2], Norbert Langer[3], and Regina Schulte-Ladbeck[4]

[1] Space Telescope Science Institute
[2] ST-ECF/European Southern Observatory
[3] Institut für Theoretische Physik und Astrophysik, Potsdam
[4] University of Pittsburgh

Abstract. We present Hα observations of the inner region of the HR Carinae nebula. The inner nebula looks clumpy and material appears to be present only to the east of the central star. A jet-like filament stretches away from the central star to the southeast.

HR Car, one of the few known galactic Luminous Blue Variables (LBVs), was originally reported to be surrounded by an "arc-shaped" nebula with a jet-like feature by Hutsemékers & Van Drom (1991). Later imaging revealed a bipolar morphology. Filamentary structure is evident over the $19'' \times 37''$ extent of the bipolar lobes with the SE lobe most prominent (Weis et al. 1997, Nota et al. 1997). Spectroscopic observations of the SE lobe show blue-shifted radial velocities up to 100 km s^{-1} while the NW lobe shows red-shifted radial velocities of the same magnitude. The bipolar nebula is oriented at 125-135°E of N in the plane of the sky–it is tilted between 40°and 60°into the plane of the sky. The innermost part of the nebula was first identified by Clampin et al. (1995) as the "waist" of the bipolar nebula–similar to the corresponding feature found in η Car. Figure 1 (left) shows the nebula as seen from the ground with coronography (Clampin et al. 1995). Figure 2 (left) shows the model of the bipolar nebula including the equatorial waist (adapted from Nota et al. 1997).

Hα images of HR Car were obtained during 1996 using the HST instrument configuration of WFPC2+F656N. Observations were separated in time such that the nominal roll of HST had changed by 22°. Multiple exposures at each epoch were combined to remove cosmic rays. The complex background caused by the bright central source was first modeled and then subtracted. Field stars were used to rotate and align the images from the two epochs. By combining two images with different orientations we were able to remove the HST secondary spider shadows from the images and were forced to mask out only the central region of the image where the CCD was saturated.

¿From the most recent HST observations of HR Car, the inner nebula appears clumpy. Figure 1 (right) shows the innermost $17''$ of the HR Car nebula as seen by WFPC2–the central star is centered in the circular masked region. We have measured four of the largest clumps of material in the waist;

the clumps have masses of: A–0.07, B–0.02, C–0.01, D–0.04 $M_{\odot}$. The four clumps account for about half of the total mass, 0.40 $M\odot$, of the inner nebula based on integrated $H\alpha$ measurements. Figure 2 (right) shows the individual clumps. The apparent "clump" marked with an asterisk is actually a previously identified star (Hutsemékers & Van Drom 1991).

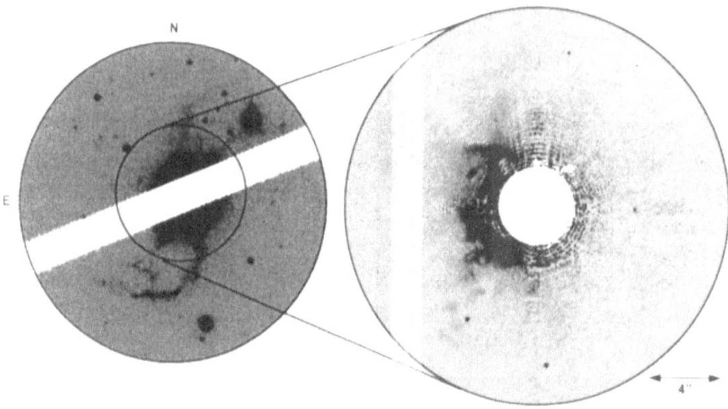

Fig. 1. Inner HR Car nebula: (left) ground-based image of HR Car bipolar nebula. (right) HST WFPCII image of HR Car inner nebula.

The inner nebula is arranged asymmetrically around the central star–material is visible only to the east of the central star. The visible material appears roughly elliptical in shape shape with a size of 3″ × 6″ and an orientation roughly E of N. The asymmetric distribution of material near the central star is curious but not inconsistent with it being the waist of the bipolar nebula. If the visible portion represents only part of the waist, an elliptical fit to the structure of the complete waist agrees with the equatorial plane being tilted into the plane of the sky about 58°–consistent with the range of tilts for the lobes of the bipolar nebula (see figure 2). In addition, the orientation of the elliptical fit on the sky (30°) is in gross agreement with the polarization results of Clampin et al. (1995).

The expansion time scale of the inner nebula is comparable to that of bipolar nebula. The nebular expansion velocity is 100 km s^{-1} yielding a dynamical time scale of 5 × 10^3 years for the bipolar nebula (Nota et al. 1997). The dynamical time scale for the clumps in the inner nebula (about 3″ away from the central star) is about 1.6 × 10^3 using an average velocity of 50 km s^{-1} taken from previous spectroscopic observations of the nebula (Nota et al. 1997). The agreement of these two time scales is, in turn, consistent with the identification of the inner nebula as the waist of the bipolar nebula.

A jet-like filament (JLF) is evident emanating from the direction of the central object. The filament extends to at least 8″ from the central star and has an orientation of 140°E of N–both the JLF and the bipolar nebula are arrayed in the same direction in the plane of the sky. This filament points directly at the rather unusual "fork-shaped" filament associated with the SE lobe of the bipolar nebula–apparently bisecting the tines of the fork. Previous spectroscopic measures that coincide with the location of the JLF show high velocity blue-shifted components along the filament of 50-100 km s^{-1} up to 10 ″ away from central object (Nota et al. 1997). Figure 2 shows the location of the JLF in the sky as well as in the model of the nebula.

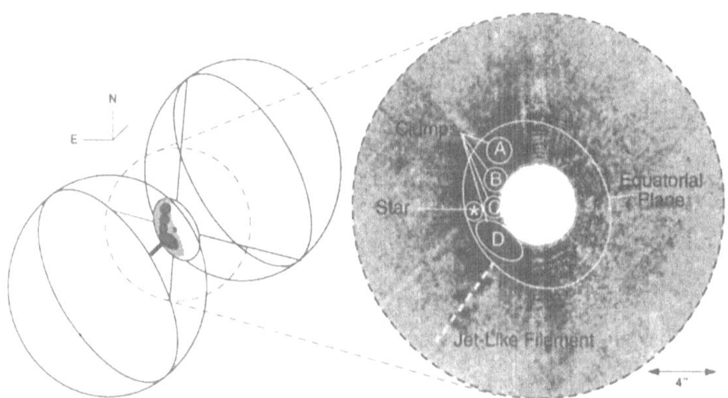

Fig. 2. HR Car Nebula Model: (left) Schematic of bipolar nebula. (right) Identification of major clumps and jet-like filament in HR-CAR waist.

References

Clampin, M., Schulte-Ladbeck, R.E., Nota, A., Robberto, M., Paresce, F., Clayton, G.C. (1995): AJ **110**, 251

Hutsemékers, D., Van Drom, E. (1991): A&Ap **248**, 141

Nota, A., Livio, M., Clampin, M.(1995): ApJ **448**, 788

Nota, A., Smith, L., Pasquali, A., Clampin, M., Stroud, M. (1997): ApJ **486**, 338

Weis, K., Duschl, W.J., Bomans, D.J., Chu, Y.-H., Joner, M.D. (1997): A&Ap **320**, 568

Discussion

N. Trams: The TIMMI images of HR Car (Voors et al. 1997) I showed in my talk yesterday show not only the small inner nebula, but also evidence for emission from the other side of the central star, which would be coming from the backside of the "waist" of the bipolar nebula.

R. Waters: The HST images nicely confirm the presence of the asymmetric inner nebula, which was discovered by Voors et al (1997). The IR images show the innermost part of the nebula ($< 2''$) and clearly demonstrate that the orientation of the inner nebula is similar to that of the outer nebula.

Peredur Williams and Stephen White

Observing Hot Stars in all Four Stokes Parameters

Thomas Eversberg[1], Anthony F.J. Moffat[1], Michael Debruyne[2], John B. Rice[3], Nikolai Piskunov[2,4], Pierre Bastien[1], William H. Wehlau[2], and Olivier Chesneau[5]

[1] Université de Montréal, Canada
[2] University of Western Ontario, Canada
[3] Brandon University, Canada
[4] Uppsala University, Sweden
[5] Observatoire de la Côte d'Azur, France

Abstract. We introduce a new polarimeter unit which, mounted at the Cassegrain focus of any telescope and fiber-connected to a fixed CCD spectrograph, is able to measure all Stokes parameters I, Q, U and V photon-noise limited across spectral lines of bright stellar targets and other point sources in a quasi-simultaneous manner. We briefly outline the technical design of the polarimeter unit and the linear algebraic Mueller calculus for obtaining polarization parameters of any point source. In addition, practical limitations of the optical elements are discussed. We present first results obtained with our spectropolarimeter for three prototype hot-star.

1 The polarimeter unit

The newly built William-Wehlau-Spectropolarimeter (WWS)[1], a collaboration between the University of Western Ontario, Brandon University and Université de Montréal, consists of two rotatable quarter-wave plates (QWP, as retarder) and a Wollaston prism (as beam-splitter and double-beam polarizer), leading light from stellar point sources via a double fiber-feed into a fixed CCD slit spectrograph.
Figure 1 shows the basic design of the polarimeter unit.
The WWS is designed to be mounted at the ≈ f/8 Cassegrain focus of medium or large size telescopes. The principle light track:

Input aperture (pinhole) → Collimator → Glan-Taylor prism (removeable, produces 100% linearly polarized light) → Two rotatable QWP's (achromatic, PC controlled) → Wollaston prism (beamsplitter) → Camera → Twin optical fibers → Spectrograph slit → CCD.

[1] The original PI for this instrument, William H. Wehlau, unfortunately passed away in February 1995. We wish to express our deepest respect for his high level of competence and leadership during the planning and early construction stages of the spectropolarimeter that now carries his name.

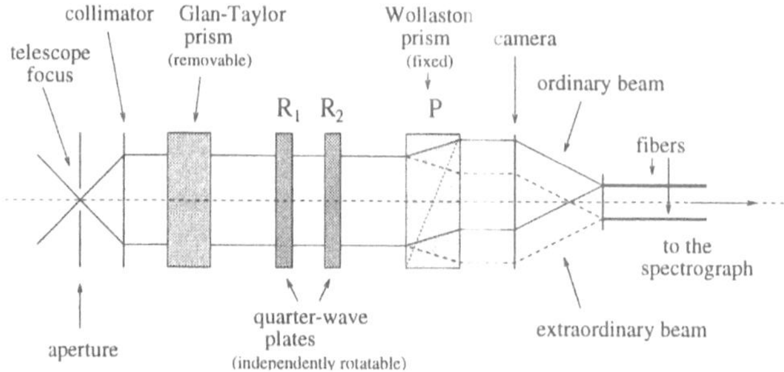

Fig. 1. Simple sketch of the William-Wehlau-Spectropolarimeter.

The independent rotation of the QWP's relative to the Wollaston prism determines the polarization components that are being measured. The spectra of the light from each fiber are imaged so that they are parallel and adjacent, with sufficient space between them on the CCD detector of the spectrograph. The two components can be combined over an observing sequence to obtain all four Stokes' parameters, that fully define the polarization of the light. When the light is dispersed by the spectrograph, the polarization of the starlight may be calculated as the wavelength-dependent quantities $I(\lambda), Q(\lambda), U(\lambda)$ and $V(\lambda)$ via the Mueller calculus. For a full description of the instrument, the Mueller calculus and the data reduction, we refer to Eversberg (1998a, PASP, in press). We refer to Serkowski (1974, in: *Methods of Exp. Phys*, Vol. 12, Chap. 8) and the references therein as a basic source for the description of the Mueller calculus and different optical devices in this nomenclatura.

2 Technical problems and first results

We report some basic problems with our instrument:

Minor: SLIGHT DEPENDENCE OF RETARDATION τ ON QWP POSITION ANGLE ϕ. The QWP frames (worm-gears) as well as some QWP crystal layers are tilted with respect to the instrumental optical axis. This introduces a wobbling effect which can be corrected with a parametric fit for $\tau(\phi)$. This is clearly a manufacturing problem and airspaced QWP's are recommended to remedy this problem.

Major: Stochastic broadband polarization variability Q, U and V due to small variations in overall illumination of the two fibers, whose spatial surface sensitivities are never perfectly uniform (see also Donati et al. 1997, MNRAS, 291, 658, who report the same problem). As a consequence we can neither measure absolute values for Q, U and V nor the polarization angle at the sky. Smoothness of fiber apertures can possibly remedy this behavior.

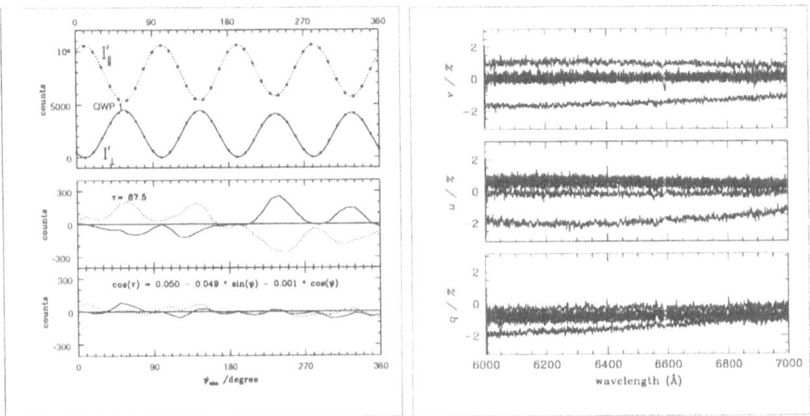

Fig. 2. Left: Top panel: Position angle dependence of the retardation $\tau(\phi)$ with 100% linear input polarization (Glan-Taylor prism). **Center panel:** χ^2 fit with fixed $\tau = 87.5°$. **Bottom panel:** Parametric fit with $\tau(\phi)$ as indicated. **Right:** Seven consecutive sequences of γ Cas during one night at Mont Mégantic Observatory.

The position angle dependence of the retardation $\tau(\phi)$ and a parametric fit is shown in Fig. 2 (left). An example for consecutive sequences of Stokes Q, U and V is also given in Fig.2 (right).
As a consequence, we cannot estimate the exact continuum polarization. For this reason we simply subtracted a fit to the original individual
Stokes Q, U and V spectra in each sequence and thus neglected broadband polarization, when combining sequences to get rectified mean $q(\lambda)$, $u(\lambda)$ and $v(\lambda)$ spectra.
We confirm previous results by Poeckert & Marlborough (1977, ApJ, 218, 220) for the Be star γ Cas, find a feature in Stokes v consistent with the measured magnetic field of $B_e = 28\,000$ G, in the Ap star 53 Cam (Borra & Landstreet 1977, ApJ, 212,141) and report newly observed line effects in q and u in the WR+O star γ^2 Vel (Eversberg et al. 1998b, submitted to ApJ).

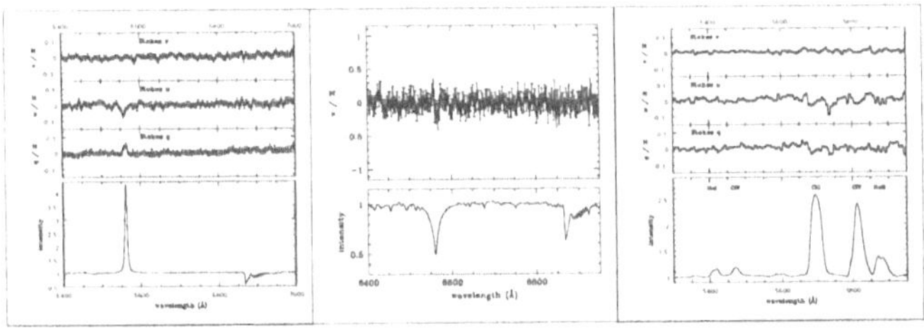

Fig. 3. Left: Stokes q, u and v for the Be star γ Cas. **Center:** Stokes v for the Ap star 53 Cam. **Right:** Stokes q, u and v for the WR+O binary γ^2 Vel. Nightly average intensity spectra are at the bottom of each panel.

Discussion

H. Henrichs: How does your instrument compare with the MUSICOS polarimeter in terms of efficiency and throughput?
T. Eversberg: You have to remember that MUSICOS uses an echelle spectrograph. Considering this, the efficiency and throughput is comparable. Of course, we lose light with the fibers, but to use a polarimeter at the cassegrain focus and measure in high resolution mode we need them to feed a coudé spectrograph.

F. Vakili: The next step to spectropolarimetric observations is to add spatial resolution. With this in mind, we have equipped our GI2T optical interferometer with polarimetric equipment.
T. Eversberg: Exactly. This is already shown by the Bjorkman/Quirrenbach paper and I think that this should be developed further.

R. Ignace: In your test observations of 53 Cam, the V feature appears to be a marginal detection to my eye (i.e., the error bars are large compared to the feature). If you did not know a magnetic field was present, would you claim to have detected one from your data? And to what confidence level?
T. Eversberg: As far as I remember, the result is from a 5-hour exposure and for this a high level of confidence is not reachable. John Rice modelled our observations, but I do not know the result so far. In general, as I mentioned in my talk, we need big telescopes for detecting circular polarisation in an accurate way.

Thomas Eversberg and Gautier Mathys

Inverse Spectropolarimetric Modelling of Hot Star Wind Structure and Variability

John C. Brown[1], Richard Ignace[1], and M. Piana[2]

[1] Dept of Physics and Astronomy, University of Glasgow, G12 8QQ, UK
[2] INFN, Universita di Genova, Genova, Italy

1 Introduction

Merits and limitations are discussed of using an inferential inverse, as opposed to the usual model-fitting, approach to diagnose stellar wind structure. We aspire to encourage the stellar wind community to use inversion rather than forward modelling by making it clear what inversion means, when and why it is valuable, and by giving examples of successful applications.

A sub-discipline of astrophysics advances beyond the discovery era as the quality of data moves from single numbers (flux, colour, size) to well measured functions or data strings $g(y)$ (e.g., light curves or spectra where g is flux and y is time or wavelength). As the precision $\delta g/g$ and the resolution $\delta y/y$ improve, we progress from data gathering and qualitative description to quantitative modelling in terms of some relevant source model function $f(x)$, describing source "structure" in some sense. In general $g(y)$ and $f(x)$ do not correspond one-to-one but rather f maps to g in a "convolved" (sometimes complex and non-linear) way through the radiation processes, while f itself may be a combination of important source properties. Here we consider only the simplest situation (though our arguments can be generalised) where the relationship is of the linear integral form

$$g(y) = \mathcal{K}[f(x);y] = \int_{D(y)} f(x)K(x,y)dx, \tag{1}$$

where the operator function $\mathcal{K}$ represents the emission physics which we assume known. The diagnostic problem is to determine as much as possible about $f(x)$ within the noise and resolution limits of the data $g(y)$ and the smearing effect of $K(x,y)$.

2 Forward and Inverse Approaches

The forward approach to (1) is to formulate some parametric empirical or (deductive) physical "model" for $f(x)$ and adjust parameters to yield the best fit to $g(y)$. This is the approach mostly, and often solely, used in astrophysics. Its principal merits are simplicity and the ability to deal readily with problems where $\mathcal{K}$ is complex or even where (1) is nonlinear, such as in full radiative transfer. When data are of moderate accuracy and resolution, it is the sensible approach. It suffers, however, from insufficiently recognised dangers of non-uniqueness of the $f(x)$ it yields. Though a "best" fit parametric f can always be found, a wide range of f's may yield a statistically acceptable fit, a possibility which should always be assessed by finding the confidence band of f within the errors. Moreover, the parametric form adopted for f may be restrictive, thereby giving a narrow parameter confidence interval on $f(x)$ that seems very good, whereas a wholly different parametric form of f might fit the data equally well. In Popperian terms, the data is not providing a critical test of the model hypothesis, goodness of fit being highly misleading (Craig and Brown 1986).

In the inverse approach, no parametric restriction is placed on $f(x)$ but rather we invert (1) to infer a model f inductively from g or, formally

$$f(x) = \mathcal{K}^{-1}[g(y); x], \tag{2}$$

though an explicit form for $\mathcal{K}^{-1}$ may not exist. In practice g is measured as a discrete data vector $\mathbf{g}$ and f is similarly represented by $\mathbf{f}$ while $\mathcal{K}$ is replaced by a matrix $\mathbf{K}$ so that (1) and (2) become

$$\mathbf{g} = \mathbf{K}\mathbf{f};\ \mathbf{f} = \mathbf{K}^{-1}\mathbf{g}. \tag{3}$$

This inverse formulation is valuable first because (2) yields an unrestricted solution for $f(x)$ from $g(y)$ alone, based solely on data rather than assumptions (other than in the adopted form for the radiation process K). Second, the non-uniqueness in f becomes fully transparent since, if g is not model sensitive, $\mathcal{K}$ is correspondingly near singular and $\mathcal{K}^{-1}$ produces large magnification of errors in g when data are inverted to yield f, whose spread is then explicit.

These advantages are only achievable when data are of sufficient resolution and accuracy for (2) and (3) to be really meaningful. This situation is reached only at quite advanced stages of astrophysical studies. For example, the inversion of the bremsstrahlung spectral problem by Brown (1971) (when data had typically 10% accuracy and 30% resolution) only recently became tenable via large, high resolution, Ge detectors with accuracy and resolution of order 1% (cf., Thompson et al 1991). The magnified errors in f caused by $\mathcal{K}^{-1}$ must be suppressed by incorporating some *a priori* smoothing condition on the inversion (cf., Craig and Brown 1986), but this is under our explicit control and not hidden in parametric assumptions. The main limitation is that simplifying assumptions may have to be made to formulate the inverse problem explicitly.

3 Examples

Here we simply list a few examples of stellar wind structure problems in which the inverse approach has been shown to be valuable.

a) Polarisation and Stellar Occultation of Non-axisymmetric Disks: Fox and Brown (1991) showed how the 2-D density structure of rotating disk features such as CIRs could be inferred from time variability of the Stokes Parameters of scattered stellar continuum emission.

b) Spectropolarimetric Line Profiles of Disk Scattered Stellar Lines: Brown and Wood (1994) inverted the equations for the profiles of scattered stellar lines to yield the velocity structure of rotating or expanding disks. The complicating effect of thermal smearing of these profiles (Wood and Bjorkman 1995) can be dealt with as an extra convolution.

c) Polarimetry and Spectrometry of Episodic Mass Loss: Brown and Wood (1992), Calvini et al (1995), and Piana et al (1995) derived variations in rate and oblateness of stellar mass loss transients by deconvolving these from light curves of continuum scattering polarisation and absorption line strength.

d) Mean Wind Velocity Structure $v(r)$ from Line Profiles: Shapes and intensities of wind emission profiles depend on $v(r)$. Brown et al (1997) and Ignace et al (1998a,b) obtained line profile inversions in several cases, including stellar occultation which improves the inversion.

With ever better stellar wind data in the temporal and spectral domains, the value of inverse diagnostic techniques over model fitting has reached the stage where they should be more fully exploited.

Acknowledgements–This work is supported by PPARC Standard and Visitor Grants and by a NATO Collaborative Research Grant.

References

Brown, J.C. (1971): Solar Phys. **18**, 489–502

Brown, J.C., Wood, K. (1992): Astron. Astrophys. **265**, 663–668

Brown, J.C., Wood, K. (1994): Astron. Astrophys. **290**, 634–638

Brown, J.C., Richardson, L.L., Ignace, R., Cassinelli, J.P. (1997): Astron. Astrophys. **325**, 677–684

Calvini, P., Bertero, M., Brown, J.C (1995): Astron. Astrophys. **309**, 235–244

Craig, I.J.D., Brown, J.C. (1986): *Inverse Problems in Astronomy* (Hilger, Bristol)

Fox, G.K., Brown, J.C. (1991): Astrophys.J. **375**, 300–313

Ignace, R., Brown, J.C., Richardson, L.L., Cassinelli, J.P. (1998a): Astron. Astrophys. **330**, 253–264

Ignace, R., Brown, J.C., Milne, J.E., Cassinelli, J.P. (1998b): Astron. Astrophys. **337**, 223–232

Piana, M., Brown, J.C., Calvini, P. (1995): Inverse Problems **11**, 961–973

Thompson, A.M., Brown, J.C., Craig, I.J.D., Fulber, C. (1992): Astron. Astrophys. **265**, 278–288

Wood, K., Bjorkman, J.E. (1995): Astrophys. J., **443**, 348–362

Discussion

A. Sapar: The inverse solution of the radiative transfer equation in its integral form belongs to the class of ill-posed problems in mathematics and its application is in the present case limited to optically thin media. What are the maximum values of optical depth for which the method can be applied successfully?

J. Brown: Most inverse problems in radiative transfer are (because of the $e^{-\tau}$ factor) like the Laplace transform, whose inversion is indeed very ill-posed. Even severely ill-posed problems can be solved by regularisation methods, so the issue is not really that of a maximum optical depth but rather what is the best resolution achievable for the source function. For the Laplace problem this is only 2 or 3 points per decade for typical data accuracies.

John Brown and Huib Henrichs (Joseph Cassinelli in the background)

Physics of Radiatively Driven Winds by High Angular Resolution Observations (HARO)

Ph. Stee, D. Bonneau, D. Mourard, and F. Vakili

Equipe GI2T, Département Fresnel
CNRS-UMR 6528
Observatoire de la Côte d'Azur
2130 route de l'Observatoire
Caussols
06460 St Vallier de Thiey, France

Abstract. Numerous models have been developed during the last two decades which try to fit a small number of "classical" observables as closely as possible (i.e Hα line profile and continuum energy distribution or polarization data). Nevertheless, little has been done to include High Angular Resolution Observations (HARO) in simulations in spite of the fact that such data can strongly constrain radiative wind models. In the following, we shall review recent results coming from HARO of active B stars and we shall focus on our radiative wind model for active hot stars which integrates these measurements.

In order to interpret HARO, we have built a latitude dependent radiative wind model which produces both spectroscopic, photometric, and interferometric synthetic data that have been compared to our interferometric observations. The line force in our model is the same as that used by Friend & Abbott (1986), but we introduce a varying contribution of thin and thick lines from pole to equator by adopting latitude dependent radiative parameters. We have performed a numerical calculation for parameters characteristic of the Be star γ Cas. Our main results are summarized in Table 1 (from Stee 1996). This model indicates that a radiative wind, driven mainly by optically thin lines at the equator, is a likely scenario for γ Cas. This is discussed in greater depth in Stee et al. (1995). We have also studied the rotational component of the wind in the envelope of γ Cas and we have found that it must be close to Keplerian rotation (Stee, 1996). A global picture of γ Cas circumstellar environment is shown in figure 1.

We have carried the first spectrally-resolved observations of γ Cas in the He I λ6678, and Hβ emission lines using the GI2T (Stee et al., 1998). We concluded that the Hβ emitting region must be smaller than 8.5 stellar radius and close to 2.3 stellar radii in He I λ6678. This is, for He I λ6678, smaller than the nearby continuum extent. These results confirmed the γ Cas basic parameters for this star obtained by Stee et al., 1995 (see Table 1). A picture of γ Cas as a function of line and wavelengths is shown in figure 2.

Table 1. Parameters and results for γ Cassiopeiae (from Stee 1996)

Parameters	
Spectral type	B0.5IVe
Effective temperature	25000 K
Mass	16 $M_\odot$
Radius	10 $R_\odot$
Stellar angular diameter	0.45 mas
Luminosity	3.510^4 $L_\odot$
Vsin i	230 km s^{-1}
Inclination angle i	45°
Results	
Polar terminal velocity	2016 km s^{-1}
Polar mass flux	$1.7 \cdot 10^{-9}$ $M_\odot$ yr^{-1} sr^{-1}
Equatorial terminal velocity	200 km s^{-1}
Equatorial mass flux	$5.1 \cdot 10^{-8}$ $M_\odot$ yr^{-1} sr^{-1}
Mass loss rate	$3.2 \cdot 10^{-7} M_\odot$ yr^{-1}
Hα major axis	17 stellar radii
Hα oblateness	0.72
Hα extension	4 mas
Mass of the disk	$6.4 \cdot 10^{-8}$ $M_\odot$

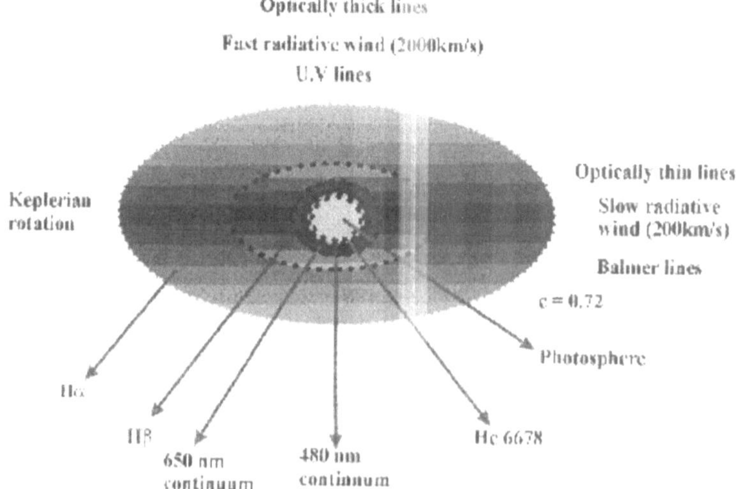

Fig. 1. A picture of γ Cas circumstellar environment from a comparison between our radiative wind model and high angular resolution observations.

Finally our results show that a lemniscat-shaped circumstellar envelope (axisymmetric thin envelopes at the inner edge and thick at the outher rim) is completely inappropriate for γ Cas (see for instance hydrostatic disks models proposed by Hanuschik, 1996). Moreover, a theoritical study by Stee (1998) on B[e] supergiant disks seems to discard the "wind-compressed disk" (WCD) model proposed by Bjorkman & Cassinelli (1993) as a possible senario for the envelope of these supergiants.

γ Cas at different wavelengths

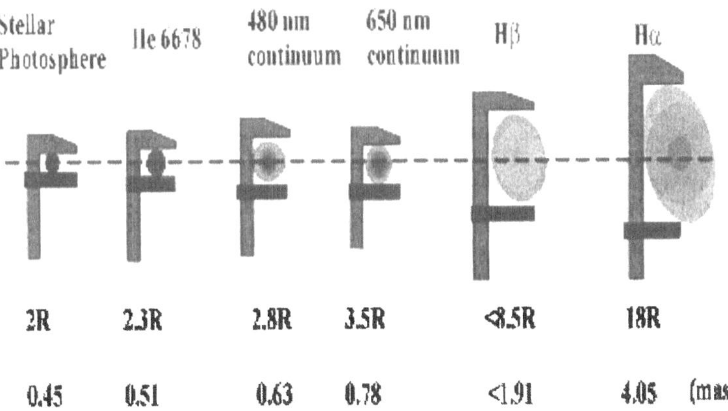

Fig. 2. A schematic view of γ Cas as a function of wavelength from our model according to GI2T observations. These envelope wavelength dependent shapes are projections onto the sky plane. The generic "slide-calliper" mimics the N-S orientation of GI2T's baseline.

References

Bjorkman, J. E., & Cassinelli, J. P. 1993, ApJ, **409**, 429

Friend, D.B., Abbott, D.C. 1986, ApJ, **311**,701

Hanuschik, R.W. 1996, A&A, **308**, 170

Stee, Ph., Araújo, F.X., Vakili, F., Mourard, D., Arnold, L., Bonneau, D., Morand, F. and Tallon-Bosc, I. 1995, A&A, **300**, 219

Stee, Ph. 1996, A&A, **311**, 945

Stee, Ph., Vakili, F., Mourard, D., Bonneau, D. 1998, A&A, **332**, 268

Stee, Ph. 1998, A&A, **336**, 980

Discussion

T. Eversberg: What is the principle idea to solve the baseline accuracy problem on very large scales like the earth's orbit?
P. Stee: The technique (which is effectively applied on ground-based interferometers: Mark III, PT I at Palomar, etc.) is a nano-metric laser-metrology, which in addition to a "dual feed" possibility (a bright reference star in the field of view) make micro-arcsecond interferometry foreseeable in the next decade.

A. Chalabaev: Your model is in contradiction with the major observational discovery made by the Petrov group, namely that the envelope of γ Cas is not centered on the B star, but on an unseen object.
P. Stee: The results presented by the Petrov group in the thesis by Sanchez are not really convincing; it is not possible to explain these observations with any model, even if γ Cas is not centered on the B star. Concerning the new results you mentioned: I am not aware of them, as they are not yet published in any refereed paper. However, having a photocenter displacement with respect to the star does not mean that the matter is not centered on the star.

Michael Marlborough, Philippe Stee and Farrokh Vakili

Session II

Theory of Non-spherical Winds

chair: B. Baschek

Wind-Compressed Disks

Jon E. Bjorkman

University of Toledo, Dept. of Physics and Astronomy, Toledo, OH 43606, USA

Abstract. We discuss the wind-compressed disk (WCD) model and its ability to produce disks in the outflows from rapidly rotating stars. In particular, we discuss the recently discovered non-radial force components that may inhibit the formation of the disk and present preliminary investigations of the ionization distribution and associated line profiles in the WCD model.

1 Introduction

Perhaps the most compelling cases for aspherical mass loss in luminous hot stars are the Be and B[e] stars. Be stars are near-main-sequence stars that have circumstellar disks as evidenced by their double-peaked Hα profiles as well as their intrinsic polarization properties (see K. Bjorkman; these proceedings). Another class of stars possessing circumstellar disks are the LMC B[e] supergiants identified by Zickgraf et al. (1985, 1986). Like Be stars, B[e] stars have a two-component outflow (a high speed, high ionization state wind in combination with a dense, low speed, low ionization state disk). In addition to Be and B[e] stars, there have been suggestions that some Wolf-Rayet stars have equatorial density enhancements (to explain their intrinsic polarization), and the bi-polar shape of the nebula surrounding the LBV η Carinae (the so-called homunculus) can also be explained by the presence of a pre-existing disk in the LBV wind (Frank et al. 1995). How such disks may form is not entirely clear, but Be stars are rapid rotators. In their study of rotating stellar winds, Bjorkman and Cassinelli (1993) have shown how rotation can naturally lead to the production of a dense equatorial disk via a mechanism they call the Wind-Compressed Disk (WCD) model.

2 Wind Compression

In a rotating two-dimensional axisymmetric model, the wind streamlines do not spiral outward on surfaces (cones) of constant latitude, as is often assumed. Instead they bend toward the equator due to the centrifugal and Coriolis forces.

Consider the forces acting on the fluid. For an axisymmetric geometry, the pressure gradient only has r- and θ-components. Although the θ-component is large at the stellar surface (to enforce hydrostatic equilibrium), it drops rapidly beyond the sonic radius. The other forces are gravity and radiation,

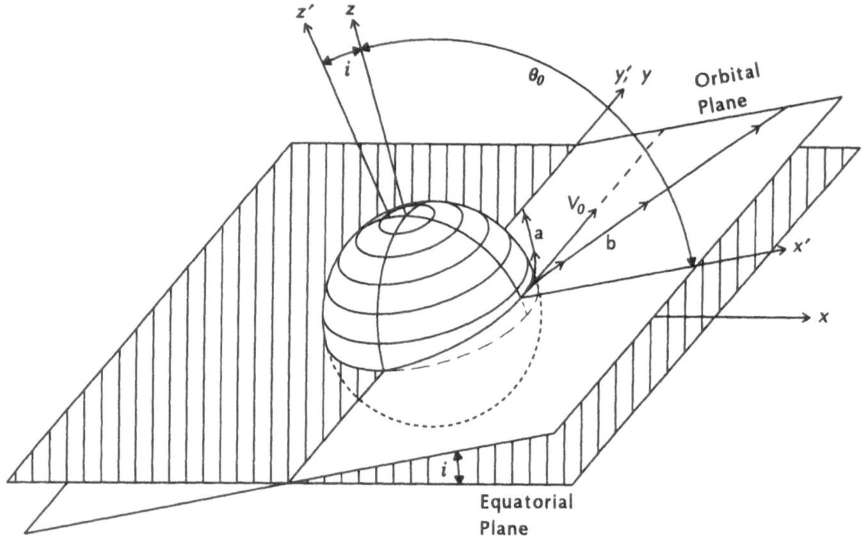

Fig. 1. Orientation of the orbital plane for a streamline originating at a polar angle θ_0. The streamline labeled (a) is a case with a high rotation rate and the streamline labeled (b) denotes a low rotation rate. (Figure from Bjorkman and Cassinelli 1993.)

which are central forces. Thus, beyond the sonic point, there are no external torques, so both the θ- and ϕ-components of the velocity are determined by angular momentum conservation and the streamlines are free particle trajectories corresponding to the external forces (radiation and gravity).

2.1 Orbital Plane

Much can be learned about the location of the streamline by utilizing the fact that gravity and radiation are central forces. Like a Keplerian orbit, the streamline lies in an orbital plane, containing the center of the star, the initial location, and velocity vector, V_0 (see fig. 1). To find the streamline trajectories, we simply rotate the Friend and Abbott (1986) 1-d solution in the equatorial plane up to the initial latitude of the streamline and adjust the rotation velocity by $V_{\rm rot} \to V_{\rm rot} \sin\theta_0$.

Figure 1 shows two trajectories labeled (a) and (b), that correspond to different initial conditions. Trajectory (a) has a slow initial acceleration and occurs when there is a large rotation rate. Trajectory (b) has a fast initial acceleration and occurs when there is a slow rotation rate. Note that as trajectory (a) orbits around the star, it has a decreasing altitude, z, and eventually crosses the equator.

2.2 Disk Formation

Unlike a non-rotating stellar wind (where the streamlines are radial) the fluid elements in a rotating wind tend to orbit the star. Thus material from high latitudes orbits toward the equator where it collides with material from the opposite hemisphere. Since the flow velocities are supersonic, this collision results in a pair of shocks above and below the equatorial plane. Between the shocks, the shock compression produces a thin, dense, wind-compressed disk. Thus the primary effect of rotation is not to change the total mass loss rate from the star; instead it *redistributes* the mass loss such that most of the mass lost from the stellar surface is funneled into the small solid angle of the equatorial disk (typically the opening angle, HWHM, of the disk is only about 3°), which results in the equatorial disk being about 100 times more dense than the polar outflow.

To form the shock-compressed disk, the star must rotate fast enough that the streamlines attempt to cross the equator. Thus there is a minimum stellar rotation speed, above which a disk forms, and the rotation threshold depends on how rapidly the wind is accelerated. Below this rotation threshold there is still some compression toward the equator, but it only produces a mild density enhancement (typically less than a factor of 10). To distinguish these two classes of equatorial density enhancements, we call the rapidly rotating shock compression disks "WCDs", while the cases with rotation speeds below the disk formation threshold, we call "wind compressed zone" (WCZ) models.

2.3 Bistability Mechanism

Although rotation typically does not appreciably change the mass flux from the stellar surface, there is an important exception that occurs for B[e] stars. This exception is the bistability mechanism discovered by Pauldrach & Puls (1990) in their study of the mass loss from P Cyg (see Lamers; these proceedings). They were studying the effect of lowering the surface gravity (i.e., increasing the stellar radius) on the mass loss rate from the star. They found for an effective temperature of 19000 K, that when the surface gravity dropped below $\log g_{\mathrm{eff}} = 1.6$, there was a sudden increase in the mass flux (about a factor of 3) and a corresponding decrease in the wind terminal speed. This was attributed to the wind becoming optically thick in the photoionizing Lyman continuum, which produces a shift in the ionization balance that changes the number of optically thick driving lines in the wind.

This discontinuity in the terminal speed has now been observed in B supergiants by Lamers, Snow, & Lindholm (1995). They find that earlier than $\log T_{\mathrm{eff}} = 4.3$ the data are consistent with $v_\infty/v_{\mathrm{esc}} = 2.7$, and later than that $v_\infty/v_{\mathrm{esc}}$ drops to 1.3. They also find that there may be a second bistability jump at $\log T_{\mathrm{eff}} = 4.0$ where $v_\infty/v_{\mathrm{esc}}$ becomes as small as 0.7.

Since B[e] stars have effective temperatures near the bistability jump, Lamers & Pauldrach (1991) investigated how it may effect the outflow from

B[e] stars. They supposed that the star is rapidly rotating and gravity darkened and estimated the hydrogen Lyman column density as a function of latitude. They found that the star could be optically thin in the polar regions and optically thick in the equatorial latitudes. Thus the bistability jump could occur at some mid latitude, producing denser slower outflow in the equatorial zones, which they argue could be responsible for the B[e] phenomenon. Since the bistability jump is not that large, the density contrast from pole to equator in their models is probably at most about a factor of 10.

Lamers and Pauldrach did not include the 2-D wind compression effects mentioned above. Although we will not attempt to quantitatively combine the bistability mechanism with the wind compressed disk model, it is nonetheless interesting to explore how the two models may interact with each other. If bistability occurs at low latitudes, then the outflow originating at low latitudes has a smaller-than-expected terminal speed and larger mass flux. This implies that the low latitude flow will be more slowly accelerating, which will increase the wind compression effects and lower the threshold stellar rotation speed for forming a disk. On the other hand, wind compression effects increase the low latitude densities, which increases the Lyman column density for two reasons. First, the larger density increases the recombination rates, which lowers the hydrogen ionization fraction and increases the Lyman bound-free opacity. Second, the increased optical depths in the equator prevent the photoionizing flux from reaching the equator and redirect it poleward, since the light will tend to diffuse out the path of least resistance. Decreasing the photoionizing flux also increases the Lyman bound-free opacity, so both these effects imply that the bistability mechanism will occur more easily in the presence of wind-compression effects. Since wind-compression enhances the likelihood of bistability and bistability enhances wind compression, there is now the possibility of feedback that will enhance the disk density even more than one would naively think. Whether or not such a runaway could occur requires detailed computations that we will not attempt. Here we will merely use bistability as a mechanism for motivating the possibility that the low latitude outflow may have terminal speeds as small as $v_\infty/v_{\rm esc} = 1.3$.

2.4 Disk Formation Threshold

Since the disk is formed from the low latitude outflow, we use the low latitude terminal speed, $v_\infty/v_{\rm esc} = 1.3$, to determine the disk formation threshold,

$$\left(\frac{V_{\rm rot}}{V_{\rm crit}}\right)_{\rm th} = \begin{cases} 0.3 & (\beta = 3), \\ 0.8 & (\beta = 0.8), \end{cases} \tag{1}$$

where β is the velocity law exponent (i.e., $v = v_0 + [v_\infty - v_0][1 - R/r]^\beta$). Since we do not know how rapidly a bistable wind accelerates, we have chosen both a slow ($\beta = 3$) and a fast ($\beta = 0.8$) acceleration to illustrate the possible range. Using a typical supergiant escape speed (240 km s^{-1}) to determine the critical speed $V_{\rm crit} = V_{\rm esc}/\sqrt{3}$, the required rotation speed to form a WCD is

$$(V_{\rm rot})_{\rm th} = \begin{cases} 40\,{\rm km\,s^{-1}} & (\beta = 3), \\ 110\,{\rm km\,s^{-1}} & (\beta = 0.8). \end{cases} \tag{2}$$

These values imply that it is relatively easy to form a disk, especially if the wind is slowly accelerating ($\beta = 3$).

2.5 Non-radial Forces

Recently, Owocki, Cranmer, & Gayley (1996, OCG) have questioned whether or not a WCD can form in a line driven wind as originally proposed by Bjorkman & Cassinelli. The radiation force that drives the wind has non-radial components arising from three effects: 1) The penetration probability $\beta = [1 - \exp(-\tau)]/\tau$, where the Sobolev optical depth $\tau \propto 1/|dv/dl|$. Since the velocity gradients are asymmetric with respect to the radial direction, the absorption of stellar photons produces a retarding torque that reduces the rotation of the wind. 2) Similarly, the absorption asymmetry also produces a θ-component of the force directed away from the equator. 3) A rotating star is oblate and gravity darkened. Although the pole is brighter than the equator, the increased solid angle subtended by the equator causes the net radiative flux vector to have a θ-component away from the equator. All three of these effects produce a non-radial component of the force that reduces the amount of wind compression. Although these non-radial forces are an order of magnitude less than the radial component of the radiation force, they are comparable to the *net* radial force, because the radiation force only marginally exceeds the force of gravity. Consequently, OCG find that the non-radial forces produced by these three effects inhibit the formation of a WCD. However, we should also keep in mind that non-radial forces are not necessarily bad; it depends on their direction. For example, if the terminal speed were to increase toward the equator instead of decreasing (as usual), then the direction of the non-radial forces would be toward the equator, enhancing the disk (see Puls; elsewhere these proceedings).

OCG use the Castor, Abbot, and Klein (1975, CAK) parameterization of the line driving force as modified by Abbot (1982), which uses a (photospheric) core and (optically thin) halo approximation for the photoionization balance of the wind. In OCG's calculation of the WCD inhibition, they assume that the wind is smoothly accelerating (so that the Sobolev approximation determines the radiation force) and that the CAK parameters (k, α, δ) are constant throughout the wind. As we discussed earlier, an optically thick disk redirects photoionizing radiation away from the equatorial regions producing lower ionization states. There is also observational evidence for latitudinal ionization gradients in the winds of rapidly rotating stars (Bjorkman et al. 1994), and in the next section we will see that the wind can dramatically increase its ionization state as the wind moves away from the star. This position dependence of the ionization balance implies that the CAK force parameters are not constant throughout the wind as assumed by OCG. Although changing the force parameters (k, α, δ) will not alter the direction of the non-radial force, it could seriously change its magnitude. For example, if the ionization balance shifts to drastically decrease the radiation force shortly after the wind is initiated, then gravity will dominate thereafter, and a WCD will form. Similarly if the wind driving lines become optically thin, the escape probability is isotropic, so there would no longer be any non-radial force components (unless the star is oblate). Finally if the wind is clumpy and not smoothly accelerating, the radiation force is no longer determined by the Sobolev approximation; it is determined instead by the physical geometry of the clump. The acceleration of the clump depends inversely on its mass; thus, if the clump is massive enough and forms close enough to the star (so that its outward velocity is less than the escape speed), then gravity would dominate, causing the clump to orbit/fall into the disk according to the WCD mechanism. The above issues indicate that, to determine when and where WCDs form, it is essential that we perform detailed calculations of the wind dynamics and ionization to determine how the CAK parameters (k, α, δ) change with location in the wind.

3 WCD Ionization

To begin a preliminary investigation of the ionization distribution within the WCD model, we employed MacFarlane's NLTE wind ionization code, which is a spherically symmetric model using the Sobolev approximation for the line transfer (MacFarlane et al. 1993). For this initial attempt to calculate the 2-D ionization distribution, we used a piecewise spherical approximation; i.e., a series of spherically symmetric models, one for each latitude.

3.1 Ionization Distribution

Figure 2 shows the ionization fraction of N V, C IV, and Si IV throughout the wind of a B2.5 IV star rotating at 80% critical with an X-ray emission measure of 10^{53} cm^{-3}, which corresponds roughly to $L_X \sim 10^{30}$ erg s^{-1}, the maximum B star X-ray luminosity observed by ROSAT (Cohen et al. 1997). Note that as the wind moves away from the star, Si IV is rapidly ionized (between 1 and 2 stellar radii). At large radii, Si VI and Si VII are the dominant states. Similarly, there is a latitudinal ionization gradient. Since Si IV is below the dominant state, it is enhanced in the denser equatorial regions. Conversely N V, which is above the dominant state, is enhanced at the poles and destroyed near the equator. Finally although C IV is mostly produced close to the star (from 1.1 to 2 stellar radii), it does persist to large radii (albeit at somewhat low levels). It is also enhanced near the equator, but destroyed in the disk.

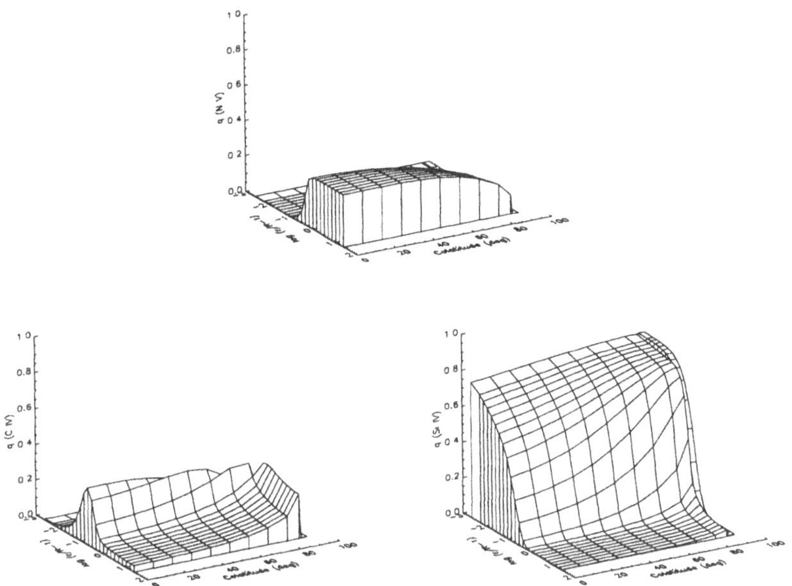

Fig. 2. WCD Ionization Fractions. Shown are the ionization fractions, q, of N V (top), C IV (bottom left), and Si IV (bottom right) as a function of position, $\log(r/R-1)$, and co-latitude, θ.

3.2 Line Profiles

Figure 3 shows the UV resonance line profiles associated with the WCD ionization fractions shown in Figure 2. These profiles were calculated using a Monte Carlo simulation that properly accounts for non-monotonic velocity fields and any associated non-local coupling of the diffuse radiation between common point surfaces in the WCD model. It also includes a realistic line blanketed stellar photosphere for the lower boundary. Note that the Monte Carlo simulation includes stellar rotation and that the small spectral features in the calculated profiles are rotationally broadened photospheric features and are not "statistical noise" in the simulation.

The solid profiles are for edge-on ($i = 90°$) observers, while the dashed profiles are for pole-on ($i = 0°$) observers. Note that the blue edge of C IV and N V is quite sharp (because the ions persist to large radii) and that the polar edge velocity is larger than the equatorial. This is because the terminal speed of the wind decreases toward low latitudes. On the other hand, Si IV has a shallow return to the continuum without a sharp blue edge and no emission. Surprisingly, the polar edge velocity of Si IV is *smaller* than the equatorial edge velocity (opposite to the trend with C IV and N V). This peculiar behavior is a result of two factors. First, Si IV is destroyed close to

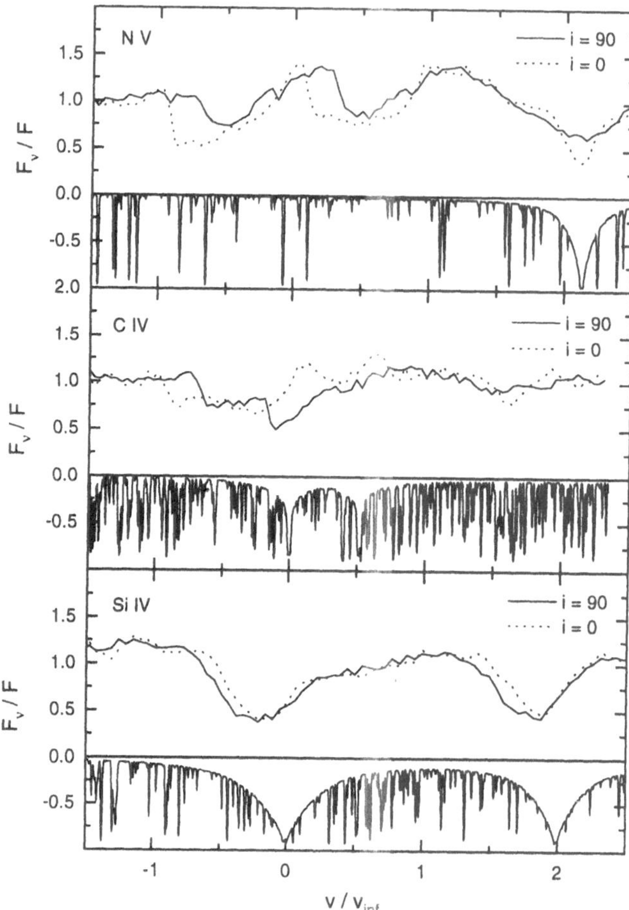

Fig. 3. WCD Line profiles. Shown are the resonance line doublet profiles of N V (top), C IV (middle), and Si IV (bottom) for a pole-on observer (dotted line) and an edge-on observer (solid line). Shown below each profile is the input stellar photospheric spectrum.

the star, which explains the shallow return to the continuum without a sharp blue edge. It also explains the lack of emission as a result of occultation by the star. Second, Si IV is below the dominant ionization stage and is enhanced near the equator. Thus Si IV persists to larger radii in the equator than in the pole. As a result it can absorb to higher velocities in the equator than in the pole despite the fact that the wind terminal speed is lower in the equator.

4 Conclusions

The ionization distribution clearly shows a large sensitivity to location in the wind. In particular the rapid ionization destroying Si IV close to the star while creating Si VII far away implies that the CAK line parameters are very likely to change quite dramatically. In general we expect that the large increase in ionization state implies that most strong lines will exist in the EUV where there is very little stellar flux. The remaining lines in the Balmer continuum are likely to be weak optically thin lines. Thus we speculate that the line driving will shift from strong lines to weak lines causing a decrease in the CAK parameter α quite close to the stellar surface. Whether this could remove the disk inhibition by non-radial forces remains to be seen.

References

Abbott, D.C. 1982, ApJ, 259, 282
Bjorkman, J. E., & Cassinelli, J. P. 1993, ApJ, 409, 429
Bjorkman, J.E., Ignace, R., Tripp, T.M., & Cassinelli, J.P. 1994, ApJ, 435, 416
Castor, J. I., Abbott, D. C., & Klein, R. I. 1975, ApJ, 195, 157 (CAK)
Cohen, D.H., Cassinelli, J.P., & MacFarlane, J.J. 1997, ApJ, 487, 867
Frank, A., Balick, B., Davidson, K. 1995, ApJ, 441, L77
Friend, D. B., & Abbott, D. C. 1986, ApJ, 311, 701
Lamers, H.J.G.L.M., & Pauldrach, A.W.A. 1991, A&A, 244, L5
Lamers, H.J.G.L.M., Snow, T.P., & Lindholm, D.M. 1995, ApJ, 455, 269
MacFarlane, J.J., et al. 1993, ApJ, 379, 659
Owocki, S.P., Cranmer, S.R., & Gayley, K.G. 1996, ApJ, 472, L115 (OCG)
Pauldrach, A.W.A., & Puls, J. 1990, A&A, 237, 409
Zickgraf, F.-J., Wolf, B., Stahl, O., Leitherer, C., & Klare, G. 1985, A&A, 143, 421
Zickgraf, F.-J., Wolf, B., Stahl, O., Leitherer, C., & Appenzeller, I. 1986, A&A, 163, 119

Discussion

A. Feldmeier: You mentioned that the azimuthal velocity law in the disk is non-Keplerian. Can you actually specify $v_\phi(\mathrm{r})$? Shouldn't it drop faster than $1/\mathrm{r}$?

J. Bjorkman: It depends on how the low specific angular momentum material mixes as it is added to the disk. Numerical simulations by Owocki show that v_ϕ in the equator does in fact fall as $1/\mathrm{r}$, so this mixing must not be large.

A. Maeder: The simple application of the wind momentum – luminosity relation, $\dot{\mathrm{M}}\, v_\infty = \mathrm{L/c}$, suggests an enhanced equatorial mass loss rather than the opposite. This is because the local luminosity on the right side goes like the effective gravity, while the terminal velocity goes like the square root of g_{eff}.

J. Bjorkman: As you suggest, using the von Zeipel theorem to obtain the radiation flux vs. latitude can cause the polar mass loss rate to be larger than the equatorial one. Stan Owocki has studied this effect and the results depend on the value of the CAK parameter, which determines the scaling of the mass flux vs. radiative flux. He finds that indeed the polar mass loss rate can exceed the equatorial one.

H. Henrichs: Since Be-star disks are time-variable, your new considerations that make the disk exist or not strongly improve the chances that the WCD model is applicable to such systems. May I encourage you to consider the implications of these "other effects", in particular what the relevant time scales are to build up or destroy a disk.

J. Bjorkman: Since the model is a time-independent model, any time dependence must be put in by hand. For example, one can vary the input parameters to get time variability. The time scale for variability is the radial flow time scale in the disk which is of the order of a few days. Any longer time scale would have to arise from the input parameters.

Alexander Fullerton and Derck Massa (Jon Bjorkman in the background)

Non-spherical Radiation-Driven Wind Models

Joachim Puls[1], Peter Petrenz[1], and Stanley P. Owocki[2]

[1] Universitätssternwarte, Scheinerstraße 1, D-81679 München, Germany
[2] Bartol Research Institute of the University of Delaware, Newark, DE 19716, USA

Abstract. The present state of modelling radiatively driven stellar winds from rapidly rotating stars is reviewed. Various processes affecting the actual, still controversial wind structure are highlighted, in particular non-radial line-forces and gravity darkening, and useful scaling relations are provided. The importance of accounting for *consistent* NLTE line-forces depending both on the actual density structure and radiation field (as function of latitude and radius) is stressed, and some independent test calculations confirming earlier numerical results are reported.

1 Geometry and basic assumptions

In the following, we consider a spherical co-ordinate system with unit vectors ($\mathbf{e}_\theta$, $\mathbf{e}_\phi$, $\mathbf{e}_r$), where the polar unit vector $\mathbf{e}_\theta$ is directed *towards the equator* (right-handedness). The co-ordinate θ denotes the co-latitute, i.e., $\theta = 0$ at the pole. Our basic assumption concerns the symmetry of the system, which we require to be *axisymmetric*. Thus, all derivatives with respect to ϕ vanish identically in the equations of continuity and momentum (comprising gravity and radiation forces) and the hydrodynmical variables $(\rho, v_\theta, v_\phi, v_r, p)$ become a function of (θ, r, t) only. The plasma is considered as isothermal at temperature $T_{\rm eff}$, which closes the system (5 variables, 5 equations) if the appropriate boundary conditions including stellar rotation are specified.

2 1-D solution in equatorial plane

The basic effects of rotation on a radiation driven wind can be studied most simply by considering only particles in the equatorial plane (cf. Friend & Abbott 1986 and Pauldrach, Puls & Kudritzki (PPK) 1986). With the assumption of a *purely radial* line acceleration, the angular momentum remains conserved (only central forces), and the rotational speed is given by $v_\phi(r) = v_{\rm rot}(R_*)R_*/r$. Thus, the usual equation of motion is modified by the centrifugal acceleration only, which leads to an effective gravity of $GM(1 - \Gamma)/r^2(1 - \Omega^2 R_*/r)$, where Γ accounts for the Thomson-acceleration and Ω is the ratio of rotational speed at the surface $v_{\rm rot}(R_*)$ to the break-up velocity $v_{\rm crit} = v_{\rm esc}/\sqrt{2}$, with $v_{\rm esc}$ the photospheric escape velocity.

Without rotation, the mass-loss rate of radiation driven stellar winds scales as (cf. PPK, Kudritzki et al. 1989, Puls et al. 1996)

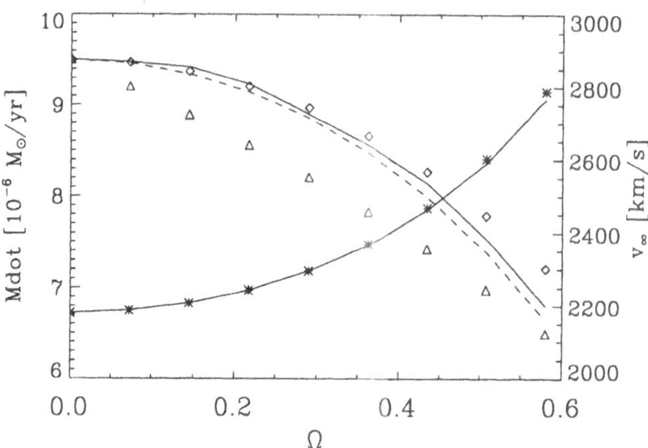

Fig. 1. Influence of rotation on $\dot{M}$ and v_∞ for particles in the equatorial plane. Stellar model: "generic OfV star" (cf. PPK) with $v_{\rm crit} = 690$ km/s. Fully drawn: exact solutions; asterisks: scaling relation for $\dot{M}$ (Eq. 2); dashed: new scaling relation for v_∞ (Eq. 4, $\beta = 0.8$); diamonds: simple scaling relation for v_∞ (Eq. 3); triangles: scaling relation for v_∞ by Friend & Abbott (1986)

$$\dot{M} \propto (kL)^{1/\alpha'} \big(M(1-\Gamma)\big)^{1-1/\alpha'}, \; \alpha' = \alpha - \delta \tag{1}$$

with luminosity L and the usual force-multiplier parameters (k, α, δ). Since the critical point of the equation of motion (determining $\dot{M}$) remains close to the stellar surface if the so-called "finite disk correction factor" is included into the radiation force (PPK), i.e., $R_*/r_{\rm crit} \approx 1$, the only difference induced by rotation is the modification of the effective mass by roughly a factor of $(1 - \Omega^2)$. Thus, we expect the mass-loss rate to scale with

$$\dot{M}(\Omega) = \dot{M}(0)(1 - \Omega^2)^{1-1/\alpha'} \tag{2}$$

A comparison with the "exact" solution obtained from the code developed by PPK confirms the validity of this expression (Fig. 1). Since the scaling law for the terminal velocity (without rotation) suggests a proportionality $v_\infty \propto v_{\rm esc} \propto (M(1-\Gamma))^{1/2}$, the most simple idea to include rotation is

$$v_\infty(\Omega) = v_\infty(0)(1 - \Omega^2)^{\frac{1}{2}} \tag{3}$$

A slightly more refined investigation accounting for the change of $\dot{M}$ and the decreasing influence of the centrifugal support in the outer wind leads to a more complicated expression, where $x = v_\infty(\Omega)/v_\infty(0)$ results from the solution of the non-linear equation

$$x^2 = (1-\Omega^2)^{1-\alpha'} x^{2\alpha-\delta}\left(1+\frac{v_{\rm esc}^2}{2\beta v_\infty^2(0)}\right) - \frac{v_{\rm esc}^2}{2\beta v_\infty^2(0)}, \tag{4}$$

which collapses to Eq. 3 for small values of $v_{\rm esc}/v_\infty(0)$ and δ. ($\beta \approx 0.7\ldots1.3$ is the exponent of the typical velocity law valid for radiation driven winds). A comparison of both relations with the exact solution as well as with the expression provided by Friend & Abbott (1986) is given in Fig. 1.

3 Wind compressed disks and zones

The details of rotationally induced wind compressed disks (WCDs) and wind compressed zones (WCZs) (Bjorkman & Cassinelli 1993) are reviewed by J. Bjorkman elsewhere in this volume. The basic idea follows again from the assumption of *purely radial* line forces. Thus, the specific angular momentum is conserved not only for particles in the equatorial plane, but for all particles, and their motion is restricted to the orbital plane to which they belong (tilted by an angle of co-latitude θ_0 from which they start). Furthermore, pressure forces are negligible in the supersonic region, and the free flow of the particles can be simulated by a corresponding 1-D treatment as above, however, of course, with a centrifugal acceleration modified for the actual azimuthal velocity in the orbital plane, $v_\phi(r) = v_\phi(\Theta=\pi/2, r)\sin\theta_0$. Thus, the mass-loss rates and terminal velocities for particles starting at θ_0 follow the same scaling relations as above, however with Ω^2 replaced by $(\Omega\sin\theta_0)^2$: $\dot M$ increases and v_∞ decreases towards the equator! In the course of the particles' motion, their azimuthal angle $\Phi'(r)$ in the orbital plane is changing, and they are deflected towards the equator. If Φ' becomes larger/equal $\pi/2$, the particles would cross the equator, where they collide supersonically with particles from the other hemisphere: finally, a wind compressed disk is formed by this process. (For further consequences, cf. Bjorkman, this volume.) If we assume a β-law for the radial velocity component, $v_r(r) = v_\infty(1-b/r)^\beta$ with b derived from the minimum velocity at $r = R_*$, we find for the asymptotic azimuthal angle

$$\Phi'(r\to\infty) = \frac{1}{\beta-1}\frac{R_*}{b}\frac{v_{\rm rot}(R_*)\sin\theta_0}{v_\infty(\theta_0)}\left(1-\frac{b}{R_*}\right)^{1-\beta}, \tag{5}$$

which immediately suggests that disk formation or a significant compression towards the equator can occur only under the following conditions (Φ' has to be large): large $v_{\rm rot}$, small v_∞, *flat velocity field*, i.e., large β. It is important to realize that models with a *typical* velocity law ($\beta = 0.7\ldots0.8$) give rise to only weak compression factors unless the ratio $v_\infty/v_{\rm rot}$ is extremely low.

Using the same assumptions as outlined above (radial line force), Owocki, Cranmer & Blondin (1994) confirmed the analytical WCD model by performing time-dependent hydrodynamical simulations. Their results were, except from some interesting details, in good agreement with the predictions by Bjorkman & Cassinelli. In order to obtain a compressed disk, however, they had to use a fairly small value of $\alpha' = 0.35$, ($\alpha = 0.51, \delta = 0.16$), since for larger values the radial velocity gradient became too large to lead to disk formation, in agreement with the above constraints.

4 Non-radial line forces

All results derived so far rely on the assumption of purely radial line-forces, implying the conservation of angular momentum. The line acceleration, however, is a vector quantity, defined per line by (in the Sobolev limit and accounting for an optically thin continuum)

$$\mathbf{g}^{\mathrm{rad}}(\mathbf{r}) = \frac{\bar{\chi}_L(\mathbf{r})}{c\rho(\mathbf{r})} \int_{\omega_c} I_*(\mathbf{n})\mathbf{n}\frac{1-\mathrm{e}^{-\tau_S(\mathbf{r},\mathbf{n})}}{\tau_S(\mathbf{r},\mathbf{n})}\mathrm{d}\omega\,;\; \tau_S(\mathbf{r},\mathbf{n}) = \frac{\bar{\chi}_L(\mathbf{r})\lambda}{|\mathbf{n}\cdot\boldsymbol{\nabla}(\mathbf{n}\mathbf{v}(\mathbf{r}))|}, \quad (6)$$

where $\bar{\chi}_L$ is the frequency integrated line-opacity, ω_c the solid angle subtended by the stellar disk, I_* the stellar intensity, τ_S the Sobolev optical depth and the denominator of τ_S contains the directional derivative of the local velocity field. From Eq. 6, it is obvious that non-radial components of $\mathbf{g}^{\mathrm{rad}}$ are present under two conditions: either the stellar intensity depends on direction (next section), or τ_S depends on direction. Note, that in the latter case for $I_* =$ const only optically thick lines induce non-radial components, which are then *proportional to asymmetries* in the derivative. Speaking in physical terms, the resulting accelerations depend on the differences in both intensities and optical depths along the various contributing lines of sight.

Let us first consider the azimuthal component g_Φ. For this quantity, the directional derivative can be written as $f_r + \sin\varphi\, f(\frac{\partial(v_\Phi/r)}{\partial r})$, where f_r is the usual symmetric radial component, and $\sin\varphi$ is proportional to the azimuthal component of $\mathbf{n}$ (defined in the same sense as Φ). Thus, the direction of the resulting azimuthal acceleration depends only on the sign of $\partial(v_\Phi/r)/\partial r$, which is negative in all cases of rotational laws slower than solid rotation. As a result, even for a symmetric core intensity (i.e., only radial fluxes), optically thick lines induce an acceleration *antiparallel* to $\mathbf{e}_\Phi$, i.e., a spin-down of the wind and loss of angular momentum. This effect was firstly investigated by Grinin (1978). For a further discussion, see Owocki et al. (1997) and Owocki, this volume.

With respect to g_Θ, the argumention is analogous. The decisive term in the velocity gradient inducing the required asymmetry is given here by $r\frac{\partial(v_\Theta/r)}{\partial r} + \frac{\partial v_r}{r\partial\Theta}$, where the second term is the important one. As long as the radial velocity law at the equator is slower than in polar regions – an almost inevitable consequence of rotation due to the reduced escape velocity $\propto (1 - \Omega^2\sin^2\Theta)^{1/2}$ –, this gradient is negative, implying a larger total directional derivative (including the symmetric radial component) towards the pole than towards the equator: Again, g_Θ is anti-parallel to $\mathbf{e}_\Theta$! Note, that this direction could be only reversed if the radial expansion at the pole were *slower* than at the equator.

Finally, by summing up the contribution of all lines (via folding with the line-strength distribution function), a polewards acceleration is created which is sufficient to stop the equatorwards motion predicted by WCD/WCZ-models and actually reverses its direction. This result (for details see the "stopping length analysis" by Owocki et al. 1997) is the so-called "inhibition effect" (Owocki et al. 1996, Bjorkman, this volume), which inhibits the formation of a wind-compressed disk due to the reversal of the (supersonic) velocities v_Θ, which are directed towards the equator only under the assumption of purely radial line-forces.

5 Distortion of the surface and gravity darkening

Because of the centrifugal acceleration $\propto \sin\Theta$, the surface of the star becomes distorted. The degree of deformation can be calculated by using an appropriate Roche potential (Collins 1963, Collins & Harrington 1966). Some applications are given by Cranmer & Owocki (1995) and Petrenz & Puls (1996). Moreover, von Zeipel's theorem may be used (for a discussion, see A. Maeder, this volume), which relates the flux and thus the effective temperature with the effective gravity as function of Θ. In consequence, the poles become hotter and the equator cooler than the average $T_{\rm eff}$ appropriate for the stellar luminosity. This change in the radiation field has a number of consequences: at first, the ionization/excitation balance becomes a function of co-latitude due to the varying radiation field (next section). Second, the polar wind plasma is irradiated by a larger flux compared to the equatorial one, which enhances the radiative acceleration and thus the expected mass-loss from this region. For a wind with typical exponent $\alpha' = 0.5$ illuminated by a uniformly bright disk, the mass-flux varies as

$$\dot{M}(\Theta) \propto (1 - \Omega^2 \sin^2\Theta)^{-1} \qquad \alpha' = 0.5, \text{ uniform disk} \tag{7}$$

(cf. Eqs. 1, 2), i.e., increases from pole to equator. If one accounts for gravity darkening, however, $L^{1/\alpha'} \rightarrow (F(\Theta)R_*^2(\Theta))^{1/\alpha'} \sim (g_{\rm eff}(\Theta)R_*^2(\Theta))^{1/\alpha'}$, an *additional* dependence on $[M_{\rm eff}(1 - \Omega^2 \sin^2\Theta)]^{1/\alpha'}$ is created. In total, we find

$$\dot{M}(\Theta) \propto (1 - \Omega^2 \sin^2\Theta)^{+1} \qquad \text{with gravity darkening} \tag{8}$$

(cf. Owocki et al. 1997, 1997a). Thus, the inclusion of gravity darkening (increased polar radiation flux) can reverse the situation and leads to a larger polar mass-flux.

Finally, the absorbing lines at mid-latitudes are irradiated by an asymmetric intensity, where the polar contribution is stronger and hotter than the equatorial one. This effect partially counteracts the optical depth effect

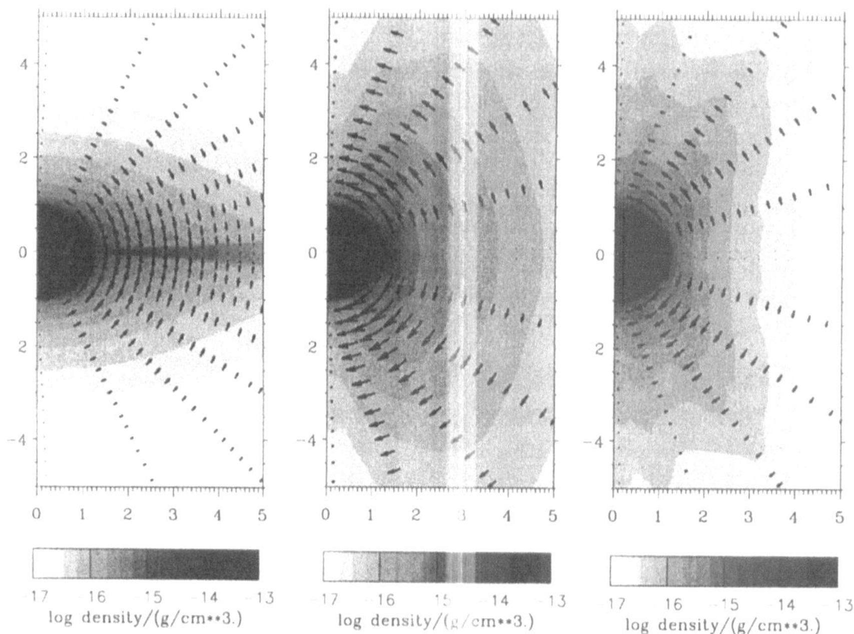

Fig. 2. Radiation driven wind models for a rapidly rotating B-star (see text). Model paramters (mean values): $T_{\rm eff} = 20,000$ K, $\log g = 4.1$, $R_* = 4\,R_\odot$, $v_{\rm esc} = 690$ km/s, $v_{\rm rot} = 350$ km/s, line force parameters: $k = 0.61$, $\alpha = 0.51$, $\delta = 0.16 \rightarrow \dot{M} = 3 \cdot 10^{-10} M_\odot/{\rm yr}$.
Hydrodynamics: ZEUS-2D (Stone & Norman 1992). Lower boundary: rotationally distorted, stair-casing (Owocki et al. 1994) plus "no slip boundary condition".

(cf. Eq. 6) and leads to a meridional acceleration g_Θ being smaller than in the uniformly bright disk case, however still sufficiently large to induce a polewards directed velocity.

As a brief summary and as an independent confirmation of the various numerical models published so far, Fig. 2 shows three different wind models for a rapidly rotating B-star calculated by one of us (P.P.) using the ZEUS-2D-code (Stone & Norman 1992), accounting for rotational distortion of the surface (for details, see caption). The first one (left) displays the formation of a wind-compressed disk, if only radial line-forces are accounted for. The second one (middle) demonstrates the inhibition effect, if non-radial forces are included and a uniformly bright stellar disk is assumed. This model shows the largest mass-flux still at the equator (Eq. 7). Model 3 (right) accounts additionally for gravity darkening: Now, the mass-flux is highest at the poles (Eq. 8), and the polar velocities (arrows, on scale) are smaller than for Model 2, in agreement with our reasoning from above.

6 Outlook – *Consistent* NLTE line forces

In all calculations performed so far, the total line force was calculated within the force-multiplier concept, where the corresponding parameters were assumed *to be constant throughout the wind*, i.e., independent both of r and Θ and taken from tables designed to cope with the 1-D situation. One has to remember, however, that these parameters are (or more precisely: should be) the result of consistent NLTE-calculations, which are affected by at least two processes: i) The Θ-dependence of both density and illuminating radiation field can have a significant effect on the ionization stratification, which in turn leads to a variation of the force-multiplier parameters with respect to co-latitude. As an example, Fig. 3 displays the ionizing radiation field for the rapidly rotating B-star wind model discussed above, when gravity darkening is included. The differences with respect to Θ are largest ($\approx$ 5000 K) close to the surface, where the ionization is controlled by the local radiation field $\propto T_{\text{eff}}(\Theta)$. At larger distances from the surface, the plasma "sees" radiation from all angles between pole and equator, so that the radiation temperatures converge to similar values for $r \to \infty$. ii) Second, the assumption of constant force-multiplier parameters as function of r is rather questionable even for a uniformly bright disk (cf. Kudritzki et al. 1997). Fig. 4 displays the behaviour of α for a spherical wind in dependence on the optical depth variable $t \propto \rho/(\mathrm{d}v/\mathrm{d}r)$ and different "mean densities" $\log n_e/W$. The overall trend of α becoming smaller for decreasing t reflects the decreasing influence of optically thick lines on the acceleration, if the wind plasma becomes thinner in the course of its outward trajectory. The reader may note, however, that the specific behaviour of $\alpha(t)$ varies strongly with spectral type and mean wind density (i.e., in concert with the dominant ions): one finds also cases with α increasing with decreasing t at first, before this behaviour is finally reversed.

Presently, incorporation of these effects into a consistent 2-D description is under way in our working group. A number of simulations for rapidly rotating O-star winds with significant mass-loss have been performed, to demonstrate operation of the basic algorithms. Before specific results can be published, however, some final test calculations have to be undertaken. Especially in those cases as displayed in Fig. 4, when α is decreasing with t, a number of interesting effects are to be expected, since then the acceleration of the outer wind (low α) can become too small for the mass-loss rate created at the wind base (larger α).– Hopefully, the final simulations will help to answer the question concerning the "real" wind structure of rapidly rotating winds as function of spectral type and luminosity class. Note, however, that magnetic fields *if* present with sufficient strength (cf. Mathys, this volume) could again alter the conclusions.

One of our final goals is the incorporation of continuum optical depth effects which in the B-supergiant domain can induce the so-called "bi-stability" mechanism (Pauldrach & Puls 1990). The decisive quantity which controls this behaviour is the optical depth in the Lyman continuum, which – for

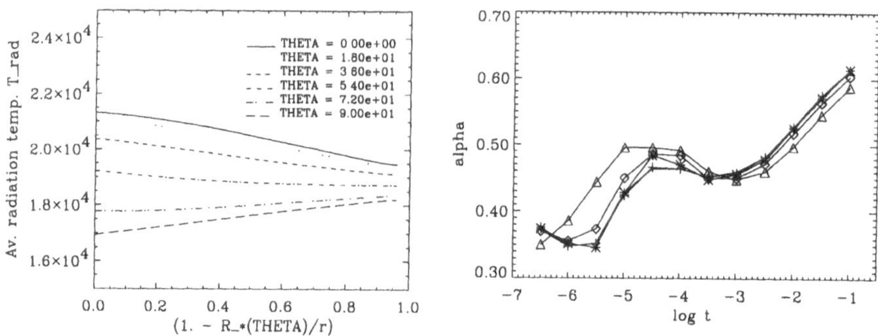

Fig. 3. Ionizing radiation temperature for a rapidly rotating B-star wind model (cf. Fig. 2) including gravity darkening, averaged over illuminating disk, as function of Θ and height over surface.

Fig. 4. NLTE-force multiplier parameter α as function of optical depth variable $t \propto \rho/(\mathrm{d}v/\mathrm{d}r)$ for a one-dimensional wind with $T_{\mathrm{eff}} = 20000$ K and different "densities" $\log n_e/W = 11.5$ (asterisks), $11.0, 10.5, 10.0, 9.5$ (triangles). Illuminating photospheric spectrum from Kurucz fluxes (Kudritzki & Springmann, priv. comm.).

rotating stars and accouting for gravity darkening – depends sensitively on rotationally induced variations of wind density and radiation temperature and thus becomes a strongly varying function of Θ (Lamers & Pauldrach 1991; Lamers, this volume, however also Owocki et al. 1997). Due to this behaviour, the bi-stability effect is thought to be responsible for the B[e] phenomenon (Lamers & Pauldrach 1991) if the star rotates close to break-up and has a rather large (undisturbed) polar mass-loss. It has to be seen if the observed bimodal B[e] wind structure (e.g. Zickgraf et al., this volume) can be actually explained by simulations accounting for consistent line-forces.

References

Bjorkman, J.E., Cassinelli, J.P., 1993, ApJ 409, 429
Collins, G.W.II, 1963, ApJ 138, 1134
Collins, G.W.II, Harrington, J.P., 1966, ApJ 146, 152
Cranmer, S.R., Owocki, S.P., 1995, ApJ 440, 308
Friend, D.B., Abbott, D.C., 1986, ApJ 311, 701
Grinin, A., 1978, Sov. Astr. 14, 113
Kudritzki, R.P., Pauldrach, A., Puls, J., Abbott, D.C., 1989, A&A 219, 205
Kudritzki, R.P., Springmann, U., Puls, J., et al., 1997, PASPC 131, 299
Lamers, H.J.G.L.M., Pauldrach A.W.A., 1991, A&A 244, L5
Owocki, S.P., Cranmer, S.R., Blondin, J.M., 1994, ApJ 424, 887
Owocki S.P., Cranmer S.R., Gayley, K.G., 1996, ApJ 472, L1150
Owocki, S.P., Cranmer, S.R., Gayley, K.G., 1997. In: Proceedings of Workshop on *B[e] Stars*, held in Paris, France, June, 1997, ed. A. Hubert, Kluwer, in press.

Owocki, S.P., Gayley, K.G., Cranmer, S.R., 1997a, PASPC 131, 237
Pauldrach, A.W.A., Puls, J., Kudritzki, R.P., 1986, A&A 164, 86 (PPK)
Pauldrach, A.W.A., Puls, J., 1990, A&A 237, 409
Petrenz, P., Puls, J., 1996, A&A 312, 195
Puls, J., Kudritzki, R.-P., Herrero, A., et. al., 1996, A&A, 305, 171
Stone, J.M., Norman, M.L., 1992, ApJS 80, 753

Discussion

N. Langer: The von Zeipel theorem is based on assumptions which do not hold for the luminous stars you consider. It should not be applied to these stars.
J. Puls: We are well aware of this problem. In our present philosophy, however, we regard the temperature difference we obtain from the von Zeipel theorem as a sort of maximum effect which has to be investigated.
A. Maeder: Corrections to the von Zeipel theorem due to the non-cylindrical rotation and the fact that real stars are not barotropic are leading to still higher polar flux; thus this would even reinforce the polar mass flux.
J. Puls: So, we even have to correct our previous maximum temperature difference argument. In this case, the resultant structure would be even more elongated in the polar direction.

Joachim Puls, Gloria Koenigsberger and Wolfgang Glatzel

Radiation-Driven Disk Winds

Janet E. Drew, Daniel Proga, and René D. Oudmaijer

Department of Physics, Blackett Laboratory,
Imperial College of Science Technology & Medicine,
Prince Consort Road, London SW7 2BZ, United Kingdom

Abstract. Recent work on numerical modelling of radiation-driven disk winds is summarised and discussed in the context of extreme B emission line stars.

1 Motivation

Bright disks are found in many astrophysical settings: around young stars, in AGN, in interacting binaries and, indeed, they probably occur in the extreme B emission line stars typically classified as B[e] stars. Such disks may shine either by virtue of radiative energy generation due to accretion, or as a consequence of reprocessing light from a luminous central object that falls upon them. In either case, radiative momentum is available to direct a flow away from the disk surface. Just as luminous spherical stars can power significant mass loss via radiation pressure, provided the stellar luminosity exceeds $\sim 0.001 L_E$ (where L_E is the Eddington luminosity), so too may a circumstellar disk. The challenge, that has only recently been met, is to demonstrate how this is realised within the more complex disk geometry.

Our work in this area is now described in three published papers: namely, Proga, Stone & Drew (1998); Drew, Proga & Stone (1998) and Oudmaijer et al. (1998). In this short contribution we draw attention to those aspects of this work most relevant to the study of luminous hot stars. It complements the efforts in recent years to explain the *origin* of circumstellar disks in terms of either wind compression (see Bjorkman, this volume) or 'bistability' (see Lamers this volume). Our philosophy is that we do not attempt at this stage to explain both circumstellar disks and the flows associated with them – instead we assume that an optically-thick, geometrically-thin disk exists and then numerically model the radiation-driven outflow that may arise from it. Our models will accordingly apply directly where circumstellar disks are nearly Keplerian, and a product of non-radiative dynamical processes (e.g. accretion, or excretion as a result of pulsation or prior stellar dynamical evolution).

2 Method

Over the past decade or so, there have been several onslaughts on the disk wind problem. The majority have been semi-analytic treatments (e.g. Vitello

& Shlosman 1988, Murray et al. 1995) that achieved little predictive power. Our approach owes most to the pioneering work of Icke (1981). We have adopted a numerical approach similar to his in order that approximation of the truly multi-dimensional character of the problem be kept to a minimum. We have moved on from Icke's electron-scattering models in that we treat line-driving using a CAK-like parameterisation of the radiative acceleration (see Proga et al. 1998 for details). The hydrodynamical code we have used is a version of ZEUS-2D (Stone & Norman 1992) in which the third spatial dimension is accounted for by axisymmetry and conservation of angular momentum.

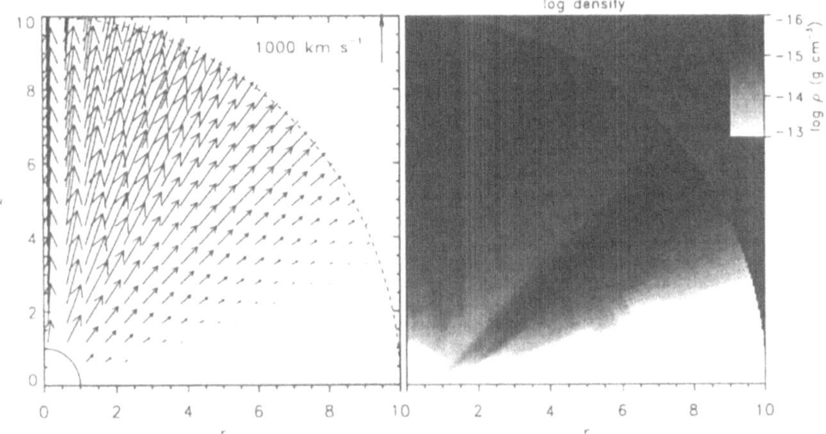

Fig. 1. The velocity field (left) and grey-scale density map (right) for a model of an evolved B star with a Keplerian circumstellar disk, described in the text. The rotational axis of the disk is along the left hand side vertical frame in each panel, while the disk midplane is along the bottom horizontal frame. The vertical and horizontal spatial scales are both expressed in units of the stellar radius. To emphasise the density changes in the outflow, we set the greyscale upper limit to saturate at 10^{-13} g cm^{-3}. Note that the left panel omits the rotational component of the velocity.

In our more recent models, calculated for two early B-star examples (Drew et al. 1998, Oudmaijer et al. 1998), the surfaces of both the circumstellar disk and star are included in the calculations as mass-losing and radiant boundaries. Purely to ease gridding of the computational domain, the star is assumed to be non-rotating (this is a convenient rather than a shaping feature of the models). However, an assumption that is essential to the outcome is that the circumstellar disk is optically-thick – only because of this is the disk able to reprocess the starlight incident upon it and become a dynamically significant source of radiation. Gayley (this volume) presents an independent discussion of how the CAK force responds to an irradiated surface. In advance of the appropriate full disk atmosphere calculations, we make the simplest assumption that the disk re-emits starlight falling upon it isotropically (cf. Kenyon & Hartmann 1987).

3 Results

Figure 1 shows the calculated density and velocity field obtained for the case of mass loss driven from a disk and evolved B star (taken from Oudmaijer et al. 1998). Note that the disk surface flux due to reprocessing of starlight falls off as R^{-3} in essentially the same manner as in a self-luminous accretion disk. This trend may be compared with the R^{-2} overall decline in gravity. Hence regardless of whether the disk shines by its own light or starlight, radiation is most effective at driving mass loss from the inner disk. This shows up in the illustrated model as a peak in mass flux (ρv_r) along streamlines emanating from within two stellar radii at a polar angle of 70°.

In brief, we find that a radiation-driven circumstellar disk wind is typically a slower, denser flow than the flow originating directly from the stellar surface. This difference can be seen as a consequence of the changed inner boundary condition – the stellar wind component is launched from the sub-critically rotating photosphere, while the disk wind is launched from Keplerian orbits in the equatorial plane beyond the stellar surface. A new result, that is at once both reassuring and very striking, is that the integrated disk wind mass loss rate can be shown to be within a factor of two of the CAK prediction for its spherically-symmetric equivalent (Proga, in preparation). Observationally the significance of such disk winds would be that low equatorial wind speeds can be easily explained, as might high pole/equator density contrast (depending on how the details of spectral line transfer work out). The results presented in Oudmaijer et al. (1998) are, in effect, a possible numerical realisation of Zickgraf et al.'s (1985) conceptual model for B[e] stars. A next step for us will be to discover whether these models can be converted into synthetic line profiles reminiscent of real systems. Work is in progress.

References

Drew J.E., Proga D., Stone J.M. 1998, MNRAS, 296, L6
Icke V., 1981, ApJ, 247, 152
Kenyon S.J., Hartmann L. 1987, ApJ 323, 714
Murray N., Chiang J., Grossman S. A., Voit G. M., 1995, ApJ, 451, 498
Oudmaijer R. D., Proga D., Drew J. E., de Winter D., 1998, MNRAS, in press
Proga D., Stone J. M., Drew J.E. 1998, MNRAS, 295, 595
Stone J. M., Norman M. L., 1992, ApJS, 80, 753
Vitello P. A. J., Shlosman I., 1988, ApJ, 327, 680
Zickgraf F.-J., Wolf B., Stahl O., Leitherer C., Klare G. 1985, A&A 143, 421

Discussion

A. Feldmeier: I am wondering why your model prefers a wind from inner disk regions. For outer disk regions, the increase in the flux with height due to illumination by inner hot regions should be more pronounced, and should help the gas to overcome the linear increase of gravity with height. Furthermore, both g and F_{disk} go essentially as $\sim 1/r^2$.

D. Proga: In the ideal α-disk considered in the models, the effective temperature declines as (disk radius)$^{-3/4}$. Hence the luminosity per unit area falls off as the cube of the radius – more quickly, therefore, than gravity. Illumination by inner hotter regions is not enough of an effect to overcome this.

Janet Drew and Daniel Proga

Radiative Fluxes and Forces in Non-spherical Winds

Rainer Wehrse[1,3] and Guido Kanschat[2,3]

[1] Institut f. Theoret. Astrophysik, Tiergartenstraße 15, D-69121 Heidelberg, Germany
[2] Institut f. Angewandte Mathematik, Im Neuenheimer Feld 294, D-69120 Heidelberg, Germany
[3] Interdisziplinäres Zentrum f. Wissenschaftliches Rechnen, Im Neuenheimer Feld 368, D 69120 Heidelberg, Germany

Abstract. For the modelling of the radiation fields in stellar winds and the resulting forces new efficient algorithms are presented. In the first one, the radiative transfer equation for moving 3D media is solved analytically with the assumption that the source function is known, eg. from the solution of the NLTE rate equations. For a wind with inhomogeneities an a-posteriori error controlled finite element code is described that takes scattering explicitly into account. Finally,we present possibilities for the accurate inclusion of an arbitrary number of spectral lines in a deterministic and in a stochastic way.

1 Introduction

It is now well known for more than twenty years that radiative transfer is of crucial importance for stellar winds from hot stars (Lucy and Solomon, 1970; Castor, Abbot and Klein, 1975) since it essentially determines the thermodynamical state of the matter involved and the force that acts on every volume element. Unfortunately, the radiative transfer equation differs from the hydrodynamic equations that describe the winds in that it is usually an integral-differential equation and therefore may involve the coupling of very distant regions. Hence special techniques for the solution have to be invoked, in particular if many lines are present and/or the geometry is not simple. Since 'classical methods' (cf. Kalkofen, 1987) either involve many severe simplifications or are by far too expensive computationally new types of solutions of the transfer equation are required.

In this contribution some analytical and numerical solutions of the transfer equation in moving media will be reviewed that are relevant to stellar winds and that have recently been obtained in Heidelberg in an interdisciplinary effort.

In the next section the transfer equation is given and interpreted as an equation in configuration$\otimes$frequency$\otimes$direction space. This allows a very convenient specific formulation e.g. for rotating winds not only of the transfer equation itself but also of the solutions if the source if given. In section 3 a finite element algorithm is described that allows the accurate solution for coherently scattering media with arbitrary density and temperature distributions. Section 4 is devoted to the inclusion of many lines and the corresponding evaluation of frequency integrated quantities relevant for hydrodynamics. We close with brief discussion and outlook.

2 The Radiative Transfer Equation and its Analytical Solution for Rotating Winds

We base our discussion on the transfer equation for unpolarized time independent radiation (cf. Oxenius, 1986, for a derivation and its inherent limitations)

$$\frac{dI(\mathbf{s})}{ds} = -\chi(\mathbf{s})(I(\mathbf{s}) - S(\mathbf{s})) \tag{1}$$

with I = specific intensity, $\mathbf{s}$ = vector in ray direction $\mathbf{n}$ =, ν = frequency, χ = extinction coefficient, S = source function. If the medium is static the coordinate system for $\mathbf{s}$ can be positioned in such a way that $\mathbf{s}$ reduces essentially to a scalar s. Frequency and angles can be considered then just as parameters.

In order to take velocities $\beta = (\beta_x(\mathbf{x}, \beta_y(\mathbf{x}\beta_z(\mathbf{x}))$ ($\mathbf{x}$ = geometrical coordinates (x, y, z)) we apply a Lorentz transformation to Eq. 1 (cf. Baschek, et al., 1997a). The resulting general expression is very complicated but it becomes very convenient if advection/aberration terms as well as terms involving β^n with $n > 1$ are neglected. For the relativistically invariant intensity in terms of the logarithmic wavelength

$$\xi = \ln \lambda \tag{2}$$

it reads

$$\begin{aligned} \mathbf{n} \cdot \nabla I(\mathbf{x}, \mathbf{n}, \xi) + \mathbf{n} \cdot \nabla(\mathbf{n} \cdot \beta(\mathbf{x}))I(\mathbf{x}, \mathbf{n}, \xi) \\ = -\chi(\mathbf{x}, \xi)(I(\mathbf{x}, \mathbf{n}, \xi) - S(\mathbf{x}, \xi)) \end{aligned} \tag{3}$$

The solution of Eq. 3 can either be derived from the standard solution of Eq. 1

$$I(\mathbf{s}) = \exp(-\tau(\mathbf{0}, \mathbf{s}))I(\mathbf{0}) + \int_{\mathbf{0}}^{\mathbf{s}} \exp(-\tau(\mathbf{s}', \mathbf{s}))\chi(\mathbf{s}')S(\mathbf{s}')d\mathbf{s}' \tag{4}$$

with

$$\tau(\mathbf{s}', \mathbf{s}) = \int_{\mathbf{s}'}^{\mathbf{s}} \chi(\mathbf{s}'')d\mathbf{s}'' \tag{5}$$

and $I(\mathbf{0})$ is the incident intensity and the subsequent application of the appropriate Lorentz transformation to this solution or the solution can be obtained directly from Eq. 3 by means of characteristics. In both cases the solution is given by Eq. 4 but all integrals have to be considered as *path integrals* (cf. Wehrse and Baschek, 1998) along the curve $\mathcal{C}(s) = (\mathbf{x}_0 + \mathbf{n}s, \mathbf{n} \cdot \beta(\mathbf{x}_0 + \mathbf{n}s))$.

It is now straightforward to use the velocity field for a rotating wind using a Cartesian coordinate system

$$v(x,y,z) = v_{exp}(\sqrt{x^2+y^2+z^2})(x,y,z)^t/\sqrt{x^2+y^2+z^2} + v_{rot}(\sqrt{x^2+y^2+z^2})(x,y,0)^t/\sqrt{x^2+y^2}. \quad (6)$$

The results will be presented and discussed in detail in a separate paper (Wehrse and Müller, 1998).

3 Numerical Solution of the Transfer Equation for Arbitrary 3D Geometries

Whenever the wind is clumpy a **finite element** algorithm is advantageous since it takes the inhomogeneities into account in a natural way. Kanschat (1996) has developed a corresponding code for the transfer equation 3 with

$$S = \epsilon B + \frac{1-\epsilon}{4\pi} \int_{4\pi} p(\mathbf{n}, \mathbf{n}' I(\mathbf{x}, \mathbf{n}')d\omega'$$

(B =Planck function, p =phase function, ω =solid angle). The code employs unstructured grids that are adaptively refined by means of an a-posteriori error estimate which is obtained from the solution of the dual problem. In this way it is guaranteed that the numerical solution of the transfer equation does not deviate from the exact analytical one by more than a prescribed value and simultaneously the computation time is minimized. An additional reduction of the CPU time is achieved by the use of a variant of the conjugate gradient method in place of the usual Jacobi iteration (Λ or *approximate* Λ iteration in the astronomical literature): since in this scheme the unknowns are eliminated according to the absolute value of corresponding eigenvalue starting with the largest one, convergence problems for optically thick media are completely avoided. The code is written in C++ in order to reduce the book-keeping concerning the interaction of the various cells and is designed for multiple instruction parallel! machines but runs also on workstations. In spite of the high numerical efficiency the machine requirements are quite high since always all radiative couplings have to evaluated. Therefore, the CPU time and memory vary strongly with required accuracy and the dimension and the complexity of the medium. Typical values are in the range of a few Gigabyte and several hours on an IBM SP2.

Fig. 1 shows, as an example, the radiative accelerations in an inhomogeneous medium that is illuminated from below. Note that the forces seem to compress the regions of higher density. Consequences for stellar winds appear to be obvious but firm conclusions cannot be drawn before this transfer code is coupled with a hydrodynamic one.

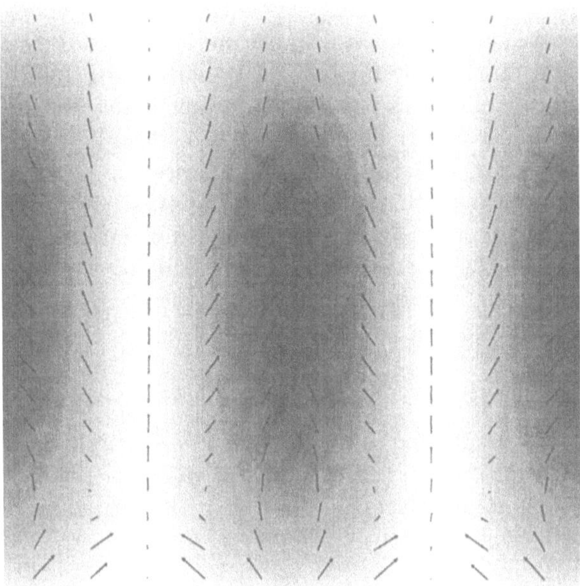

Fig. 1. Example for the complex behavior of radiative forces in an inhomogeneous, static scattering 2D medium with small thermal emission that is illuminated from below (Kanschat, 1996). Only one frequency is taken into account. The shading indicates the density distribution and the arrows indicate strengths and direction of the radiative acceleration. Differential motion, wavelength dependent extinction and thermal emission would introduce additional complications.

4 Many Lines

Since it seems that the winds from hot stars are not accelerated by just a few lines but by the combined action of all lines (cf. Kudritzki's contribution in this volume) it is important to include in simulations the line absorption as completely as possible; in particular, weak lines should be represented adequately. Unfortunately, even very fast present-day computers do not allow such a representation if the well known solutions of the transfer equation are employed.

In our search for alternative methods it turned out (Baschek et al., 1997b) that the solution of eq. 3 for a layered moving medium can easily be expressed in terms of the *spectral thickness*

$$\psi(\xi) = \int_{\xi_0}^{\xi} \chi(\xi')d\chi'. \tag{7}$$

The spectral thickness is much better suited for the evaluation of integrals over wavelength than the extinction coefficient itself since it is much smoother due to integration; in fact; in most cases it can be well approximated by a piece-wise linear function. Since, in addition, it was found that the integral over the Planck function over an arbitrary wavelength interval can conveniently be expressed in terms of poly-logarithmic functions, the computation times for total fluxes and radiative accelerations could be reduced by more than 5 dex with a loss in accuracy of only 1 percent.

As an alternative to this deterministic approach we also developed a method in which the line positions are assumed to follow a Poisson process with mean density $\rho(\xi)$(Wehrse et al., 1998). The line strengths and broadening parameters may be correlated with wavelength and may obey some suitable distribution, as e.g.

$$\rho(\xi, \vartheta) = \rho_0 f(\vartheta) \tag{8}$$

where f is a suitable function to describe the line strengths and widths (combined in the parameter ϑ). The expectation values of the specific intensity for a shell of geometrical thickness $z << R$ with constant acceleration w is then given by

$$\langle I \rangle = \int_0^\infty I\mathbf{P}(dI) \tag{9}$$

$$= \langle I_0 \rangle - \int_{\xi - wz}^{\xi} \left\langle \exp\left(-\frac{1}{w}\int_\eta^\xi \chi(\zeta)d\zeta\right)\right\rangle \frac{d}{d\eta} S(z - \frac{\xi-\eta}{w}, \eta)d\eta$$
$$+S(z,\xi) - S(0, \xi - wz)\left\langle \exp\left(-\frac{1}{w}\int_{xi-wz}^{\xi} \chi(\zeta)d\zeta\right)\right\rangle \tag{10}$$

where the crucial term

$$\left\langle \exp\left(-\frac{1}{w}\int_\eta^\xi \chi(\zeta)d\zeta\right)\right\rangle = \exp\left(-\chi_c(\xi-\eta)/w\right) \Omega(\xi,\eta) \tag{11}$$

with

$$\Omega(\xi,\eta) = \exp\left(\int_S \rho(\xi',\vartheta)\left\{\exp\left(-\frac{1}{w}\int_\eta^\xi \chi_l(\xi',\vartheta,\zeta-\xi')d\zeta\right) - 1\right\} d\xi' d\vartheta\right). \tag{12}$$

In these expressions we have splitted the extinction coefficient in an continuum (subscript c) and a line contribution (subscript l). S s the combined set of wavelengths and line parameters; the three arguments of the line absorption coefficient χ indicate the actual wavelength, the line parameters, and the center of line. The general formula for Ω given here looks rather complicated; in actual cases, however, one integration can often be performed analytically so that quite handy expressions can be obtained. Fig. 2 gives as an example the expectation value of the intensity emerging from a rotating stellar wind under the assumption that the invidual lines have a δ-function profile.

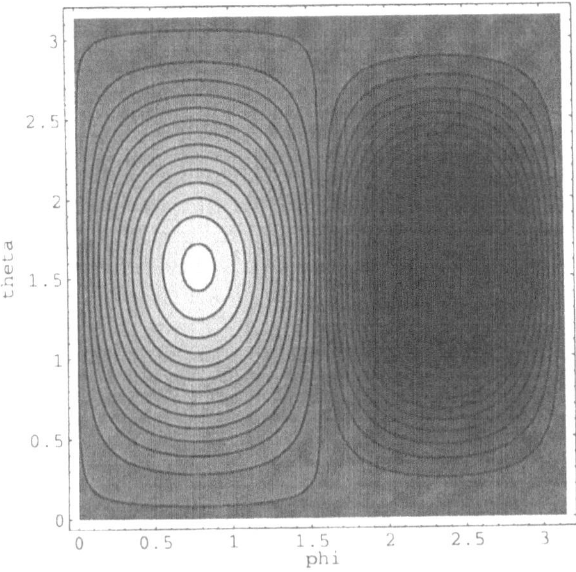

Fig. 2. Typical expectation value of the comoving frame specific intensity from an rotating wind as a function of latitude θ and longitude ϕ. Spectral lines are included in a stochastic description, see text.

5 Discussion and Outlook

In this contribution some powerful new methods for the modelling of radiation fields in non-spherical wind have been discussed. Up to now they have been applied to rather simple situations only in order to *understand* the way radiation fields operate in such complicated systems as non-spherical and rotating winds. The next step will be the proper determination of the opacity input data, in particular the distribution functions mentioned in the previous section.

Acknowledgment: We thank B. Baschek, G. Efimov, P. Müller, R. Rannacher, and W. v. Waldenfels for their continuous collaboration on radiative transfer problems. This work was supported by the Deutsche Forschungsgemeinschaft (Project C2 of SFB 359 and We 1148/7-1).

References

Baschek, B., Efimov, G.V., v. Waldenfels, W., Wehrse (1997a): *Astron. Astrophys.* **317** 630

Baschek, B., Grüber, C., v. Waldenfels, W., Wehrse (1997b): *Astron. Astrophys.* **320** 920

Castor, J.I., Abbott, D.C., Klein, R.I.: (1975) Astrophys. J. **195** 157

Kalkofen, W. (1987): *Numerical Radiative Transfer* (Cambridge University Press)

Kanschat, G. (1996): PhD thesis, Universität Heidelberg

Lucy, L.B., Solomon, P.M. (1970): Astrophys. J. **159**, 879

Oxenius, J. (1986): *Kinetic Theory of Particles and Photons* (Springer, Berlin, Heidelberg)

Wehrse, R., v. Waldenfels, W., Baschek, B.: (1998) J. Quant. Spectrosc. Rad. Transf., in press

Wehrse, R., Baschek, B.: (1998) Physics Reports, in press

Wehrse, R., Müller, P.: (1998) in preparation

Discussion

J. Bjorkman: One of the more difficult aspects of radiation transfer is that the opacity depends on the radiation field. In the model you presented, the opacity constants are independent of the radiation field. How would one modify your model to account for these effects?

R. Wehrse: In the analytical solutions for moving media presented here it is indeed assumed that the source function is known in advance. However, the solutions can also be used advantageously in NLTE calculations by inserting them into the rate equations.

Line-Driven Ablation by External Irradiation

Kenneth G. Gayley, Stanley P. Owocki, and Steven R. Cranmer

[1] University of Iowa, Iowa City, IA, 52242, USA
[2] Bartol Research Institute of the University of Delaware, Newark, DE 19716, USA
[3] Harvard-Smithsonian Center for Astrophysics, Cambridge, MA 02138, USA

Abstract. The Sobolev approximation for supersonic flows creates an effective opacity distribution that is nonisotropic, because the line-of-sight velocity gradient is different in different directions. To better understand the importance of this phenomenon in a simplified geometry, we consider line-driven flows in the plane-parallel zero-sound-speed limit, and solve for the wind driven by radiation with an arbitrary angular distribution. One conclusion, surprising at first glance, is that the acceleration component normal to the surface is *independent* of both the strength and angular profile of the driving radiation field. The flow *tilt* and overall *mass-loss rate* do depend on the character of the radiation field. Also interesting is that mass loss through a surface may be generated or enhanced by irradiation that originates above the surface.

1 Introduction

The theory of radiatively driven mass loss, as grounded in the line-driving formalism developed by Castor, Abbott, and Klein (1975; hereafter CAK), has proven remarkably successful in explaining the general characteristics of stellar winds from hot, luminous, OB-type stars (e.g., Kudritzki & Hummer 1990). In recent years there has developed considerable interest in applying this theory toward modeling radiatively driven mass flows in more complex geometries. Here we consider a plane-parallel atmosphere for simplicity, but include the effects of external irradiation superimposed against the intrinsic upwelling flux. In contrast with previous studies (Friend & Castor 1983; Drechsel et al. 1995), we include the component of radiation *reflected* by the illuminated surface.

2 Vertical Acceleration

The momentum equation we apply normal to the surface is simply the standard CAK expression, and it leads directly to the surprising conclusion (see Gayley, Owocki, & Cranmer 1998 for details) that the vertical acceleration is completely *independent* of the strength and angular distribution of the driving radiation field. This is a consequence of the assumption that the mass-loss rate stabilizes only once it achieves its maximum possible value, a generalization of the CAK "critical point" conditions. The overall magnitude of the driving intensity, as well as the total opacity, are then relevant only for fixing the mass-loss rate, while the angular character of the radiation also influences the azimuthal tilt of the flow.

3 Azimuthal Force Equation

To consider tilted flows, we must include the transverse from equation, which contains no gravity terms and involves different angular moments of the radiation field. Since the vertical acceleration is already known, we form the ratio of the transverse acceleration to the vertical, and after various constants in the radiative driving cancel, we are left with the essential angular moments. These moments depend on the angular anisotropy of the Sobolev opacity, so must be solved simultaneously with the wind dynamics to achieve self-consistency. The result is a nonlinear equation for the non-vertical flow direction, and its solution also determines the mass-loss rate.

4 Results for an External Point Source

The results for the simple example of irradiation from an external point source coupled with an intrinsic upwelling flux are shown in Figure (1a). An interesting conclusion from this figure is that the flow solution tends to align nearly perpendicular to the incident rays. Physically, this occurs because a wind that is perpendicular to the incident photon stream presents zero line-of-sight velocity gradient. Therefore, the downward photons slip easily through the wide frequency gaps between lines, whereas the upward reflected or intrinsic flux can couple strongly to the steep velocity gradient, supporting the mass-loss rate. This "bootstrap" quality is common in line-driven flows.

5 Potential Effects of Finite Source Size and Disk Winds

A perpendicular flow direction can only be defined for nearly parallel external irradiation, such as from a point source. If the external source is distributed, as for a nearby binary companion or the central source for a disk wind, the nearly perpendicular solution may cease to exist. Figure 1b shows the parameter regimes for the various solutions for a spread external source. Notice the tendency for the solution to bifurcate to high tilt and low mass loss.

Although plane-parallel atmosphere models differ in fundamental ways from the physics of winds in other geometries, such as disk winds, our results may provide useful insight into the complexities induced by nonisotropic opacity in a broader context. For example, the blank region in Fig. (1b) indicates that no ablation occurs for strongly external illumination from a

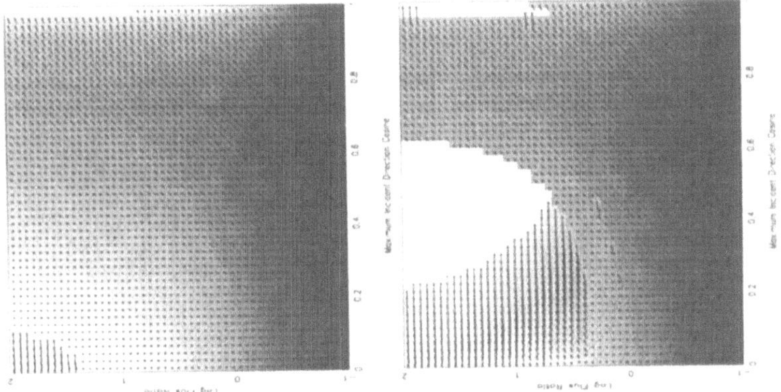

Fig. 1. a) Left: mass-loss-rate (gray scale) and flow direction (arrows) for outflow driven in part by irradiation from an external point source. The plot depicts a two-dimensional parameter space, with the external-to-intrinsic flux ratio at the bottom and the direction cosine to the external point source at the right. The flow direction is defined relative to an imaginary stellar surface at the right of the figure. b) Right: similar format to (a), except that the external source is spread from 0 to the maximum direction cosine indicated at the right.

distributed source with $\mu_{max} > 0.34$, since then there is simply too much downward pressure through which the wind cannot negotiate a path. This suggests it might prove difficult to use irradiation from a central source to drive a dynamically self-consistent wind from a disk inside about 3 times the source radius.

6 Conclusions

We stress that the two key elements that lead to surprising results in our models are the non-isotropic quality of the opacity, and the requirement for self-consistency between this opacity and the flow dynamics. Models which neglect either of these aspects may miss important characteristics of the ensuing flow.

This work was supported by NASA grants NAG5-4065, NAGW-2624, and NGT5-40024.

References

Castor, J. I., Abbott, D. C., & Klein, R. I. 1975, ApJ, 195, 157 (CAK)
Gayley, K. G., Owocki, S. P., & Cranmer, S. R. 1998, ApJ, in press
Drechsel, H., Haas, S., Lorenz, R., & Gayler, S. 1995, A&A, 294, 723
Friend, D. B., & Castor, J. I. 1983, ApJ, 272, 259
Kudritzki, R.-P., & Hummer, D. G. 1990, ARA&A, 28, 303

Discussion

A. Sapar: According to Sobolev theory, photons absorbed in a thin accelerating layer do not scatter monochromatically, but survive complete frequency redistribution and scatter in equal amounts to both sides of the scattering layer, thus giving to it their momentum before scattering. Backscattering effects can play a role only in the case of partial redistribution; but this is not the Sobolev scheme.
K. Gayley: The statement is correct, but the implication is not. Our models are purely within the Sobolev scheme and require no backscattering. All forces are due to the interaction of the non-isotropic Sobolev opacity with the incident continuum. The key is that the flow solution is dynamically self-consistent.

Bodo Baschek and Rainer Wehrse

Extremely Luminous Atmospheres

Nir J. Shaviv

Theoretical Astrophysics 130-33, Caltech, Pasadena, CA 91125

Abstract. We present the effects that inhomogeneities have on radiating atmospheres. It is shown that nonuniformities in a medium induce a reduction of the effective opacity which subsequently increases the Eddington Luminosity. The most striking effect however that arises from the dependence of the opacity on the inhomogeneities, is the possibility of a phase transition, where the atmosphere energetically favors exciting horizontally propagating waves due to large fluxes.

Atmospheres with a large radiative flux are extremely interesting as they are important for the behavior of objects such as luminous stars, novae or accretion disks. Moreover, they exhibit many effects in radiative hydrodynamics. We summarize here two of them. We first show that the effective bulk opacity of an inhomogeneous medium is changed and that this can subsequently induce a phase transition in a very luminous atmosphere.

The important effect that arises when a system becomes inhomogeneous is the change of its effective opacity. Shaviv (1998a) has shown that the effective opacity relevant for the calculation of the average radiative force is not necessarily a simple mass or volume weighted average of the effective opacity. Instead, the opacity per unit mass is generally given by:

$$\kappa_m^{\mathit{eff}} = \langle H \kappa_m \rho \rangle \,/\, (\langle H \rangle \langle \rho \rangle). \tag{1}$$

It is found by comparing the total radiative force on the system with the total flux. For small amplitude perturbations of the form $\delta\rho/\rho = \delta \cos(kx - \omega t)$, the effective opacity becomes[1]: $\kappa_m^{\mathit{eff}} \approx \kappa_m^{(0)} / \left(1 + \nu\delta^2\right)$, where ν is a constant that depends on the wave type and on the form of the opacity. If for example $\kappa_m \propto \rho^\alpha T^\beta$ and $T \propto \rho^\mu$ then optically thin and thick waves respectively have (Shaviv 1998b):

$$\nu_{thin} = -\left[(\alpha(\alpha+1) + \beta\mu^2(\beta-1) + 2(\alpha+1)\beta\mu\right]/4. \tag{2}$$

$$\nu_{thick} = \left[(\alpha+2)(\alpha+1) + \beta\mu^2(\beta+1) + 2(\alpha+1)\beta\mu\right]/4. \tag{3}$$

Evidently, the bulk opacity can under various circumstances be reduced. We now show that this induces a phase transition. To see it, we calculate the atmosphere's energy when a perturbation is added to it. If we find that the homogeneous equilibrium is not a minimum of the energy when small adiabatic perturbations are added, then it will be unstable as the latter will grow. The system will then have a new equilibrium state.

[1] The expression is accurate only for small amplitudes; since we wish to qualitatively analyze also large ones and strong fluxes, we choose an expression that correctly gives the quantitative behavior, namely, that $\kappa_{\mathit{eff}} \to 0$ for $\delta \gg 1$.

For simplicity, we assume that the relative perturbation is not a function of height and that the atmosphere resides on a rigid surface. Although generally not the case, these assumptions simplify the derivation as the specific energies become independent of height and of global energy changes in the system (e.g. a star) as a whole when the opacity is changed.

The basic equation is the hydrostatic equation with the radiation forces:

$$\frac{dp}{dz} = c_T^2 \frac{d\rho}{dz} = (-g + g_{rad})\rho \quad \text{with} \quad g_{rad} = g_{rad}^{(0)} \frac{\kappa_{eff}}{\kappa_0}. \tag{4}$$

Here c_T the isothermal speed of sound and $g_{rad}^{(0)}$ is the unperturbed acceleration due to radiation. The latter is changed from its unperturbed value when the opacity has corrections due to inhomogeneities. After integration, one finds that $\rho = \rho_0 \exp\left(-g_{eff}/c_T^2 z\right)$, with $g_{eff} = g - g_{rad}$. Even if the atmosphere is perturbed, the total mass of it per unit area – Σ should be constrained to remain the same. Using this condition we find that $\rho_0 = \Sigma g_{eff}/c_T^2$.

Two terms contribute to the total energy when a wave of amplitude δ is excited. The first is the acoustic energy in the wave per unit area. It is $A = (\epsilon/2)\Sigma\delta^2 c_T^2$, where ϵ is a constant that depends on the type of wave. The second is the potential energy. It is composed of the interaction with both the gravitational and radiation fields. For an optically thin atmosphere, one finds after proper integration that (Shaviv, 1998b):

$$U + A - U(\Gamma_0 = 0, \delta = 0) = \Sigma c_T^2 \left[-\ln\left(\frac{1 - \Gamma_0 + \nu\delta^2}{1 + \nu\delta^2}\right) + \frac{\epsilon}{2}\delta^2 \right] \tag{5}$$

where we have introduced Γ_0 – the ratio between the radiation pressure gradient and the unperturbed Eddington Limit. For small amplitudes, we have:

$$U + A - U(\Gamma_0 = 0, \delta = 0) \approx \Sigma c_{T^2} \left(\frac{\epsilon}{2} - \frac{\nu\Gamma_0}{(1 - \Gamma_0)}\right)\delta^2 \tag{6}$$

Evidently, the total energy of the system is quadratic in the amplitude of the perturbations. However, the coefficient of the second order term can be either positive or negative depending on the value of Γ_0, if ν is positive. For small values of Γ_0, the sign is clearly positive and a zero amplitude wave is clearly the least energetic, however, above the critical value of:

$$\Gamma_{crit} = 1/(1 + 2(\nu/\epsilon)), \tag{7}$$

it is apparent that by exciting a small amplitude wave, the total energy of the system will be lowered and thus more favorable! Moreover, the most favorable excitation or the one that is excited first when the flux is increased, is the one that has the lowest Γ_{crit}. In other words, the least stable mode is the one that has the lowest ϵ to ν ratio – a larger opacity change with a lower energy excitation cost.

We also see that one cannot find the equilibrium from the linear or small amplitude analysis. Although eq. 5 is quantitatively valid only for small amplitudes ($\delta^2 \ll 1$), we can use its qualitative behavior to understand what happens at large amplitudes as well. This behavior can be seen in figure 1.

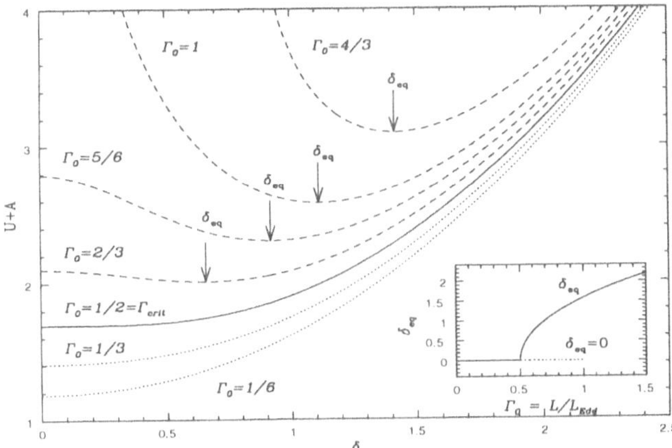

Fig. 1. The total energy of an atmosphere for different Γ_0 (fluxes) when g, c_T, ϵ and $2\nu = 1$ (according to eq. 5). For values of Γ_0 that are less than the critical value, the minimum of the total energy is at the origin. When Γ_0 is larger, the minimum energy is obtained for a finite amplitude wave. The inset depicts the equilibrium value as a function of Γ_0. It is of course *accurate* only for small amplitudes. The dotted line represents the unstable equilibrium of $\delta_{eq} = 0$ that exists for $\Gamma_{crit} < \Gamma_0 < 1$.

When Γ_0 is smaller than the critical value, the equilibrium amplitude is 0. When Γ_0 is larger, the total energy has a minimum for a finite amplitude of δ given by:

$$\delta_{eq}^2 \left(\Gamma_0 > \Gamma_{crit}\right) = \left(\Gamma_0 - 2 + \sqrt{\Gamma_0^2 + 8\Gamma_0(\nu/\epsilon)}\right) \Big/ \nu. \tag{8}$$

It approaches 0 for $\Gamma_0 \to \Gamma_{crit}^+$. Namely, it is similar to a second order phase transition: The order parameter δ is continuous but its derivatives are not.

We have seen two interesting characteristics of radiative hydrodynamic flows. First, inhomogeneities can decrease the effective opacity of a medium, and second, that this change can reduce the potential energy of the system and induce a phase transition. The effects have interesting implications to luminous objects. They affect phenomena like wind acceleration, change characteristics such as the Eddington luminosity and induce both spatial and temporal variability. A numerical simulation exhibits the qualitative results found here. A more general treatment can be found in Shaviv (1998b).

References

Shaviv, N. J., (1998a): *The Eddington luminosity limit for multi-phased media.* The Astrophys. J., **494**, L193.

Shaviv, N. J., (1998b): *A phase transition of luminous atmospheres.* Submitted to the Astrophys. J.

Discussion

S. Owocki: In your simulation, you eventually find only one horizontal structure in the periodic box. In a real star, what sets the limiting horizontal scale?
N. Shaviv: From the analytical treatment we know that different wavelengths are preferred for different opacity laws. Often, as was the case in the simulation, the preferred scale is $\sim 2\pi H_p$. Since the horizontal extent is roughly that, the periodic condition forces a wavelength that is exactly the width of the box. A larger horizontal extent gives a simulation that results in two wavelengths in the box.

S. Shore: What is the effective viscosity of your calculations? In other words, what is your effective Reynolds number?
N. Shaviv: The viscosity is of course limited by the finite resolution of the simulation (100x100). An increase in the resolution reduces the viscosity and could theoretically introduce more phenomena. This is why an analytical treatment is done as well.

Nir Shaviv and Achim Feldmeier

Disks formed by Rotation Induced Bi-stability

Henny J.G.L.M. Lamers[1,2], Jorick S. Vink[1], Alex de Koter[3], and Joseph P. Cassinelli[4]

[1] Astronomical Institute, Utrecht University, Princetonplein 5, NL-3584 CC, Utrecht, The Netherlands
[2] SRON Laboratory for Space Research, Utrecht, The Netherlands
[3] Astronomical Institute, University of Amsterdam, Kruislaan 403, NL-1098 SJ, Amsterdam, The Netherlands
[4] Dept of Astronomy, University of Wisconsin, 475 N Charterstreet, Madison, WI 53706-1582, Madison, USA

Abstract. We discuss the evidence for the existence of bi-stable stellar winds of early type stars, both theoretically and observationally. The ratio between the terminal wind velocity and the escape velocity drops steeply from about 2.6 for stars with $T_{\rm eff} > 21\,000$ K to about 1.3 at $T_{\rm eff} < 21\,000$ K. This is the *bi-stability jump*, which is due to a change in the ionization of the wind and in the wind driving lines. The mass loss rate increases across the jump by about a factor 2 to 5 from the hotter to the cooler stars. The mass flux from rapidly rotating stars can also show the bi-stability jump at some lattitude between the pole and the equator, with a slow high density wind in the equatorial region and a faster low density wind from the poles. This might explain the disks of rapidly rotating B[e] stars, formed by the *Rotation Induced Bi-stability* mechanism. We discuss the RIB mechanism and its properties. We also describe some future improvements of the model.

1 Introduction

The concept of bi-stability of the winds of early type stars was first described by Pauldrach and Puls (1990, hereafter P&P) based on their calculations of the line driven wind of P Cygni. They showed that the wind of a star with $T_{\rm eff} = 19\,300$ K can be either dense and slow if the wind is driven by lines in the Balmer continuum, or fast and less dense if the driving is due to lines in the Lyman continuum. The change occurs over a small range of $T_{\rm eff}$ or gravity because the optical depth in the Lyman continuum is very sensitive to T and ρ in the wind near $T_{\rm eff} \simeq 20\,000$ K. The change from the fast to the slow wind solution is called *the bi-stability jump.* Lamers et al. (1995) found in a study of the terminal velocities of stellar winds that the bi-stability jump also occurs for normal OB supergiants near about type B1, where the ratio $v_\infty / v_{\rm esc}$ drops steeply from about 2.6 for the hotter stars to 1.3 for the cooler stars. Lamers and Pauldrach (1991, hereafter L&P) showed that the same bi-stability may occur in the wind of a rapidly rotating early-B star between the hotter pole and the cooler equator. This will lead to a fast low density wind from the pole and a slow dense wind from the equator, i.e. a disk like structure. This mechanism is called the *Rotation Induced Bi-stability* or *RIB*.

2 The bi-stability of the winds of normal stars.

Fig. 1a shows the ratio v_∞/v_{esc} from Lamers et al. (1995) as a function of T_{eff} for normal OB supergiants. There is a clear jump near $T_{eff} \simeq 21\,000$ or spectral type B1. (A similar jump in v_∞/v_{esc} may be present near 10 000 K). Fig. 1b shows the same data but now in terms of v_∞ versus v_{esc} in different T_{eff}-intervals. The jump coincides with a drastic change in the ionization of the wind: stars on the "hot side" of the jump have strong CIV and weak CII wind lines, and the situation is reversed for stars on the "cool side" of the jump (see Fig. 8 of Lamers et al. 1995).

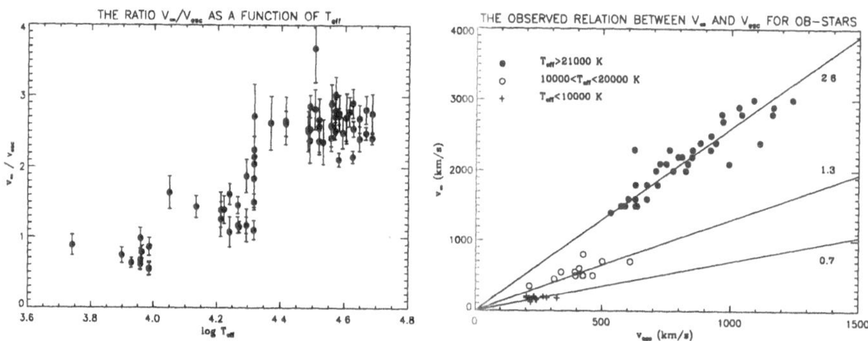

Fig. 1. Left: the observed ratio $v_\infty/\ v_{esc}$ for supergiants as a function of T_{eff}. Notice the jump near $T_{eff} \simeq 21\,000$ K, and possibly also near $T_{eff} \simeq 10\,000$ K. Right: the same data plotted as v_∞ versus v_{esc} (from Lamers et al., 1995).

The bi-stability model predicts that the mass loss rate should also show a jump, with $\dot{M}$ increasing by about a factor 2 to 5 when v_∞ decreases (P&P and L&P). The counteracting behaviour of $\dot{M}$ and v_∞ indicate that the wind momentum loss $\dot{M}v_\infty$ is approximately invariant across the bi-stability jump. This may be expected as the ratio between the radiative momentum loss, L/c, and the wind momentum loss hardly changes across the jump. It should be said that observational support for this statement is presently meagre, mainly because of the difficulty in deriving mass loss rates from UV lines with uncertain degrees of ionization.

Vink et al. (1999) have calculated the mass loss rates as a function of T_{eff} for stars with $\log L/L_\odot = 5.0$, using de Koter's ISA-WIND and MC-WIND codes (de Koter et al. 1993 and 1997). The first program solves the statistical equilibrium and radiative transfer equations for extended atmospheres and winds, with H, He, C, N, O and Si in non-LTE. The mass loss rate $\dot{M}$, and the velocity law, specified by β and v_∞, are input parameters. The second program calculates the total radiation pressure due to the continuum and lines, including about 10^5 of the strongest lines, mainly of the iron group elements. The mass loss rate is found by requiring that the radiative momentum input into the wind is equal to the wind momentum (see Abbott & Lucy, 1985 and Lucy & Abbott, 1993 for details). Fig. 2a shows the mass loss rates as a function of $T_{\rm eff}$ for wind models with values of $v_\infty/v_{\rm esc}$= 2.6, 2.0 and 1.3 respectively. From the observations we know that the ratio $v_\infty/v_{\rm esc}$ decreases steeply from 2.6 to 1.3 near 21 000 K. Adopting these ratios we find the variation in mass loss rate as shown in Fig. 2b. This figure shows a steep increase in $\dot{M}$ of about a factor five between $T_{\rm eff}$= 26 000 and 20 000 K. This is due to a change in the driving lines in this temperature range (Vink et al. 1999).

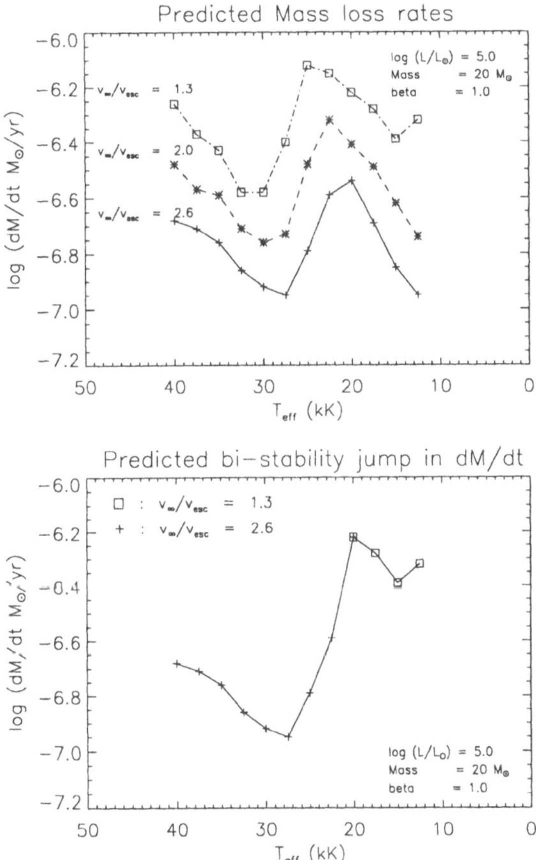

Fig. 2. Top: the predicted mass loss rates as function of $T_{\rm eff}$ for a grid of models with three values of $v_\infty/v_{\rm esc}$. Bottom: the mass loss rate as a function of $T_{\rm eff}$, for winds with the observed ratio $v_\infty/v_{\rm esc}$=2.6 if $T_{\rm eff} > 21\ 000$ and 1.3 if $T_{\rm eff} < 21\ 000$ K (from Vink et al., 1999).

3 Rotation-Induced Bi-stability

P&P and L&P showed that line driven winds from early-B stars can change from low density and high velocity to high density and low velocity if the optical depth of the wind in the Lyman continuum becomes larger than about unity. The basic idea of the RIB model is that the same transition can occur in the wind of a rotating early B star between the pole and the equator. This is because the reduction of the effective gravity and $T_{\rm eff}$ between the pole and the equator due to rotation enhances the optical depth in the wind from the pole to the equator. If the optical depth changes from a value smaller than unity at the pole, to a value larger than unity at the equator, the bi-stability jump will occur at some intermediate latitude. If this happens, the equatorial wind will be slow and dense, whereas the polar wind will be fast and tenuous. This is very similar to the observed characteristics of B stars with outflowing disks. An attractive feature of the RIB mechanism is that it is expected to work only for early B stars, because the models and the observations show that the bi-stability can occur around $T_{\rm eff}$ about 20 000 K. (The exact value of $T_{\rm eff}$, where the bi-stability jump occurs, depends on luminosity, gravity and rotational velocity.)

Figure (3) illustrates the nature of the wind predicted by the RIB model for B[e] supergiants. This picture is quite similar to the empirical model of these stars derived by Zickgraf et al. (1986).

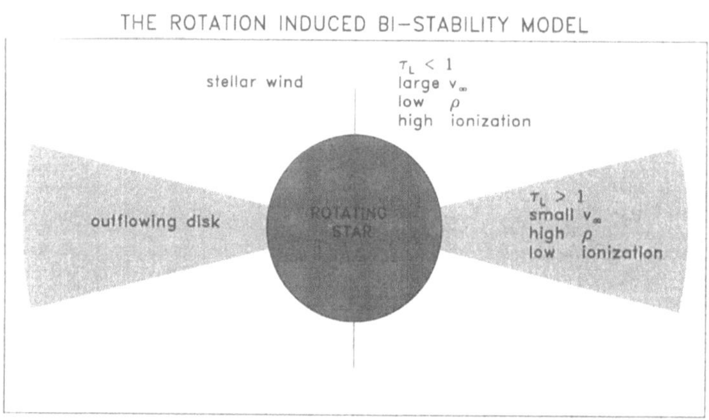

Fig. 3. A schematic picture of a rapidly rotating B[e] supergiant formed by the bi-stability mechanism. The wind is optically thin, ($\tau_L < 1$) in the polar regions, and so the wind there has a high velocity, high ionization state and low density. The wind is thick ($\tau_L > 3$) in the equatorial region so the wind there has a low velocity, lower ionization, and a higher density. (From Lamers and Pauldrach, 1991)

Since the bi-stability depends on the optical depth of the wind in the Lyman continuum, we consider the various effects of stellar rotation on the wind parameters and optical depth, following L&P.

The net surface gravity, $g_{\rm net}$, is given as a function of the polar angle θ on the star by

$$g_{\rm net}(\theta) = \frac{GM_*(1-\Gamma_e)}{R(\theta)^2}(1-\omega^2 \sin^2\theta) \tag{1}$$

with $\theta = 0$ and π at the poles and $\pi/2$ at the equator and with $\omega < 1$ defined by

$$\omega \equiv \frac{v_{\rm rot}}{v_{\rm crit}} = v_{\rm rot}\sqrt{\frac{R_*}{GM_*(1-\Gamma_e)}} = \Omega\sqrt{\frac{{R_*}^3}{GM_*(1-\Gamma_e)}} \tag{2}$$

We ignore the rotational distortion of the star, so $R(\theta) = R_*$. According to the von Zeipel theorem the radiative flux at the stellar surface is proportional to the local gravity, so

$$T_{\rm eff}{}^4(\theta) \sim g_{\rm net}(\theta) \sim (1-\omega^2\sin^2\theta) \tag{3}$$

The changes in gravity and radiation temperature as a function of θ affect the wind in several ways.

(a) The wind speed in the radiation driven wind theory is proportional to the escape speed. Consequently, the wind speed at θ is

$$v_\infty(\theta) \sim v_{\rm esc}(\theta) \sim \sqrt{R_* g_{\rm net}(\theta)} \sim (1-\omega^2\sin^2\theta)^{0.5} \tag{4}$$

So the wind speed decreases from the pole to the equator.

(b) In the piecewise spherical model that we consider here, the mass flux of a radiation driven wind, F_m, depends on the local radiative flux, i.e. on $T_{\rm eff}(\theta)^4$, and on the local effective gravity $g_{\rm eff}$. From the mass loss calculations of Vink et al. (1999) for $\log L/L_\odot = 5.0$ and for different masses and effective temperatures we find that in the range of 20 000 $\leq T_{\rm eff} \leq$ 30 000 K the mass flux can be written as

$$F_m \sim \left\{T_{\rm eff}{}^4(\theta)\right\}^{1.66} g_{\rm net}^{-1.2} \sim (1-\omega^2\sin^2\theta)^{+0.5} \tag{5}$$

This means that the mass flux decreases slightly from the pole to the equator.

(c) The Lyman continuum flux is very sensitive to the $T_{\rm eff}$. The brightness temperature in the Lyman continuum scales as $T_L \sim T_{\rm eff}{}^{1.6}$ for early-B stars (Kurucz, 1979). The fraction of neutral H is inversely proportional to the photo-ionization rate $R_i = 4\pi\int_{\nu_0}^{\infty}(J_\nu/h\nu)a_\nu d\nu$, which varies as $R_i \sim T_L^{9.7}$ in the range of 15 000 $< T_L <$ 20 000 K, corresponding to about 20 000 $< T_{\rm eff} <$ 25 000 K. So we find that the ionization rate of neutral H is

$$R_i \sim T_{\rm eff}{}^{15.5} \sim (1-\omega^2\sin^2\theta)^{3.9} \tag{6}$$

(d) The optical depth in the Lyman continuum through the wind, τ_L, is proportional to the column density of neutral H. As this is determined by ionization and recombination, τ_L is proportional to $\rho^2\alpha/R_i$, where $\alpha \sim T^{-0.5}$

is the recombination rate. Accounting for the dependence of the optical depth on both the ionization and recombination rate and on the column density, we find that the optical depth of the wind in the Lyman continuum varies with polar angle approximately as

$$\tau_L \sim \left\{ \frac{F_m(\theta)}{v_\infty(\theta)} \right\}^2 \frac{\alpha}{R_i} \sim \left\{ \frac{F_m(\theta)}{v_\infty(\theta)} \right\}^2 T_{\text{eff}}{}^{-16.0} \sim (1 - \omega^2 \sin^2 \theta)^{-4.1} \tag{7}$$

This relation implies that an early B star that is rotating will have a larger wind optical depth in the Lyman continuum in the equatorial region than near the polar region. The difference will be a factor of 16 for a star that is rotating with $\omega = 0.7$ and a factor of 3 if $\omega = 0.5$.

In this estimate we have assumed that the velocity law $v(r)/v_\infty$ is constant for all polar angles. If the velocity law is slower (larger β) for lower temperatures, as suggested by the study of velocity laws of B-stars (Rivinius et al. 1997), the optical depth will depend even stronger on θ than given by eq. (7).

P&P and L&P have shown that the bi-stability jump occurs when the optical depth of the wind in the Lyman continuum at a wavelength of 600 Å reaches a value of $\tau_L \simeq 1$ (see Fig. 1 of L&P), where τ_L is the optical depth of the wind from infinity to the sonic point.

This implies that the half opening angle is given by the condition

$$(1 - \omega^2 \sin^2 \theta)^{-4.1} \simeq 1/\tau_L(\text{pole}) \tag{8}$$

or

$$\theta \simeq \arcsin \left\{ \frac{1}{\omega} \left[1 - \tau_L(\text{pole})^{1/4.1} \right]^{0.5} \right\} \tag{9}$$

Fig. 4 shows the half opening angle of the RIB-disk of a B-type supergiant of T_{eff}=20 000 K, $M_* = 35 M_\odot$, $L = 5 \times 10^5 L_\odot$ as a function of ω for different polar mass loss rates defined by $\dot{M}_{\text{pole}} = 4\pi R_*^2 F_m(\text{pole})$. The polar optical depth of the wind in the Lyman continuum at 600 Å is about 0.05 if $\dot{M}_{\text{pole}} = 2 \times 10^{-6} M_\odot \, \text{yr}^{-1}$. This is a typical value for the polar mass loss rate of B[e] stars (Zickgraf et al., 1986).

4 Conclusions and future work

The RIB model uses the properties of rotating stars such as the von Zeipel theorem to show that the bi-stability jump of a rotating star may occur between the pole and the equator. This will lead to a high density and low velocity wind from the equator and a low density and high velocity wind from the pole. The density contrast is about a factor of ten. This contrast is similar to what is observed for the B[e] supergiants. So the RIB mechanism might explain the winds and disks of B[e] stars if the stars are rapidly rotating. An

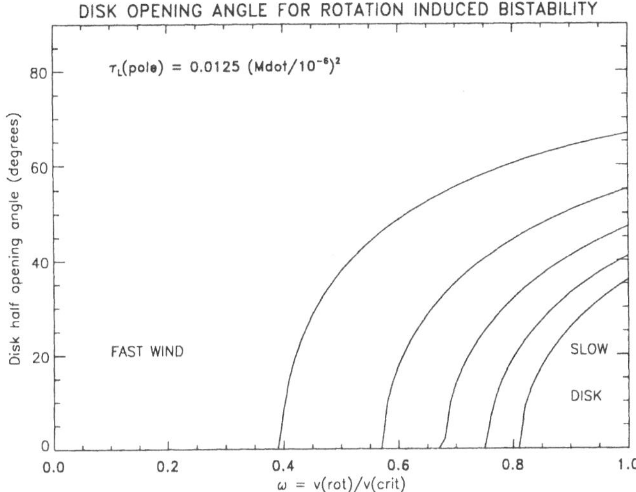

Fig. 4. The half opening angle of the disk of a star with $T_{\rm eff}$=20 000 K, $M = 35M_\odot$, $L = 5 \times 10^5 L_\odot$ as a function of ω for different polar mass loss rates: $\log(\dot{M}_{\rm pole}) =$ -5.2 (upper curve), -5.4, -5.6, -5.8 and -6.0 (lower curve) respectively. The observed polar mass loss rate of B[e] stars is $\log(\dot{M}_{\rm pole}) \simeq -5.7$.

attractive feature of the RIB mechanism is that it is expected to work only for the early-B stars, which agrees with the spectral types of the B[e] stars!

The mechanism described by L&P and in this paper is very descriptive and simplified. The major simplification is the piecewise spherical concept, where we use the information about the dependence of the mass flux on the local quantities $T_{\rm eff}(\theta)$ and $g_{\rm eff}(\theta)$ derived from the observations and theory of winds from normal, slowly rotating stars. We are presently studying the formation of disks by the RIB mechanism in a better approximation where the radiation pressure is calculated from the proper distribution of the radiation over the surface of the star (Pelupessy and Lamers, in preparation).

The disks produced by the RIB mechanism are rather wide, with full opening angles on the order of 90 degrees or more. Possibly the wind compression effect (Bjorkman and Cassinelli, 1993 and Bjorkman, these proceedings) will compress the disks to smaller opening angles. The differences between the RIB and the Wind Compressed Disk model (WCD) are:

(a) In the RIB model the disk is due to the higher equatorial mass flux, whereas in the WCD model the disk is due to the compression of the wind towards the equator.

(b) The RIB model can give at most a factor of ten density contrast from equator to pole and works for B supergiants with high mass loss rates. The WCD model can produce a higher density contrast, by an orbital effect that is independent of mass loss rate.

The combination of the bi-stability mechanism and the WCD mechanism will result in even higher density contrasts: the bi-stability producing a higher mass flux from the equator, and the WCD effect concentrating the flow towards the equator.

The formation of disks by the RIB mechanism depends on the stellar rotation. Maheswaran and Cassinelli (1994) have calculated the rotation history of massive stars as a function of mass loss rate for different mass loss mechanisms. They have shown that if there is no significant increase in angular velocity inward from the surface during the main sequence and the early supergiant phase, then the equatorial rotation rate will be extremely small before the star reaches the B[e] phase and the star will be an almost non-rotator during all subsequent evolution phases. The existence of B[e] stars shows that this is not the case and the stars survive the main sequence phase and the early supergiant phases with sufficient angular momentum in the envelope. Hegel and Langer (see Langer, these proceedings) have shown that massive stars returning from the red supergiant phase by contraction will rotate almost rigidly, because of the short convection turn-over time. This will result in a speeding up of the star when it moves to the left. So we expect that a massive star can have a B[e] disk in two phases: the most rapidly rotating stars may immediately become B[e] stars when they evolve off the main sequence. Mildly rotating stars can devellop a B[e] disk when they return from the RSG phase with a significantly smaller mass and angular momentum than during the first crossing.

References

Abbott, D.C., Lucy, L.B. 1985, *Ap.J* **288**, 679

Bjorkman, J.E., Cassinelli, J.P. 1993, *Ap.J* **409**, 429

de Koter, A., Schmutz, W., Lamers, H.J.G.L.M. 1993, *A&A* **277**, 561

de Koter, A. Heap, S.R., Hubeny, I. 1997, *Ap.J.* **477**, 792

Kurucz, R.L. 1979, *Ap.J.Supl* **40**, 1

Lamers, H.J.G.L.M., Pauldrach, A.W.A. 1991, *A&A* **244**, L5 (L&P)

Lamers, H.J.G.L.M., Snow, T.P, Lindholm, D.M. 1995, *Ap.J.* **455**, 269

Langer, N. 1999 (these proceedings)

Lucy, L.B., Abbott, D.C. 1993 *Ap.J.* **405**, 738

Mahareswaran, M., Cassinelli, J. 1994, *Ap.J.* **421**, 718

Pauldrach, A.W.A., Puls, J. 1990, *A&A* **237**, 409 (P&P)

Pelupessy, I., Lamers, H.J.G.L.M. 1999 (in preparation)

Rivinius, Th. et al. 1997, *A&A* **318**, 819

Vink, J.S., de Koter, A., Lamers, H.J.G.L.M. 1999 (in preparation)

Zickgraf, F.J. et al. 1986, *A&A*, **163**, 119

Zickgraf, F.J. 1999 (these proceedings)

Discussion

M. Friedjung: What is the wind velocity? Taking S 22, which I studied some years ago, you see narrow, optically thick Fe II lines in the optical without P Cygni components. On the other hand, the Balmer lines have P Cygni components at about 100 km/s. This seems also to be the case for the optically much thicker UV Fe sc ii lines. If the disk is edge on, you must have a very slow non-classical wind in the plane of the disk.

R. Ignace: Using the von Zeipel theorem, Owocki et al. indicated that $\dot{M} \sim g_{\rm eff}$, in contrast to $g_{\rm eff}^{-1.5}$ that you use. This will make $\tau_{\rm L}(\Theta)$ a less-sensitive function of latitude, and the disk opening angle may not be such a strong function of the rotation.
S. Owocki: I agree. Henny seems to be overlooking the flux dependence of the mass-loss rate variation. This will work against the equatorial increase in density for the bi-stability jump.

A. Maeder: Your bi-stability model predicts that there should be a discontinuity in the mass-loss rates (for slowly rotating stars) around 20 000 K. Do we observe it?
H. Lamers: This requires a mass-loss study of early B stars based on P Cygni profiles. It has not been done yet, probably because we do not know the ionisation structure of the winds of B supergiants.

E. Verdugo: What does the jump between the B supergiants and the A supergiants mean? Is it real?
H. Lamers: The figure showing v_∞ versus $v_{\rm eff}$ indeed suggests a second bi-stability jump near 10 000 K. However, this depends on the observations of only a few stars, with $T_{\rm eff} \sim 10\,000$ K. So at the moment we can say that $v_\infty/v_{\rm esc}$ decreases as $T_{\rm eff}$ decreases between B and A stars, but whether it does so discontinuously remains to be confirmed.

G. Koenigsberger: If a star's wind is exactly at the bi-stability limit, can it flip from one state to another?
H. Lamers: Yes. This was proposed for P Cygni by Pauldrach and Puls (1989). They suggest the following scenario: if the star is on the cool side of the bi-stability limit, the high mass loss rate will result in strong wind blanketing. This will heat up the photosphere and move the star to the hot side of the bi-stability limit. The decreased mass-loss rate will then reduce the blanketing, the atmosphere will cool and the star will cross the bi-stability jump again. They estimate the fluctuation time, i.e., the time to build up and destroy the blanketing layers, to be of the order of one month or so for P Cygni.

K. Gayley: In the absence of bi-stability, the gravity darkening would increase the mass loss through the polar region. Are you saying that in spite of that, bi-stability will reverse that effect? Will mass loss then peak at an intermediate angle where the transition occurs?

H. Lamers: There are two opposing effect: (a) the increase of T_{eff} to the pole, which implies a higher flux ($F_{rad} \sim T^4$); and (b) the increase of g_{eff} towards the pole. The first effect results in an increase of the mass flux; the second in a decrease to the pole. If $F_m \sim g_{net} M - 3{-}1.5\ F_{rad}^{+1.5}$ (as suggested by the models) then the two effects cancel because of the von Zeipel effect: $F_{rad} \sim T^4 \sim g$. Subtle effects may affect the $\dot{M}(\Theta)$ dependance either way.

Henny Lamers and Stan Owocki

The Effects of Magnetic Fields on the Winds from Luminous Hot Stars

Joseph P. Cassinelli and Nathan A. Miller

University of Wisconsin, Astronomy Department, 475 N. Charter St. Madison WI. 53706, USA

Abstract. Here we explore three types of models in which the combination of magnetic fields with line driving forces leads to faster winds than line forces alone. The fields can change the flow geometry; they can couple with rotation and transmit angular momentum; and they can carry transverse Alfvén waves. All three can lead to a deposition of momentum beyond the critical point and accelerate winds. Some history of the attempts to use magnetic fields to understand hot star winds is discussed and several new models for magnetic rotator B[e] winds are presented.

1 Introduction

There are several lines of evidence that magnetic fields play a role in hot star winds. The co-rotating interaction region (CIR) models of Cranmer and Owocki (1996) require a longitudinal dependence of the wind properties, which can arise naturally in stars with significant magnetic fields. There are correlations between Hα and He II 4686Å that form at the base of winds and the wings of UV line profiles that form far from the star (Kaper *et al.* 1996). The X-rays from OB stars show anomalies that require a multi-component wind structure (Cohen *et al.* 1997). All of these can be taken as evidence that fields near the star affect conditions far out in the wind. In this paper, we will focus on just one question: How can a combination of magnetic wind forces and line driven wind forces lead to *faster* outflows than would arise from line forces alone? We review three basic effects of fields: a geometrical effect, a rotational effect, and a transverse Alfvén wave effect.

First, let us briefly summarize the history of magnetic modelling of the winds of luminous hot stars. In 1979, it became clear for the first time that Wolf-Rayet stars have a "wind momentum problem", i.e. the ratio of the wind momentum to the radiative momentum, $\eta = \dot{M}v_\infty/(L_*/c)$, is much larger than unity- by factors as high as $\eta = 30$ to 50. Currently it is widely thought that the problem can be explained by multiple scattering of radiation (Gayley *et al.* 1995), but when first discovered the momentum ratio seemed to pose an insurmountable problem for radiation driven wind theory. Because of this difficulty with radiation driven wind theory Hartmann and Cassinelli (1981) and Cassinelli (1982) investigated whether magnetic forces could explain the winds. Since the momentum excess in the wind was so much greater than the radiation field, the effects of radiation were ignored and the focus was on the effects of fields alone. Both magnetic rotator models and Alfvén wave driven wind models were found capable of driving the wind of a WR star having $\dot{M} = 2 \times 10^{-5}\ M_{\odot}\,\mathrm{yr}^{-1}$ and $v_{\infty} \approx 1800\ \mathrm{km\,s}^{-1}$.

The rotating models were based on the fast magnetic rotator theory of Hartmann and MacGregor (1980). To drive the WR wind it required the star to have a field of 20,000 G and be rotating at 95% critical, both a field and a rotation rate which seemed implausible.

The Alfvén wave driven wind theory developed by Hartmann and MacGregor (1982) required the same large field, a wave flux of 1.1×10^{14} ergs $\mathrm{cm}^{-2}\ \mathrm{sec}^{-1}$, assuming a damping length of 1 R_*. In this model both the field and the wave luminosity ($= 1/3\ L_*$) seemed excessive.

During the subsequent decade, Maheswaran and Cassinelli (1988, 1992) developed upper and lower limits on surface fields based on virial theorem considerations, and the submersion of surface fields by the effects of circulation currents. Models combining radiation and magnetic forces were also developed and these will be discussed below. Missing throughout the history of the subject have been observational detections of magnetic fields. This appears to be the area that should show the greatest improvement in the future.

2 Flow Tube Geometrical Effects

MacGregor (1988) investigated the effects of non-spherical expansion on radiation driven winds. Seemann (1998) has also investigated the effects of flow tube geometry for luminous magnetic rotators. In the geometrical model of MacGregor, there is a velocity change that arises from a modification of the conservation of mass equation. Normally we use $\rho(r)v(r)r^2 = constant$ to eliminate the density from the wind momentum equation. However, for the case of a non-radial outflow, MacGregor uses $\rho(r)v(r)r^2\mathrm{f}(r) = constant$, where $\mathrm{f}(r)$ is illustrated in Figure 1. The factor arises in the case of the solar wind in coronal hole regions, where the base of the tube is constricted by neighboring closed field regions.

The factor $\mathrm{f}(r)$ appears in the wind momentum equation both in the gas pressure gradient $\rho^{-1}dp/dr$ and in the CAK line driven wind acceleration term $\left(\rho^{-1}dv(r)/dr\right)^{\alpha}$.

The results of the non-radial divergence factor are as follows:
a) the flow tube geometry can lead to an *increase* of the flow speed by a factor of two to three relative to the pure CAK wind case, for example increasing the wind speed from 1500 $\mathrm{km\,s}^{-1}$to 3000 $\mathrm{km\,s}^{-1}$. This increase appears to be more than enough to account for the most extreme velocity differences needed to explain discrete absorption components (DAC's) as proposed in the CIR models.
b) the mass loss rate remained unchanged from the CAK result in MacGregor's models. This is explained by the fact that although the flow tubes mod-

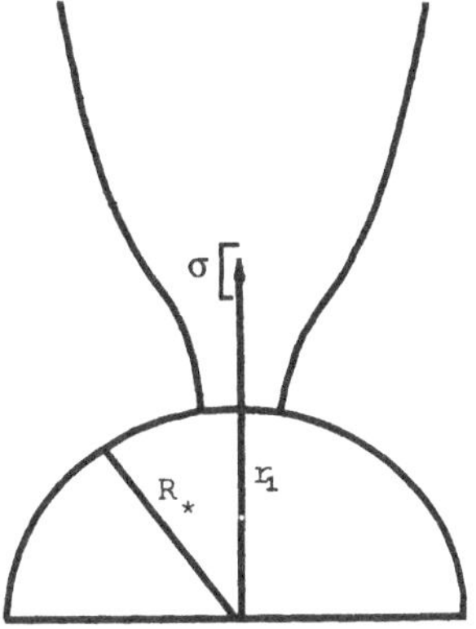

Fig. 1. The non-radial flow tube geometry used by MacGregor (1988). This particular case led to an enhancement of the wind speed by a factor of three.

ified the location of the CAK critical point, the actual velocity distribution in the subcritical point region was not changed significantly. In other words, the flow tube effect is an example of the "afterburner" effect (Cassinelli and Castor, 1973, Leer and Holzer, 1978). This states that momentum changes concentrated in the region beyond the critical point affect the terminal velocity but not the mass loss rate. If the momentum is also deposited in the subcritical region it will affect the mass loss rate as well. (see Ch 3, Lamers and Cassinelli, 1998)

3 Luminous Magnetic Rotator Theory

The basic physics explaining how an open magnetic field emerging from a rotating star will affect the mass flux, and the radial and azimuthal components of velocity was developed for the solar wind by Weber and Davis (1967). The equations describing how radiation forces can drive a flow from a luminous star were developed by Castor, Abbott, and Klein (1975, CAK). Luminous Magnetic Rotator (LMR) theory concerns the combination of the forces treated in these two papers. Cassinelli (1998) has recently reviewed the development of the theory in a sequence of papers by Friend and MacGregor (1984), Poe and Friend (1986), and Poe, Friend, and Cassinelli (1989).

There are three domains of magnetic rotator theory. A star can be a *slow magnetic rotator* (SMR), a *fast magnetic rotator* (FMR), or a *centrifugal magnetic rotator* (CMR). The defining characteristic of an SMR wind is that the magnetic field and rotation have almost no effect on the values of either the mass loss rate or wind terminal velocity. So, for an SMR, the values for $\dot{M}$ and v_∞ are completely set by the primary mechanism. However, even for an SMR wind there is a transfer of angular momentum from the star to the wind, which leads to a spin down of the star.

In the case of a FMR wind, the deposition of the momentum occurs mostly in the region beyond the inner critical point, and thus the terminal velocity is increased, but with little change in the mass loss rate.

Centrifugal magnetic rotator theory was developed by Hartmann and MacGregor (1982), regarding mass loss from rapidly rotating protostars. The magnetic field causes solid body rotation out to the inner critical point. Thus, there is a simple modification owing to centrifugal forces to the hydrostatic density distribution, and this causes $\dot{M}$ to increase. The terminal velocity of a centrifugal magnetic rotator is given to a very good approximation by the speed at the fast critical point,v_M, called the Michel velocity

$$v_M{}^3 = \left(R_*{}^4 \Omega^2 B_{r,\circ}{}^2 / \dot{M}\right) \approx v_\infty{}^3. \tag{1}$$

Note here that the terminal velocity is fully determined by the basic parameters at the base of the wind, the rotation rate, Ω (which determines the mass loss rate $\dot{M}$), and the magnetic field at the surface of the star. So unlike $\dot{M}$, the value of v_∞ depends explicitly on the field strength; given a rotation rate, the higher the field the faster the wind.

Radiation also plays a role in increasing the velocity of the wind along the FMR portion of the track in the $\dot{M}$ versus v_∞ plot. Because of the velocity gradient effect, the radiative acceleration is enhanced by the effects of the strong magnetic field. There is an increase the radiation force because in Sobolev line transfer theory the larger the velocity gradient the larger the amount of stellar radiation that is Doppler shifted into the line absorbing region per unit length in the wind. Thus, the magnetic rotator forces make the stellar radiation field more effective, per spectral line, in driving a wind.

These effects are illustrated in figures 2 and 3 for the case of a B[e] star (as originally treated by Cassinelli *et al.* 1989). We compare a non-rotating star with one rotating at 60% critical and with one that is rotating at that rate but which also has a magnetic field of 500 G. Though radiation remains the dominant force throughout the wind, the extra acceleration caused by the magnetic force desaturates the radiation force, making it more effective at driving much faster winds than would be achieved otherwise.

4 Alfvén Wave Amplified Winds from Hot Luminous Stars

There are many similarities between Alfvén wave driven winds and magnetic rotator winds. In the case of the magnetic rotators centrifugal forces and the Lorentz force associated with the curvature of the field lines transmit momentum to the wind. In the Alfvén case the origin of the force is presence of an oscillation at the base of the field lines, this then leads to the deposition of momentum both by a centrifugal force and a Lorentz force as the wave disturbance propagates through the plasma. The basic process as applied to

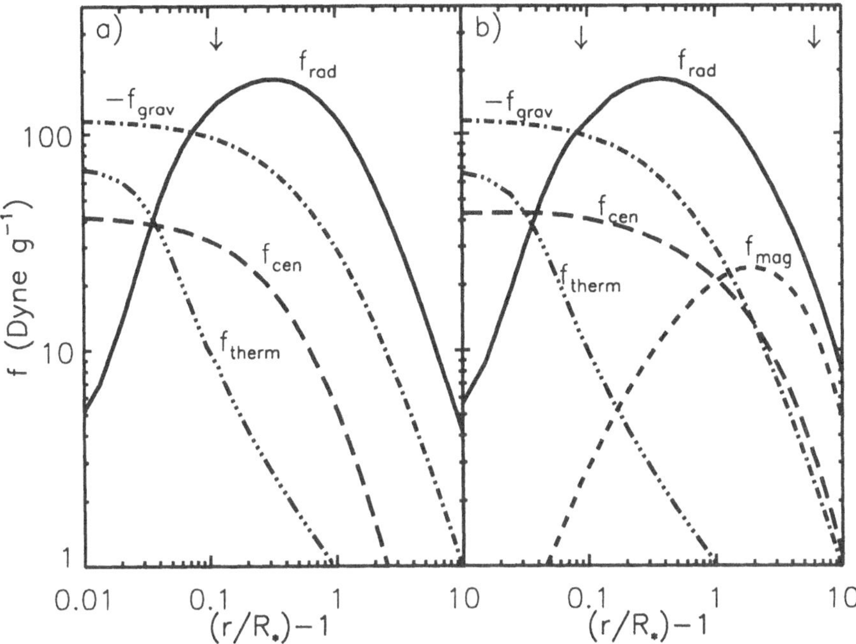

Fig. 2. The distribution, as a function of radius, of 5 forces per unit mass: thermal, magnetic, centrifugal, gravitational, and line radiation, for a B[e] star which is assumed to rotate at 60% critical speed. Panel (a) shows results for a model star with no magnetic field, resulting in a terminal velocity of 1200 km s^{-1}, and a mass-loss rate of 6.6×10^{-6} $M_\odot$ yr^{-1}. The arrow indicates the inner critical point. For Panel (b), on the right, $B_{r,o}$=500 G, and we see the added magnetic force leads to a steeper velocity gradient near the Alfvén radius resulting in a visibly increased radiation force. Both the inner and Alfvén critical points are indicated. This highly magnetic model has a terminal velocity of 2000 km s^{-1}, much faster than the panel (a) model, and a mass loss rate (7.9×10^{-6} $M_\odot$ yr^{-1}) that is only slightly larger than the panel (a) model, as is characteristic of models in the FMR regime (from Friend and MacGregor, 1984).

coronal winds is described by Holzer, Flå, and Leer (1983). In the case of the sun, Alfvén waves are considered responsible for driving the fast ($\approx$ 800 km s^{-1}) outflows from coronal hole regions. The mechanism was proposed for explaining the outflows of red giants by Hartmann and MacGregor (1982). However Holzer *et al.* (1983) criticized the model because it depended too sensitively on the wave damping length. This is because the winds of red giants are slow, while the forces tend to lead to fast winds. In the case of hot stars the flow speed produced by the Alfvén waves is not a problem because the winds tend to be fast in any case, and we are using the waves to produce a

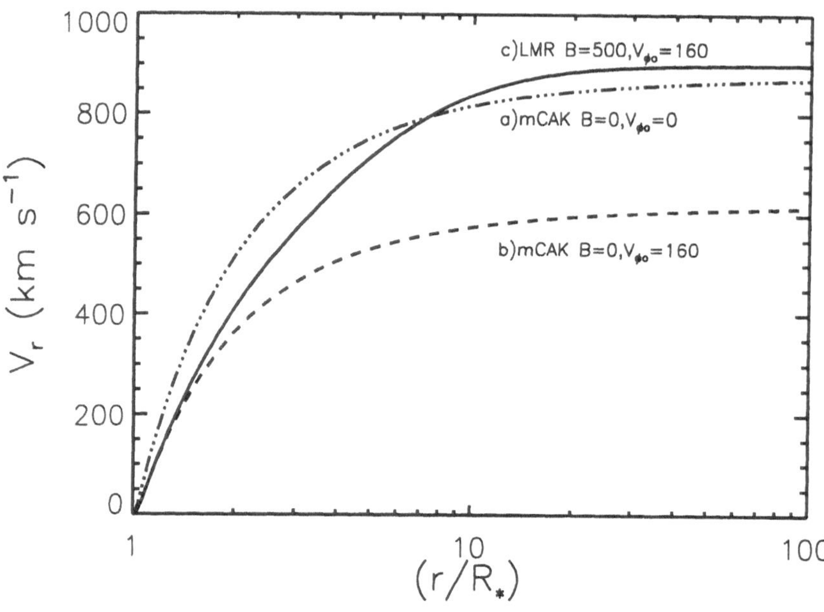

Fig. 3. The radial velocity distribution for three B[e] models: Curve a) shows the velocity distribution for a mCAK (Friend and Abbott (FA) 1986) model for a star with no rotation or magnetic field. Curve b) is for a FA model that includes rotation. Note this reduces both the velocity near the star and the terminal velocity (which scales as $v_{\rm esc}$). Curve c) is for a rotating magnetic model. Near the star the model is similar to the purely rotating model, but farther out near the Alfvén point the wind is accelerated by magnetic force effects.

wind that is even faster than can be produced by radiation alone. The Alfvén wave forces have been combined with radiation forces by Dos Santos, Jatenco-Pereira, and Opher (1993), specifically to explain the Wolf-Rayet momentum problem that was discussed in the introduction. Dos Santos *et al.* considered WR outflows with Alfvén waves and the modified CAK forces. Unfortunately they also included the effects of the non-radial spherical outflow discussed in section 2, so it is not possible to isolate the Alfvén wave effects. The Alfvén models lead to the following conclusions:

a) The mass loss rate is determined by the average of the square of the transverse wave amplitude $\langle(\delta V_\phi)^2\rangle$ This is analogous to the centrifugal magnetic rotator case in which $\dot{M}$ is determined by the (transverse) rotation velocity at the base of the magnetic tubes.

b) The terminal velocity is determined by the magnitude of the B-field, as

most of the deposition of momentum occurs beyond the inner critical point, but near the Alfvén radius. This again is analogous to the CMR case in which the magnetic field determines v_∞.
c) Dos Santos *et al.* could achieve WR wind conditions with a magnetic field of less than 1000 G. The model had sufficient momentum to achieve a momentum ratio of $\eta \approx 10$. Furthermore the model had an energy flux of less than about 1 percent of the radiative luminosity.

The combined Alfvén wave plus line driven wind model has not been pursued recently perhaps because: a) It is not clear what would produce the oscillations at the base of the winds. b) Transverse motions in general may be strongly damped by the interaction with the radiation forces (Owocki and Rybicki, 1986). Whether this is a problem or not is unclear as an unknown damping force is incorporated in the equations in any case. c) It is also not clear whether the Dos Santos *et al.* solution crossed mCAK type critical points, and this is a major problem and concern of any model that uses line driving forces.

5 Summary

Our goal has been to investigate whether the combination of radiation and stellar magnetic fields could increase the terminal velocity of the wind from a hot luminous star. We have shown that there are three ways that this can be done: 1) a non-spherical divergence of area in flow tubes defined by open field geometries, 2) fast magnetic rotator forces 3) Alfvén wave forces, at least in combination with non-spherical geometries. We have indirect evidence for the presence of fields as given by fast wind slow wind collision effects and the dependence of wind properties on the rotation of the star. What is most seriously needed is direct observational information about field strengths and field geometries. Several of us have been studying the Hanle effect (Cassinelli and Ignace 1996, Ignace Nordsieck and Cassinelli, 1997,1998). This effect requires high-resolution polarimetric observations in the far-ultraviolet, and only a rocket experiment has been approved thus far. There is no doubt that magnetic fields are capable of producing the general fast wind slow wind phenomena or rotational modulations that are inferred from observations. The problem is that there remain too few observational/diagnostic constraints to guide more detailed theoretical modelling. Here we have focussed on changes that fields can induce regarding the terminal velocity. In principle the fields can also lead to a mass-loss rate that varies across the face of a hot star. Perhaps it is possible to distinguish between the two effects, and this will provide a better understanding between forces acting primarily at the base of the wind and those operating beyond the wind critical point.

References

Castor, J.I., Abbott, D.C., and Klein, R.I. 1975, ApJ, 195, 157. (CAK)
Cassinelli, J.P., Schulte-Ladbeck, R.E., Abbott, M., and Poe, C.H. 1989, in Proc. IAU Colloq. 113, Physics of Luminous Blue Variables, Moffat, A.F.J., Lamers, H.G.J., eds., Dordrecht, Kluwer, p. 121.
Cassinelli, J.P. 1998, to appear in ESO workshop, Cyclical Variability in Stellar Winds.
Cassinelli, J.P., and Castor, J.I. 1973, ApJ, 179, 189.
Cassinelli, J.P. and Ignace, R., 1996, in Proc. 33rd Liege Intl. Astrophys. Colloq., Wolf-Rayet Stars in the Framework of Stellar Evolution, Vreux, J.M., Detal, A., Fraipont-Caro, D., Gosset, E., Rauw, G., eds., p. 531.
Cohen, D.H., Cassinelli, J.P., MacFarlane, J.J. 1997, ApJ, 487,876.
Cranmer, S.R. and Owocki, S.P. 1996, ApJ, 462, 469.
Dos Santos, L.C., Jatenco-Pereira, V., and Opher, R. 1993, ApJ, 410, 732.
Friend, D.B., and Abbott, D.C. 1986, ApJ, 311, 701.
Friend, D.B., and MacGregor, K.B. 1984, ApJ, 282, 591.
Gayley, K.G., Owocki, S.P., and Cranmer, S.R. 1995, ApJ, 442, 296.
Hartmann, L., and Cassinelli, J.P. 1981, BAAS, 13, 795.
Hartmann, L., and MacGregor, K.B. 1980, ApJ, 242, 260.
Hartmann, L., and MacGregor, K.B. 1982, ApJ, 257, 264.
Holzer, T.H., Flå, T., and Leer, E. 1983, ApJ, 275, 808.
Kaper, L., Henrichs, H.F., Nichols, J.S., Snoek, L.C., Volten, H., and Zwarthoed, G.A.A. 1996, A&AS 116,257.
Ignace, R., Nordsieck, K.H., and Cassinelli, J.P. 1997, ApJ, 486, 550.
Ignace, R., Nordsieck, K.H., and Cassinelli, J.P. 1998, ApJ, submitted.
Lamers, H. J. G. L. M. and Cassinelli, J. P. 1998: *Introduction to Stellar Winds*, (Cambridge, Cambridge), Chap. 9.
Leer, E., Holzer, T.E. 1979, Solar Phys., 63, 143.
MacGregor, K.B. 1988 ApJ, 327, 794.
Maheswaran, M., and Cassinelli, J.P. 1988 ApJ, 335, 931.
Maheswaran, M., and Cassinelli, J.P. 1992 ApJ, 386, 695.
Owocki, S.P., and Rybicki, G.B. 1986, ApJ, 309, 1270.
Poe, C.H., and Friend, D.B. 1986, ApJ, 311, 317.
Poe, C.H., Friend, D.B., and Cassinelli, J.P. 1989, ApJ, 337, 888.
Seeman, H. 1998, Thesis, University of Bonn.
Weber, E. J., and Davis, L. Jr. 1967, ApJ, 148, 217.

Discussion

S. Owocki: In a paper with George Rybicki in 1985, we showed that horizontal velocity perturbations in a line-driven wind will be strongly damped by the line-drag effect. This suggests that radially propagating Alfvén waves, with horizontal velocity fluctuations, should be strongly damped.

J. Cassinelli: Thanks for reminding me about that. It would be good to consider specifically the case of transverse Alfvén waves.

H. Henrichs: Can you make a quantitative estimate of the magnetic field needed to provide the link between Hα and the UV fast wind?
J. Cassinelli: Sure, if you can tell me the velocity contrast needed to provide the observed DAC. That velocity difference can be used in conjunction with the $\dot{M}$ versus v_∞ plot of Luminous Magnetic Rotator theory. The larger the B field, the greater the difference in velocity between a fast magnetic rotator and the mCAK velocity associated with a non-magnetic longitude sector.

Joseph Cassinelli, Jon Bjorkman and Richard Ignace

Modeling Oblique Rotators: Magnetospheres and Winds

Steven N. Shore

Department of Physics and Astronomy, Indiana University South Bend, 1700 Mishawaka Ave., South Bend, IN 46634-7111 USA

Abstract. The upper main sequence chemically peculiar (CP) stars display evidence of trapped circumstellar gas and nonspherical outflows. These stars are also known to possess strong magnetic fields that are often highly inclined to the rotational axis. Their phenomenology can be understood by using the oblique rotator model, which has successfully accounted for the observed behavior of the cooler CP stars. This paper reviews some features of the oblique rotator model, in which the magnetic field is assumed to provide a rigid framework for the structuring of the stellar and circumstellar gas. Corotation of circumstellar plasma is enforced out to the Alfven radius in the magnetic equatorial plane, while for the hotter stars, a radiatively driven wind emerges from the magnetic polar caps. Some observable consequences of the model are discussed, especially the Hα and ultraviolet resonance line absorption and emission periodic variability that has been observed in the He-peculiar stars and nonthermal radio emission. Magnetospheres may also be present in O stars, e.g. θ^1 Ori C, and in the Herbig Ae/Be stars.

1 Introduction

The phenomenological oblique rotator model is actually as old as mathematical astronomy itself, having been introduced by Eudoxus (ca. -350) to explain the planetary motions (see Neugebauer 1983). For more recent applications, however, the history dates back to the explanation of variable stellar spectra and magnetic fields observed in the upper main sequence chemically peculiar (CP) stars (Deutsch 1958, 1970). In its most basic form, the model contains only two adjustable *geometric* parameters that permit transformation between the magnetic and observer's frames. The magnetic fields in the CP stars are frozen into the stellar envelope and not actively generated by contemporaneous dynamos. The field thus provides a rigidly rotating[1] coordinate system. The first parameter is i, the inclination of the rotational axis to the line of sight. This is trivially derived, since model atmosphere studies and parallax measurements combine to provide the stellar radius and

[1] The stability of the surface features is attested to by the longterm regularity of the photometric variations of a few Ap stars, especially α^2 CVn, which has maintained the same period since it was first observed photometrically early in this century (Pyper 1966).

thus $v_{eq} = 50.6(R_star/R_\odot)(P/\text{days})^{-1}\text{km s}^{-1}$, where P is the rotation period in days and $R_\star$ is the stellar radius.[2] The other model parameter is $\beta = \cos^{-1}\hat{\omega}\cdot\hat{\mathbf{B}}$, the obliquity of the magnetic field symmetry axis, $\hat{\mathbf{B}}$, to the rotational axis, $\hat{\omega}$. For CP stars, although the mechanism is not thoroughly understood (Moss 1990)[3], the field is inclined to the rotation axis. We can therefore write the simple coordinate transformation of rotation matrices, $\mathcal{R}$, as $\mathbf{x} = \mathcal{R}(\omega)\mathcal{R}(i)\mathcal{R}(\beta)\mathbf{x}'$. Here $\mathbf{x}'$ is the magnetic coordinate system and $\mathbf{x}$ is the observer's frame, which is centered on the visible hemisphere and over which all averages are taken including the limb darkening, $\Lambda(\theta)$. For an unresolved surface, the model treats all quantities distributed over the photosphere by averaging over the surface brightness, given by the series expansion:

$$\xi(\theta',\phi') = \sum_{lm} a_{lm} Y_{lm}(\theta'\phi') \tag{1}$$

where θ and ϕ are the meridional and azimuthal angles, respectively, and Y_{lm} is a spherical harmonic of order (l,m). Thus,$Q(\theta,\phi)$ transforms into a mean quantity that is a function only of phase:

$$< Q(\Phi) >= \int d\Omega Q(\theta,\phi)\Lambda(\theta)\xi(\theta,\phi) \tag{2}$$

where Ω is the subtended solid angle, Λ and ξ are normalized in the observer's frame, and Φ is the rotational phase. This is essentially the basis for Doppler imaging since, for a rigid rotator, each point on the surface maps to a position in the line profile whether due to velocity alone or to the Zeeman effect and velocity acting together (Deutsch 1970; Mihalas 1973; Mesessier et al. 1979; Rice et al. 1989). Doppler imaging using maximum entropy reconstructions of stellar surfaces have been successful in providing statistically robust, model-free snapshots of elemental and flux distributions (cf. Hatzes et al. 1989; Piskunov and Rice 1993 and refs. therein), although the resultant map is hard to quantitatively associate with any feature of the magnetic field geometry or polarity (see however Semel 1989).

The line of sight (longitudinal) magnetic field, B_{eff}, varies due to the rotation and the obliquity is provided by measurements of $r = B_{\text{eff},+}/B_{\text{eff},-}$, the ratio of the projected magnetic field extrema (e.g. Borra, Landstreet, and Mestel 1982). The combination of field distribution and velocity shifting of the Zeeman components produces the cross-over effect that yields important model information (e.g. Mathys 1995). Finally, restricting the distribution to a dipole in the magnetic frame, an axisymmetric distribution produces a double minimum in the light curve that corresponds to magnetic equatorial

[2] Additional support for this picture comes from measurements of the phasing and variations of pulsation of the rapidly oscillating Ap stars (roAp) (e.g. Kurtz 1990).

[3] This obliquity is also observed for virtually all planets in the Solar System, although the explanation is equally elusive there.

crossing, so that $\cos(\Delta\Phi/2) = \cot i \cot\beta$, where $\Delta\Phi$ is the phase separation of the mimima. This simplification in the geometry is especially important for diagnosing and modeling magnetospheres.

2 Magnetospheres

There are several signatures of magnetospheres in the CP stars. In the He-strong stars, Balmer line emission is modulated with the rotational phase (*e.g.* Nakajima 1985; Bolton 1994). The periodic Hα variations in the O7 V star θ^1 Ori C (Stahl et al. 1996) seem to be similar. The best studied case, HD 37479 = σ Ori E, which is not axisymmetric in *either* the magnetic or rotational frame (*e.g.* Hunger et al. 1993; Bolton 1994; Groote and Hunger 1997), also shows periodic shell-like absorption on the highest Balmer series lines when Hα is strongest. Some He-weak stars also display variable Hα emission that is modulated on the rotational phase (see Shore et al. 1998 and refs. therein) (especially HD 79158 = 36 Lyn and HD 35502). It therefore seems that many of the hotter members of the He-weak sequence of chemically peculiar stars possess magnetospheres.

The unambiguous evidence of magnetospheric plasma is the detection of enhanced, variable C IV and Si IV resonance line absorption in the CP stars (Shore 1987; Shore, Brown and Sonneborn 1987; Shore et al. 1990; Shore 1990). This behavior has also been detected in the O7 V star θ^1 Ori C (Walborn and Nichols 1994). The phase relation for the variable CP star line profiles clearly shows that the strongest absorption occurs when the magnetic equator crosses the line of sight. The profiles are only moderately variable in symmetry, indicating that the gas is constrained to corotate to large distance – the extent can be determined by the occultation of the magnetospheric gas by the stellar disk (Fig. 1). The geometry is usually dipolar, but here one star deserves special mention: HD 37776. This is the only known He-strong star with a predominantly quadrupolar surface field (Thompson and Landstreet 1985). The C IV variations, although precisely periodic over the magnetic cycle, do not correlate well with the projected line of sight field strength.

A clue to the geometry of the circumstellar plasma comes from the slowly rotating He-strong stars: the stars that show emission (HD 5 and HD 96446) have constant strong magnetic fields while those showing strong, constant C IV absorption, HD 60344 and HD 133518, show no detectable fields[4] (Shore and Brown 1990).

In addition, a number of CP stars are *nonthermal* radio emitters (Linsky, Drake, and Bastien 1992; Leone, Umana, and Trigilio 1996). Phillips and Lestrade (1988) have even used VLBI to place limits of about $6R_\star$ on the

[4] At this meeting, Mathys mentioned observations of HD 96446 that seem to present a problem for the simple unified picture I have just described. This star shows periodic variations of the magnetic field that are not consistent with the geometry implied by the emission line profile and hint at a non-dipolar geometry.

size of the emission regions in HD 37017 and HD 37479, two prototypical He-strong stars. For the He-weak and Si stars, Linsky et al. (1992) obtained many detections, while Drake, Linsky, and Bookbinder (1994) find for a sample of 23 candidates that no non-magnetic CP stars are radio sources and therefore argue that such stars lack magnetospheres. Linsky et al., concentrating exclusively on radio measurements, find a general scaling for the 6 cm luminosity, $L_{6cm} \sim \dot{M}^{2/5} B \omega^{1/3}$, where B is the surface field, $\dot{M}$ is the mass loss rate, and ω is the rotational frequency. There is considerable uncertainty in the mass loss rates for the He-weak stars compared with the He-strong stars so their scaling relation is only preliminary. One advantage of the UV resonance line measurements is that it provides direct information about the density and velocity of the material in the circumstellar plasma. The implicated mechanism for emission is gyrosynchrotron radiation (Linsky 1993). Havnes and Goertz (1984) suggested that the emission is powered by reconnection at the Alfven surface. The formation of a compact disk, of the sort discussed by J. Bjorkman and Cassinelli at this meeting, results from compression in the magnetic equatorial plane, with outflow and heating occurring due to the resulting pressure gradient from the equatorial collisions.

For cooler B stars, a centrifugally driven wind is also a possible source for circumstellar plasma (Mestel 1968) since the He-weak stars do not possess strong radiatively driven winds with which to supply the circumstellar gas. In this case, the rapidly rotating magnetic stars might then be expected to show more circumstellar gas, and there should be a strong dependence of the density of the magnetospheric gas on the rotation frequency and magnetic field strength (Mestel 1968; Nerney and Suess 1987; Shore and Brown 1990). It is therefore something of a surprise that we do not see any evidence for so simple a picture among the helium-weak and silicon stars. For the He-strong stars, the periods are pretty uniform, from about 1 to 10 days. The He-weak stars show a larger range of rotational periods with which the magnetospheres are not well corelated; e.g. HD 5737 = α Scl has a 21 day period and displays a similar variability to HD 142301, which has a period of $\approx$1.5 days. In addition, the magnetospheric C IV strength seems not to depend on magnetic field strength or oblique rotator parameters.

Theoretically, the magnetosphere is a closed surface whose extent is the Alfven radius (e.g. Dressler 1983; Michel 1991). This is related to the intrinsic properties of the star through $R_A \sim B^{1/3} (f v_\infty)^{1/6} \dot{M}^{-1/6} \omega^{-1/3} R_\star$, where $f v_\infty$ is the fraction of the terminal velocity of the wind (cf. Shore 1987, 1993). As observed for the outer planets, such as Jupiter (Dressler 1983), the circmstellar plasma is forced into approximate corotation out to R_A, and inertial effects concentrate matter into a disk in the magnetic equatorial plane. This should lead to formation of a current sheet. In light of the outflow that is expected in the hottest stars, this sheet could evolve by driven reconnection. Because its symmetry plane does not coincide with the rotational equator, the plasma is forced into a periodic oscillation on the rotational timescale. For planetary magnetospheres, which are strongly interacting with the solar wind, this produces reconnection in the magnetotail with attendant acceleration of electrons (cf. Nishida 1984). It is thus reasonable to expect reconnection to occur in CP star magnetospheres, leading to sporadic variations of the radio emission and local heating due to turbulence. In addition, the backward flowing particles will dissipate their energy in the stellar atmosphere, producing an auroral ring – a region in a planetary atmosphere that is strongly heated by particle precipitation from the current sheet – and providing the hot ions that are responsible for C IV absorption in the cooler He-weak stars.

A fundamental difference between stellar and either Solar System magnetospheres or those in cataclysmic variables is the absence of external flows in the isolated star case. The CP stars are self-structured, in the sense that the heating and shear is due to internal processes, such as stellar winds and reconnection. To date, the most comprehensive models for heating and dynamics of the trapped circumstellar plasma in magnetic CP stars has been by Babel and Montmerle (1997a, 1997b), following formalism by Havnes and Goertz (1984), who use an analogy to the terrestrial magnetosphere. Babel and Montmerle successfully explain the X-ray emission from HD 34452 as arising from shocks within the magnetosphere. A problem with this picture may be that this star does not show enhanced C IV absorption, while other stars that do show this are not X-ray emitters.

As an example of the generic behavior expected for the Hα emission, I show in Fig. 1 the variation of the line profile with phase for a simple model with a disk-like distribution for the emission $j(\theta') \sim P_2(\theta')^n$ where P_2 is the Legendre polynomial in the magnetic frame and n is a free parameter describing the flattening of the magnetosphere (Shore and Brown 1990). Figure 2 shows the integrated emission strength as a function of phase. The parameters were chosen to reproduce the variations of θ^1 Ori C (Stahl et al. 1996). The width of the line provides the Alfven radius and requires at least a kilogauss field to maintain corotation to around $4R_\star$. Such a field has not, however, been detected, although Babel and Montmerle (this meeting) argue that a much weaker field can yield the observed X-rays. The integrated flux curve is generic (Fig. 2) and any oblique magnetospheric distribution will produce the same behavior. The C IV lines, however, indicate a very thick region and again require a large field to maintain the structure. Yes, this is a puzzle.

3 Non-spherical Mass Loss

In the hottest stars, radiation pressure is expected to drive outflows. That is what much of this meeting has been about. The imposition of a strong magnetic field, however, necessarily renders the flow aspherical. The magnetosphere is a "dead zone" for outflow, so any mass loss can only occur from

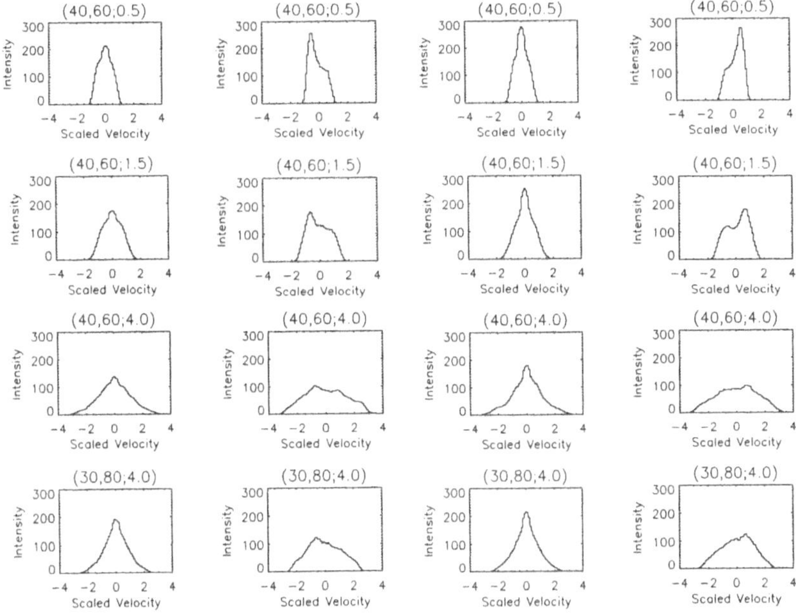

Fig. 1. Sample line profile variations, labeled by (i, β) and the outer radius (R_A-$R_\star$). Notice the relative insensitivity to oblique rotator parameters although strong dependence on radius. The profiles show phases 0.0, 0.25, 0.5, and 0.75, respectively.

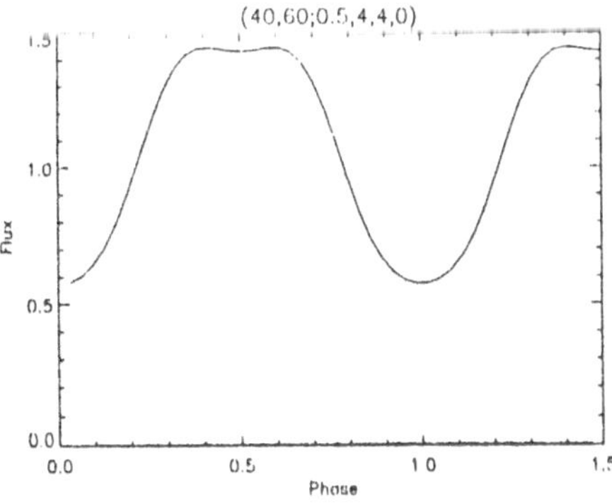

Fig. 2. Integrated line flux curve for $i = 40, \beta = 60$ (see Fig. 1). Magnetic polar traverse is at $\Phi = 0^p.5$.

latitudes whose field lines close outside the Alfven surface. This restricts the opening cone for the flow to $\Delta\theta = \sin^{-1}(R_*/R_A)^{1/2}$. Heating may be provided by plasma turbulence or by shocks. Nonthermal radio emission is observed in several stars, notably the strongest field He-strong stars (especially HD 37479 = σ Ori E), but it also extends to the cooler He-weak stars with the strongest fields. While the He-strong stars appear to show rotational emission modulation, this is not as clear for the He-weak emitters.[5]

The strongest case for a collimated outflow is HD 21699 (Brown et al. 1985). This He-weak Si star is unique in showing the strongest C IV absorption when the strongest magnetic pole is in the line of sight. In addition, the absorption is strongly asymmetric, which is not seen in any of the variable magnetospheric stars, and the terminal velocity is consistent with corotation out to R_A. This star is a member of the α Per cluster so its age is known and the star is clearly on or near the ZAMS, but not pre-main sequence.[6] Among the He-strong stars, HD 58260 and HD 96446 show only nonvariable, asymmetric emission. There is an extended red wing but the blueward side of the line is truncated at the rest wavelength. Shore and Brown (1990) schematically model this as a combination of a magnetosphere and polar outflow (see their fig. 10).

For collimated flows, the solid angle of the polar cone is relatively small so the normal P Cygni emission line wings are absent. The simplest models (Kunacz 1984; Brown et al. 1985; Shore et al. 1994) produce only absorption troughs whose radial velocity and depth change with rotational phase. The absorption disappears for transverse presentation of the flow to the line of sight. Thus, for orthogonal rotators such as the He-strong stars HD 60344 and HD 133518, there is no contribution from the wind to the line profile. The weakness of the emission wing may also explain the pbserved profiles in some of the Herbig Ae/Be, such as AB Aur.

Acknowledgments: I wish to thank the SOC for their kind invitation to this wonderful meeting and Jacques Babel[7], Joe Cassinelli, Henny Lamers, Gauthier Mathys, Stan Owocki, and Steve White for discussions during the sessions. Some of the work reported in this review is part of a larger study of the helium peculiar stars with David Bohlender, Tom Bolton, and Pierre North. IUE observations of the helium peculiar stars were supported by NASA through NAG5-2612 to IUSB.

[5] Linsky et al. find variable radio emission but the data set to date is too sparse to permit accurate phasing of the detections.

[6] Many stars in this intermediate spectral range that show anomalous C IV absorption turn out to be Herbig Ae/Be stars (cf. Imhoff 1994). These may show the combined effects of outflow and magnetospheres, but they have yet to be modeled within this picture.

[7] ... even though we disagree on the nature of θ^1 Ori C.

References

Babel, J., Montmerle, T. 1997, A&A, **323**, 121

Babel, J., Montmerle, T. 1997, ApJ, **485**, L29

Barker, P. L., Brown, D. N., Bolton, C. T., Landstreet, J. D. 1982, in *Advances in Ultraviolet Astronomy: Four Years of IUE Research*, ed. Y. Kondo, J. Mead, R. D. Chapman (NASA CP-2238), p. 589

Bolton, C.T. 1994, ApSS, **221**, 95

Brown, D.N., Shore, S.N., Sonneborn, G. 1985, AJ, **90**, 1354

Deutsch, A. 1958, in *IAU Symp. 6* ed. A. Lehnert (North-Holland: Amsterdam) p. 209

Deutsch, A. 1970, ApJ, **159**, 985

Drake, S.A., Abbott, D.A., Bastien, T.S., Bieging, J.H., Churchwell, E., Dulk, G., Linsky, J. L. 1987, ApJ, **322**, 902

Drake, S.A., Linsky, J.L., Schmitt, J. H. M. M., Russo, C. 1994, ApJ, **420**, 387

Drake, S. A., Linsky, J. L., Bookbinder, J. A. 1994, AJ, **108**, 2203

Dressler, A. ed. 1983, *The Jovian Magnetosphere* (Cambridge Univ. Press: Cambridge)

Groote, D. Hunger, K. 1982, A&A, **116**, 64

Groote, D., Hunger, K. 1997, A&A, **319**, 250

Hatzes, A., Penrod, G. D., Vogt, S. S. 1989, ApJ, **341**, 456

Havnes, O., Goertz, C.K. 1984, A&A, 138, 421

Hunger, K., Groote, D. 1993, *Peculiar versus Normal Phenomena in A-type and Related Stars* , ed. F. Castelli, M. M. Dworetsky (ASP Conf. Series: San Francisco), p. 394

Imhoff, C. 1994, in *First International Meeting on Herbig Ae/Be Stars* eds. M. Perez, E. van den Heuvel (ASP Conf. Series: San Francisco)

Kurtz, D. W. 1990, ARAA, **28**, 607

Leone, F., Umana, G., Trigilio, C. 1996, A&A, **310**, 271

Linsky, J.L. 1993, in *Peculiar versus Normal Phenomena in A-type and Related Stars* , ed. F. Castelli, M. M. Dworetsky (ASP Conf. Series: San Francisco), p. 507

Linsky, J.L., Drake, S.A., Bastien, T.S. 1992, ApJ, **393**, 341

Mathys, G. 1995, A&A, **293**, 733

Megessier, C., Khokhlova, V. L., Ryabchikova, T. A. 1979, A&A, **71**, 295

Mestel, L. 1968, MNRAS, 138, 359

Michel, F. C. 1991, *Theory of Neutron Star Magntospheres* (Univ. of Chicago Press: Chicago)

Mihalas, D. 1973, ApJ, **184**, 851

Moss, D. 1990, MNRAS, **244**, 272

Nakajima, R. 1985, ApSS, 116, 285

Nerney, S., Suess, S. 1987, ApJ, 321, 355

Neugebauer, O. 1983, *Astronomy and History: Selected Essays* (Springer-Verlag: Berlin)

Nishida, A. 1984, in *Magnetic Reconnection in Space and Laboratory Plasmas* (AGU Geophys. Mono. 30) (American Geophysical Union: Washington DC) p.159

Phillips, R. B., Lestrade, J-F, 1988, Nature, **334**, 329

Piskunov, N. E., Rice, J. B. 1993, PASP, **105**, 1415

Rice, J. B., Wehlau, W. H., Khokhlova, V. L. 1989, A&A, **208**, 179
Semel, M. 1989, A&A, **225**, 456
Shore, S.N. 1987, AJ, **94**, 731
Shore, S.N. 1993, in *Peculiar versus Normal Phenomena in A-type and Related Stars* , ed. F. Castelli, M. M. Dworetsky (ASP Conf. Series: San Francisco), p. 528
Shore, S. N., Bolton, C. T., Bohlender, D. A., North, P. 1998, *preprint*
Shore, S.N., Brown, D.N. 1990, ApJ, **365**, 665
Shore, S.N., Brown, D.N., Sonneborn, G. 1987, AJ, **94**, 737
Shore, S.N., Brown, D.N., Sonneborn, G., Landstreet, J.D., Bohlender, D.A. 1990, ApJ, **348**, 242
Stahl, O. et al. 1996, A&A, **312**, 539
Thompson, I. B., Landstreet, J. D. 1985, ApJ, **289**, L9
Walborn, N. R., Nichols, J. S. 1994, ApJL, **425**, L29

Discussion

H. Henrichs: Could it be that your θ^1 Ori C magnetic field measurements were taken around a phase when the field was small?
G. Mathys: No magnetic field was detected in θ^1 Ori C in about a dozen observations distributed in phase: this rules out the possibility of null measurements by chance.

G. Mathys: The observations of the He-strong star HD 96446 could probably be easily modelled, as you claim, if the star were an oblique rotator. However, as mentioned in my talk, the magnetic observations would in this case imply that HD 96446 should have an unrealistically small radius of less than 2 $R_\odot$. Therefore, the oblique rotator model must be questioned for this star.
S. Shore: This could be evidence of trapped polar oscillations, analogous to the roAp stars. Since HD 96446 should have i and $\beta \sim 0$ (aligned, low inclination), it could be showing this (if the star can have an overstable mode or modes). Then the same should be seen for the emission line He-strong stars; but the He-strong stars with only C IV steady absorption should not show it.

S. White: I would like to comment that we see the magnetosphere in radio observations as well: a small number of stars have been observed over several rotational periods and they show a clear signal in the circular polarisation of the underlying rotating magnetic field (work by Jeremy Lim et al., in preparation).
S. Shore: Both the He-weak and He-strong stars show this sort of emission. What we have found in the Sco OB 1 sample is that the strong radio emitters are also strong magnetosphere stars. There are lots of ways of accelerating particles in these dynamical structures, especially K-H instabilities in the boundary between the trapped and outflowing plasma. But the radio emission could also be the clue to the heating required to produce the excess ionisation in the coolest stars.

X-Ray Emission from Magnetically Confined Winds

Jacques Babel

Battieux 36, 2000 Neuchatel, Switzerland

Abstract. We consider the effect of large scale magnetic fields on the circumstellar environment of hot stars. In these stars, magnetic fields of order of 100 G lead to magnetically confined wind shocks (MCWS) and then to the existence of large X-ray emitting region. MCWS lead also to the presence of corotating cooling disks around hot stars.

We discuss the case of θ^1 Ori C, which is perhaps the hottest analog to Bp stars and consider the effect from rotation and instabilities. We finally discuss the case of the Herbig Ae-Be HD 104237 and show that MCWS might also explain the X-ray emission from this star.

1 Magnetically confined wind shocks

Magnetic fields certainly play a major role in the wind variability, also X-ray variability, of hot stars (i.e. Fullerton, Henrichs this meeting). The O7V star θ^1 Ori C is perhaps one of the most impressing case. It presents a strong periodic variability in the wind absorption and emission lines (e.g. Stahl 1996) and an X-ray variability with the same period (Gagné et al. 1997). These variations are interpreted as a possible signature of a strong magnetic field. In Babel & Montmerle (1997b), and based on the MCWS model developed by Babel & Montmerle (1997a, BM97a) for Bp stars, we modelised the wind of θ^1 Ori C in presence of a dipolar magnetic field.

The main effect of a dipolar field is to confine the wind component from the two hemispheres towards the magnetic equator, where the components collide leading to a strong shock. We modelise then the postshock region postshock region and X-ray emission from θ^1 Ori C (BM97b). We compute the X-ray emission spectra from each point of the magnetosphere, the 3-D absorption by the surrounding "cold" wind and by the interstellar media, and convolve with the detector effective area. The theoretical and observed *ROSAT* HRI count rate are shown in Fig. 1. The observed X-ray luminosity and variability are well explained by the MCWS model and by an oblique rotator provided that a dipolar magnetic field with $B_d = 270-370$ G (strength at the pole for $r = R_*$) is present. This value is much lower than the lower limit of 1.6 kG obtained by Donati & Wade (1998).

The variability is well explained by an oblique rotator model and by eclipses caused by the cooling disk and the star, and by the varying absorption from the "cold" wind. In Fig. 1, we also see that Hα emission has

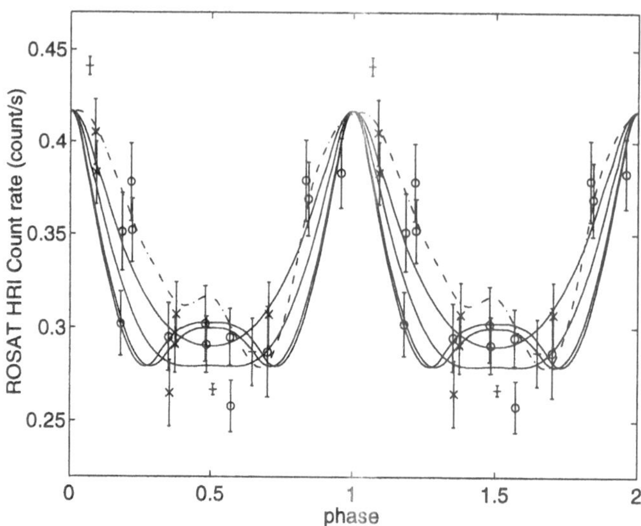

Fig. 1. Theoretical *ROSAT* HRI count rate for the MCWS model (closed case) for a magnetic field of 370 G, and for various angles i and β (see BM97b). The points are for the HRI observations by Gagné et al. Is also plotted (dotted line) the ligntcurve of Hα emission in arbitrary vertical units (Stahl et al. 1996).

a variability quite similar to the X-ray one. We also note that some features in the Hα lightcurve, like the two minima around phase 0.5, are strongly reminiscent of those predicted by the MCWS model for the X-ray lightcurve. This strongly suggests that the Hα emitting region is located close to the X-ray emitting region and thus close to the cooling disk (see also Shore in this meeting). This also indicates an inclination angle i between about 30° and 50° with $\beta + \imath \simeq 90^o$ (note the typo in BM97b).

2 Instabilities and effect from rotation on MCWS

Rotation and instabilities play also an important role in the expected X-ray emission and structure of the MCWS.

MCWS are expected to be quite instables as:

- The flow balance can make the height of the cooling disk grow so much, that the disk pressure against the wind becomes larger than the wind ram pressure. We expect then transient phases of downflows towards the stellar surface (see BM97a).
- The stagnation locus between the flows from the two hemispheres is largely instable. Indeed, in the absence of large rotation, a small asymmetry in the wind flows from the two hemispheres may prevent the building of the stagnation disk at the equator and lead also to downflows, the

wind from one hemisphere quenching the smaller wind from the other hemisphere.
- Magnetic reconnection may also lead to "flaring" events in the outer parts of the cooling disks.

This may explain the strong and time variable continuum circular polarisation observed by Donati & Wade (1998).

Rotation cause effects mostly in stars with small mass loss rates like Bp stars. In these stars, it hase two effects:

- It affects directly the wind flows at the wind base . This leads first to complex mass loss spots with various shapes like: ring, lunated or elongated spots at the stellar surface (Babel, in preparation) and thus to warped disks cooling disks around these stars.
- It modifies also directly the structure of the postshock region around these stars increasing sometimes largely the X-ray emission (BM97a). In oblique rotator models, this make also the cooling disk move out of the magnetic equator and lay between the magnetic and rotation equators.

3 Conclusion

Magnetic field of order of 100 G have very large effects on the wind from hot stars, leading to MCWS, thus to X-ray emission, and circumstellar disks. Stars like θ^1 Ori C or Bp stars are thus expected to present a large range of variability going from X-ray spectral variability to circumstellar signatures.

Recently, Donati (1998) detected a longitudinal field of order of 50 G in the Herbig Ae-Be HD 104237. This star present also an X-ray emission with a cool component around 0.3 keV and an hotter one not well constrained but above 1.6 KeV (Skinner & Yamauchi 1996). Using the wind observed parameters from HD 104237 and the value of the detected magnetic field, the MCWS model (for a dipole) explains both the X-ray luminosity and the temperature of the cool component. While this star shares little properties with Bp stars, the magnetic field being certainly much more complex and the origin of the wind being not known, the X-ray emission might be a signature of wind shocks at the top of magnetic flow tubes.

References

Babel, J., Montmerle, T. (1997a), *A&A* 323, 121
Babel, J., Montmerle, T. (1997b), *Apj* 485, L29
Donati, J.-F., Wade, G.A.:(1998), *A&A* in press
Donati, J.-F.:1998, Thèse d'habilitation, Univ. Toulouse III
Gagné, M., et al. (1997), *Apj* 478, L87
Stahl, O., et al. (1996), *A&A* 312, 539

Discussion

G. Mathys: If the 270–370 G magnetic field value required by your model refers to the polar field strength, the corresponding observable mean longitudinal magnetic field should be on the order of 100 G; the exact value depends on the geometry of the observation. Its definite detection by the kind of techniques described in my talk would accordingly require the measurement uncertainties to be ~10 times less than achieved so far. Hence your model is fully consistent with the constraints currently available.

T. Berghöfer: You mentioned our work from 1994. This paper was entitled "Are late B stars intrinsic soft X-ray emitters?". We still owe the community a final answer. The answer is "no"! It is true that we found 10 % of all B and A stars in the Rosat all-sky survey. The detection rate of Bp and Ap stars is the same. If the X-ray production is related to the magnetic field, one would expect to see higher luminosities for Bp/Ap stars, or at least a higher detection rate. Many of the detected late B/A stars show radial velocity changes and suggest X-ray emitting companions. So what?
J. Babel: As mentioned in your paper devoted to IQ Aur, we expect that most Ap stars have small mass-loss rates, which are not sufficient to lead to an X-ray luminosity detectable in the RASS. Significant winds are only expected during short phases on IQ Aur or in hotter Bp stars. For the latter, the observations of GHz radio emission due to gyrosynchrotron emission from mildly relativistic electrons at the same time as wind signatures, magnetic fields, and X-ray emission much larger than in normal B stars give us much confidence that the X-ray emission is intrinsic (statistically). I agree, however, that for an individual object like IQ Aur we cannot be completely sure that the X-ray emission is intrinsic.
Finally, as has been shown, the model we propose has a much larger application than for Ap stars, as it shows the general consequence of the magnetic confinement of the wind from hot stars on the X-ray luminosity.
S. Shore: To come to your defense, let me add to what I discuused in my talk, i.e., that there is an aspherical wind so there must be a strong shearing interface between the magnetospheric plasma and the wind flow. The result is certainly some non-thermal heating. In fact, this must be present in order to explain the C IV and other high-ionisation lines in the first place. The shear flows, tearing modes, shocks, etc. that result can accelerate the electrons for the radio sources and generate Alfvén waves for heating. Whether you can reach 10^6 K or only a few times 10^4 K is in question.

Session III

Variable Winds

chair: A.W. Fullerton, N. Markova and D. Massa

O-Star Wind Variability in the Ultraviolet and Optical Range

Lex Kaper

Astronomical Institute, University of Amsterdam, Kruislaan 403, 1098 Amsterdam, The Netherlands

Abstract. Variability is a fundamental property of O-star winds. One of the major breakthroughs in this field of research has been the recognition that wind variability is cyclical in nature. This suggests that stellar rotation is an important piece in the unsolved puzzle of the wind-variability mechanism. The current idea is that Corotating Interacting Regions (CIRs) are responsible for the observed wind variability. These conclusions are based on detailed and extensive monitoring of ultraviolet resonance lines formed throughout the stellar wind, and subordinate lines like Hα which probe the regions close to the star.

1 Introduction

The P Cygni profiles of strong ultraviolet resonance lines (e.g., of N V, Si IV, C IV) have proven to be a powerful tool in the study of the supersonically expanding winds of hot stars. The blue-shifted absorption troughs trace the wind material in front of the star, allowing a direct measurement of the velocity distribution of the accelerating plasma. The most prominent features of wind variability are the so-called *discrete absorption components* (DACs), which migrate through the profile from red to blue on a timescale of hours to days. Obviously, DACs cannot be observed in saturated P Cygni profiles; however, the steep blue edges of these profiles often show regular shifts of up to 10% in velocity. Edge variability is most probably related to the DAC behaviour, but the precise phase relation has not been unraveled yet. The P Cygni emission, centered at rest wavelength, is constant with time. This is what one would expect in the case of relatively modest variations in the stellar-wind structure. However, the intervening material has to cover a significant fraction of the stellar disk in order to give rise to a noticeable change in P Cygni absorption.

Because of their specific shape DACs are readily recognized in single snapshot spectra. In ultraviolet spectra obtained with the Ultraviolet International Explorer (IUE), Howarth & Prinja (1989) detected DACs in more than 80% of a sample of 203 galactic O stars. Taking into account that saturation of P Cygni profiles in O stars with very dense stellar winds prohibits the detection of DACs, one must conclude that the occurrence of DACs is a fundamental property of O-star winds.

The key problem is to understand why DACs start to develop and how they evolve. Extensive monitoring campaigns with the IUE satellite (e.g. Prinja et al. 1992, Howarth et al. 1993, Massa et al. 1995 (MEGA campaign), Kaper et al. 1996, 1998a) have resulted in a detailed quantitative description of wind variability in about a dozen bright O stars. For a given star, the DAC behaviour results in a characteristic variability pattern, which undergoes detailed changes from year to year.

In search for the origin of DACs, spectral lines of different strength can be used to probe the stellar wind at various depths. The UV resonance lines form throughout the stellar wind up to the outer regions where the wind has reached its terminal velocity. Strong subordinate lines (like N IV 1718 Å and Hα) are formed closer to the star where the density is higher, because the population of the lower level of the transition (which is not the ground state) is proportional to ρ^2. Monitoring of these lines would tell us whether wind variability sets in already close to the star. A complication is that both emission and absorption changes might contribute to the variability, and projection effects have to be taken into account. Whether or not the formation of DACs is triggered by variations at the photospheric level has to be investigated with help of high-resolution (both in wavelength and time) spectroscopy of deep photospheric lines. Therefore, coordinated ultraviolet and optical spectroscopic monitoring is required to study the relationship between variability in different lines in order to reveal the eventual "photospheric connection".

2 Quantitative analysis DAC behaviour

A detailed analysis of time series of ultraviolet spectra from a sample of 10 bright O stars has been presented by Kaper et al. (1996, 1998a). The migrating DACs are isolated from the underlying P Cygni profiles using a least-absorption template constructed from the datasets. Fits with exponential gaussian functions, with the known doublet separation and ratio in oscillator strength as fixed parameters, yield DAC parameters like the central velocity and column density, which are studied as a function of time. The DACs start at low velocity as broad absorption features and accelerate, while narrowing, until an asymptotic velocity ($v_{\rm asymp}$) is reached. This velocity is systematically lower (by 10-20%) than the maximum "blue edge velocity" observed in saturated P Cygni lines, which has historically been identified with the terminal velocity, v_∞, of the stellar wind. The proposed solution is that $v_{\rm asymp}$ measures the "real" v_∞ and that the additional blueshift in saturated P Cygni lines is caused by small-scale structure in the stellar wind (e.g. due to shocks resulting from the instability of radiation-driven winds, cf. Owocki, Feldmeier, this volume). In some stars, $v_{\rm asymp}$ systematically changes from event to event. Taking into account optical depth effects, this might explain the gradual changes observed in the steep edge in saturated profiles.

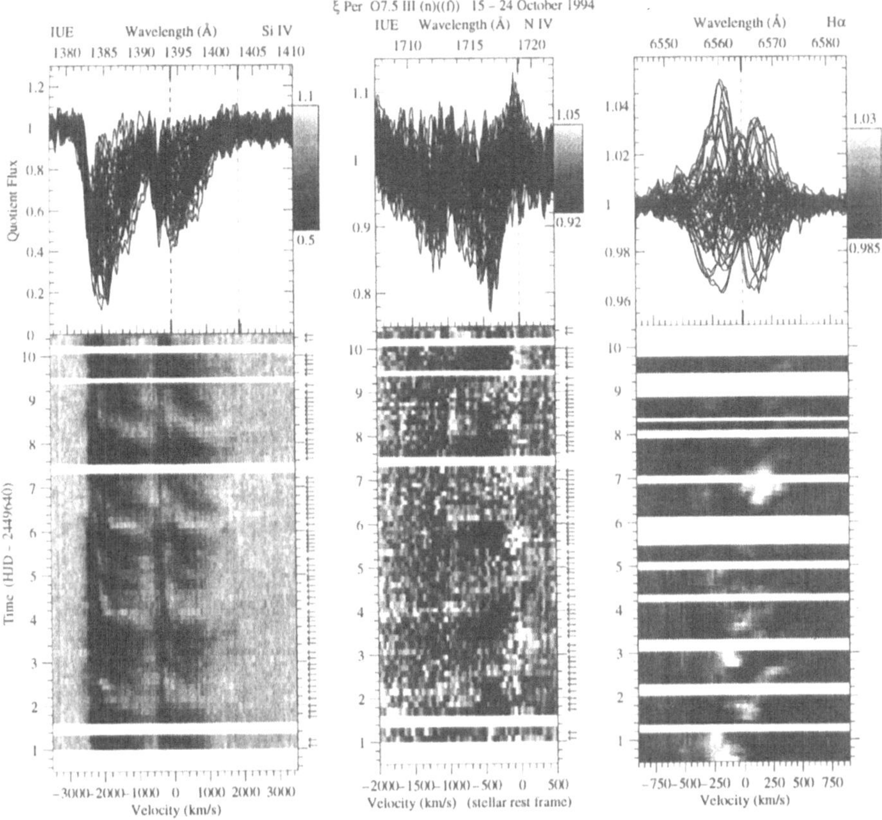

Fig. 1. Dynamic quotient spectra showing the variability of the ultraviolet Si IV resonance doublet and subordinate N IV line, and the optical Hα line in the O7.5III star ξ Per. The UV and optical data were taken simultaneously during a period of more than 10 days of continuous observations. The N IV line indicates the appearance of a strong DAC, whose subsequent evolution is registered by the Si IV doublet. Also the Hα line exhibits a clear increase in absorption at low blue-shifted velocities down in the base of the wind. Close to line center the variations are much more complicated due to projection effects, a varying amount of incipient wind emission, and photospheric variations (from Henrichs et al. 1998).

As is found to be true in several O stars, the DACs reach a maximum column density at a velocity of about 0.75 v_∞ and subsequently fade in strength. When DACs are measured in more than one resonance doublet, consistent results are obtained, supporting the interpretation that DACs are due to changes in wind density and/or velocity rather than changes in the ionization structure of the stellar wind. The change in $N_{\rm col}$ suggests that the covering factor of the absorbing material in the line of sight changes as a function of distance from the star.

A key issue is the recurrence timescale of DACs; they repeat on a timescale comparable to the estimated stellar rotation period. Fourier analyses performed on the datasets clearly reveal this periodicity (Kaper et al. 1998a). In Tab. 1 the observed "wind periods" (corresponding to the DAC recurrence timescales) are listed for a sample of O-type stars. The stars are ordered according to the estimated maximum rotation period:

$$P_{\max} = 50.6\,(v \sin i)^{-1} \left(\frac{R_\star}{R_\odot}\right) \text{days},$$

where $R_\star$ is the stellar radius. The table shows that stars with a shorter DAC cycle end up higher on the list, consistent with the interpretation that this cycle relates to the rotation period of the star.

3 Coordinated Hα monitoring

Also subordinate lines, formed relatively close to the star, show signatures of wind variability. Prinja et al. (1992) and Henrichs et al. (1994) demonstrated that for some stars DACs appear as well in the N IV line at 1718 Å, in concert with the DACs observed in the UV resonance lines (Fig. 1). Strong subordinate lines are present in the optical spectrum as well (e.g. He I 5876 Å, Hα). In his survey of a dozen OB supergiants Ebbets (1982) found dramatic changes in shape and strength of Hα, but the time sampling of his observations was too irregular to permit the timescales (1-10 days) to be estimated reliably. Kaper et al. (1997) obtained coordinated UV and Hα observations for a number of bright O-type stars and demonstrated that the Hα variability is directly linked to the DAC behaviour in the UV resonance lines. The Hα line yields the same wind period (Tab. 1) and can thus be used to measure the cyclical appearance of DACs in UV resonance lines.

An extensive campaign of coordinated UV and optical observations of the O7.5III star ξ Per was carried out in October 1994 (Henrichs et al. 1998). The wind period of 2 days is very apparent in both the resonance and the subordinate lines; the strong DACs in the Si IV doublet are accompanied by phases of enhanced blue-shifted absorption at low velocities in the N IV and Hα lines. The Hα observations were obtained at 7 different observatories spread over the northern hemisphere. The core of the Hα line shows much more complicated variations, probably because of varying amounts of wind emission (also from the regions outside the line of sight) and variations deep down in the photosphere (De Jong 1998).

4 Hα campaign

In order to test the hypothesis that O-star winds are modulated by the rotation of the underlying star, we monitored a large sample of O stars (38) for

Table 1. Measured "wind" period in O star spectra compared to the estimated maximum rotation period (based on the stellar parameters listed by Howarth & Prinja (1989) and references therein). Runaway stars are indicated by (R). References: 1) Kaper et al. 1998a; 2) Prinja et al. 1992; 3) Howarth et al. 1993; 4) Fullerton et al. 1992: from He I 5876 Å; 5) Prinja 1988; 6) Stahl et al. 1996; 7) Kaper et al. 1997; 8) Kaper et al. 1998b.

Name	Sp. Type (Walborn)	$v \sin i$ ($\mathrm{km\,s^{-1}}$)	R ($R_\odot$)	P_{max} (days)	P_{wind} (days) UV	Hα	Reference
ζ Oph (R)	O9.5 V	351	8	1.2	0.9		3
68 Cyg (R)	O7.5 III:n((f))	274	14	2.6	1.4	1.3	1,7
ξ Per (R)	O7.5 III(n)((f))	200	11	2.8	2.0	2.0	1,7
λ Cep (R)	O6 I(n)fp	214	19	4.5	1.3	1.2	1,7
ζ Pup (R)	O4 I(n)f	208	19	4.6	0.8	0.9	2,8
HD 34656	O7 II(f)	106	10	4.8	1.1		1
15 Mon	O7 V((f))	63	10	8.0	>4.5		1
63 Oph	O7.5 II((f))	80	16	10.2	>3		5
HD 135591	O7.5 III((f))	65	14	10.9		3.1	8
λ Ori A	O8 III((f))	53	12	11.5	~ 4	2.0:	1
θ^1 Ori C	O6-O4 var	50:	12:	12:	"15.4"	15.4	6
19 Cep	O9.5 Ib	75	18	12.1	4.5	~ 5	1,7
μ Nor	O9.7 Iab	85	21	12.5		6.0	8
α Cam (R)	O9.5 Ia	85	22	13.2		5.6	8
ζ Ori A	O9.7 Ib	110	31	14.3	~ 6	6	1,8
10 Lac	O9 V	32	9	15.3	~ 7		1
HD 112244 (R)	O8.5 Ib(f)	70	26	18.8		6.2	8
HD 57682 (R)	O9 IV	17	10	29.8		>6	8
HD 151804	O8 Iaf	50	35	35.4		2-3,7.3	4,8

Hα-profile variations (Kaper et al. 1998b). Most (75%) of the monitored O stars, excluding the known short-period spectroscopic binaries, exhibit line profile variations in Hα. Fig. 2 presents the amplitude of Hα variability (expressed as the intergral of the time variance spectrum (TVS), Fullerton et al. 1996) as a function of Hα equivalent width. The open symbols correspond to spectroscopic binaries, the closed symbols are single stars. The main-sequence stars (circles) do not show any indication of (wind) variability, though many of them are known to have variable DACs in their UV resonance lines. This is probably due to the low density of their winds causing any variations to remain below our detection threshold. The main-sequence binaries have a strong TVS spectrum, because the intrinsically constant line profile is moving back and forth due to orbital motion. With increasing luminosity (triangles: giants; squares: supergiants) the Hα equivalent width gets more negative (incipient emission due to stellar wind) and the variability amplitude gets larger. This indicates that the variability in the Hα line is mainly due to variations

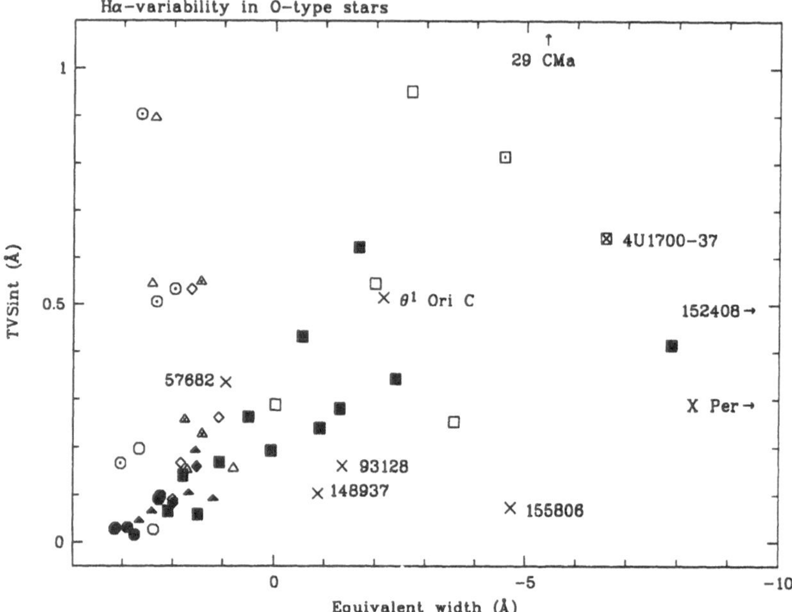

Fig. 2. The amplitude of Hα variability shown as a function of Hα equivalent width. The single main-sequence stars (closed circles) do not exhibit wind variability. The spectroscopic binaries (open symbols) have relatively large variability amplitudes, because of the radial-velocity variations resulting from orbital motion. The Hα profiles of stars with stronger stellar winds (triangles: giants; squares: supergiants) have a larger contribution of wind emission (negative EW) and a larger variability amplitude. The main-sequence stars θ^1 Ori C and HD57682 have strongly variable Hα emission. The largest variability amplitude is observed for the SB2 29 CMa which includes a strong shock resulting from the stellar-wind collision.

in the stellar wind which are more easily detected in strong stellar winds. In close, massive binary systems the collision of the two stellar winds can result in an additional variability component (Thaller 1997).

For 41% of the single Hα-variables a wind period could be determined (Tab. 1), which is in all cases consistent with (an integral fraction of) the expected stellar rotation period. For those O stars we monitored in Hα and for which a DAC recurrence timescale is known, the UV and Hα wind periods are equal (within the measurement accuracy).

5 On the origin of cyclical wind variability

Several models have been put forward to explain the observed properties of DACs. The cyclical nature of wind variability indicates that stellar rotation must play an important role. Most promising is the CIR model proposed by Mullan (1984, 1986), recently worked out by Cranmer & Owocki (1996). In this model, structure in the stellar wind is caused by the interaction of fast and slow streams that originate at neighbouring locations on the stellar surface. Due to the rotation of the star the streams are curved, causing fast wind material to collide with slow material in front. The interaction region has a spiral shape and corotates with the star, though not the wind material itself which mainly flows in the radial direction. Prinja & Howarth (1988) suggested such a spiral-shaped, large-scale structure to be present in the wind of 68 Cyg. The coordinated UV and Hα observations in Kaper et al. (1997) strongly support the CIR model. Comparison between model and observations suggests that wind streams with relatively slow speed interact with a faster ambient wind, rather than high-speed streams in a slow wind (Cranmer & Owocki 1996). Therefore, v_{asymp} might be somewhat lower than the v_∞ of the ambient wind. The systematic difference in v_{asymp} measured in some stars could be due to a difference in aspect angle of the CIR, or to a difference in contrast (velocity difference fast and slow wind), or both.

In order to work, the CIR model needs a certain structure imposed at the stellar surface to produce flows with different kinematic properties. Two candidate physical mechanisms are: (i) non-radial pulsations (NRP), or (ii) surface magnetic fields. NRP are detected in several O stars (Fullerton et al. 1996, Henrichs, this volume). De Jong et al. (1998) have recently shown that also ξ Per (Fig. 1) is a non-radial pulsator. The period and corresponding "pattern speed" of the NRP are, however, not consistent with the observed wind cycle of 2 or 4 days. Our working hypothesis is that the outflow properties at the base of the wind are influenced by a magnetic field anchored in the star. The presence of surface magnetic fields is difficult to prove with direct observational methods (Mathys, this volume). Spectropolarimetric observations to detect a surface magnetic field in ξ Per were not conclusive (Henrichs et al. 1998) and resulted in an upper limit of 70 G. The search for the photospheric connection is the subject of current investigations.

References

Cranmer, S.R., Owocki, S.P. 1996, ApJ 462, 469

De Jong, J.A. 1998, in Proc. ESO workshop on Cyclical Variability in Stellar Winds, Eds. Kaper & Fullerton, p. 336

De Jong, J.A., Henrichs, H.F., Schrijvers, C., et al. 1998, submitted to A&A

Ebbets, D. 1982, ApJS 48, 399

Fullerton, A.W., Gies, D.R., Bolton, C.T. 1992, ApJ 390, 650

Fullerton, A.W., Gies, D.R., Bolton, C.T. 1996, ApJS 103, 475

Henrichs, H.F., Kaper, L., Nichols, J. 1994, A&A 285, 565

Henrichs, H.F., De Jong, J.A., Nichols, J., et al. 1998, in Proc. "Ultraviolet astrophysics beyond the IUE final archive", Eds. Wamsteker, González Riestra, Harris (ESA SP-413), p. 157

Howarth, I.D., Prinja R. 1989, ApJS 69, 527
Howarth, I.D., Bolton, C.T., Crowe, R.A., et al., 1993, ApJ 417, 338
Kaper, L., Henrichs, H.F., Nichols, J., et al. 1996, A&A Supp. Ser. 116, 1
Kaper, L., Henrichs, H.F., Fullerton, A.W., et al., 1997, A&A 327, 281
Kaper, L., Henrichs, H.F., Nichols, J., Telting, J.H. 1998a, submitted to A&A
Kaper, L., Fullerton, A.W., Baade, D., et al. 1998b, in preparation
Massa, D., Fullerton, A.W., Nichols, J.S., et al. 1995, ApJ, 452, L53
Mullan, D.J. 1984, ApJ 283, 303
Mullan, D.J. 1986, A&A 165, 157
Prinja, R.K. 1988, MNRAS 231, 21P
Prinja, R.K., Howarth, I.D. 1988, MNRAS 233, 123
Prinja, R.K, Balona, L.A., Bolton, C.T. et al. 1992, ApJ 390, 266
Stahl, O., Kaufer, A., Rivinius, Th., et al. 1996, A&A 312, 539
Thaller, M. 1997, ApJ 487, 380

Discussion

Eversberg: Could the different terminal velocities for DACs be due to temperature differences on the stellar surface?
Kaper: It might well be that the velocity contrast between fast and slow streams in the wind is different at different areas above the stellar surface, resulting in different values for v_{asymp}. In the models of Cranmer & Owocki (1996), the difference in outflow properties is due to a difference in luminosity (bright spot).

Moffat: Do you have any constraints on the lifetime of the spot or whatever it is on the surface that causes DACs?
Kaper: Our long-term studies show that the characteristic DAC pattern is clearly identifiable in all our datasets obtained over a period of several years. Detailed differences are found from year to year, suggesting a long-term variation on the order of months.

Verdugo: You've said that the mechanism producing the DACs is close to the stellar surface in the region where Hα is formed. Then, how do you explain that stars which don't show any variability in the Hα line present DACs in their spectra?
Kaper: I think this is due to the low density of the winds of main-sequence stars. The Hα line strength is proportional to ρ^2, while the strength of the resonance lines showing the DACs is proportional to ρ.

X-Ray Evidence for Wind Instabilities

Joseph J. MacFarlane, Joseph P. Cassinelli, and D.H. Cohen

University of Wisconsin - Madison

Hot stars are known to emit X-rays with $L_X/L_{bol} \sim 10^{-7}$ for O stars, falling to $\sim 10^{-9}$ for B3 stars. These stars also lose mass at large rates through their high-speed winds. Over the years, several types of production mechanisms have been proposed to explain the X-ray emission from O stars, with source locations ranging from very near the stellar surface to very far from the star. A coronal X-ray source was originally proposed (Cassinelli and Olson 1979) to explain the presence of anomalously high ionization stages observed as P Cygni line profiles in the UV spectra of O stars. At the other extreme, Chlebowski (1989) suggested that the X-rays of O stars originate far from the star, and are produced by the interaction of the stellar wind with circumstellar matter. A model in which shocks forming due to instabilities in the line-driven winds of O stars was proposed by Lucy (1982), and studied in detail by Owocki et al. (1988), Cooper (1994), and Feldmeier (1996). In this case, the X-ray emission originates in a large number of shock-heated regions distributed throughout the wind. The shocked-wind model has also been shown to be consistent with the X-ray emission from early-B stars, such as τ Sco (MacFarlane and Cassinelli 1989). However, it appears difficult for shocked wind models to explain the X-ray emission from B3 and later stars because of their presumed low mass loss rates (Cohen et al. 1997).

Here, we discuss evidence that suggests that the X-rays we observe from hot stars originate from shock-heated plasma created by radiatively-driven wind instabilities. The observational evidence for this comes from observations at a variety of wavelengths, including: moderate-resolution X-ray spectra, EUV line emission, and UV P-Cygni profiles. The lack of significant bound-free absorption due to an overlying cool wind (Cassinelli et al. 1981; Corcoran et al. 1993) suggests that a significant fraction of the X-rays must be emitted from regions significantly above the base of O star winds. A detailed analysis of the dependence of the wind ionization distribution and resulting O VI P Cygni profile on the X-ray source distribution for ζ Pup was carried out by MacFarlane et al. (1994). They found that the UV O VI profile was not consistent with: (1) models in which the X-ray source was located at a radius much above $\sim$ several stellar radii; and (2) coronal models, unless the mass loss rate for ζ Pup is a factor of $\sim$ 3 to 5 lower than the value deduced from radio observations. More recently, EUVE observations of the B2 II star ϵ CMa (Cassinelli et al. 1995), along with a combined analysis of its ROSAT and EUVE data (Cohen et al. 1996), indicate the presence of a moderate amount of wind attenuation for this relatively low mass-loss rate star, as well as a multitemperature plasma X-ray source. Thus, at a variety of wavelengths, observations and analysis provide significant evidence for a shocked-wind X-ray source for O and early-B stars.

Discussion

S. Owocki: In your model of ζ Pup, although you may need to have the X-rays begin close to the star, the X-rays one sees come from quite far away, e.g., $r \geq 10 R_{\star}$. This suggests that the radial fall-off of your assumed X-ray source should also be an important parameter. Thus, if you vary this, you might be able to match the observed X-rays with a source that has less effect on the inner ionisation.

A. Moffat: In your model, you assume homogeneity of the wind. However, we know that hot-star winds are highly clumped (e.g., Eversberg et al. 1997 on ζ Pup). This might have a strong effect on your models. In fact, the clumps may themselves be the source of the X-ray flux via shocks.
J. MacFarlane: At present, this is a shortfall in our modelling. It would indeed be good to do calculations to study the effect of clumping.

Thomas Berghöfer and Ronny Blomme

X-Ray Variability of the O Star ζ Puppis

Thomas W. Berghöfer[1,2]

[1] Space Sciences Laboratory, University of California, Berkeley, CA 94720, USA
[2] Hamburger Sternwarte, Gojenbergsweg 112, D-21029 Hamburg, Germany

1 X-ray emission of O stars

X-ray surveys carried out with the *Einstein Observatory* (Chlebowski et al. 1989) and ROSAT (Berghöfer et al. 1996) have shown that all O stars are soft X-ray emitters. Since O star winds are opaque at soft X-ray energies the stars or their photospheres cannot be the origin of the observed X-ray emission, thus, this emission must be produced in their stellar winds. Obviously, the X-ray emission is connected to dynamical processes present in the winds of O stars; steady-state computations for O star winds which are able to explain many of the observational features cannot predict any X-ray emission.

Lucy & White (1980) suggested the presence of hot gas in the stellar winds which is produced in shocks developing from the growth of instabilities in the winds; supersonic wind flows in O stars are known to be intrinsically unstable. Numerical simulations confirmed this scenario. However, so far these simulations are limited to one or two dimensions and are not able to explain for instance the observed X-ray luminosity of O stars.

Further support for a model of shock-heated gas in the winds of O stars came from the analysis of the ROSAT PSPC spectrum of ζ Pup (O4Iaf). This spectrum demonstrates that the X-ray emission is "self-absorbed" by the stellar wind. Based on a detailed modelling of the X-ray opacity and an assumed "uniform" distribution of X-ray sources in the stellar wind, Hillier et al. (1993) were able to explain the observed spectrum of ζ Pup.

On the other hand, however, shocks should statistically occur and the total number of visible shocks and their X-ray output should be time variable. X-ray variability studies provide the only method to study the dynamical processes involved in the X-ray production in O star winds and to derive important input parameters needed for model simulations (e.g., occurrence rates and cooling times of shocks). As a result, long term variability studies by Berghöfer & Schmitt (1994a, 1994b, 1995) demonstrated that X-ray variability on all observable time scales is generally not common for O stars. So far, only for the O supergiant ζ Orionis a moderate increase in X-ray count rate has been found (Berghöfer & Schmitt 1994c). Here we present results obtained for the prototype O star ζ Pup.

2 X-ray and wind variability of ζ Pup

ζ Pup is one of the most studied O stars in the sky. Variability has been reported for this star on different time scales. A 5 d period has been attributed to rotational modulation. Several authors found evidence for periodic line profile variations on time scales between 2 and 8 h at different epochs. These are attributed to non-radial pulsations (NRP) of the photosphere. A reoccurrence time of 19.2 h for the so-called discrete absorption components (DAC) in UV wind lines has been detected during the IUE MEGA campaign. Simultaneous ROSAT and Hα observations provided evidence for correlated variability in the X-ray and Hα emission of ζ Pup (Berghöfer et al. 1996). The period of 16.7 h detected in both data sets has been attributed to a periodic modulation of the wind density at the base of the wind which extends out to the X-ray emitting regions in the wind.

To further investigate these small but significant periodic variations in the X-ray flux of ζ Pup we obtained a long observation with the BeppoSAX X-ray satellite. In order to demonstrate the long term stability of the X-ray flux of ζ Pup we show in Fig. 1 the long-term X-ray light curve of ζ Pup in the energy band 0.9–2.0 keV. Together with the BeppoSAX LECS observation we show data points obtained with the *Einstein Observatory*, ROSAT PSPC, and ASCA SIS; all count rates were converted to ROSAT PSPC counts/s taking into account the effective areas of the different detectors. As can be seen the X-ray flux of ζ Pup is constant over a time scale of about 16 years, the deviations from the best fit constant model (dashed line in Fig. 1) are small.

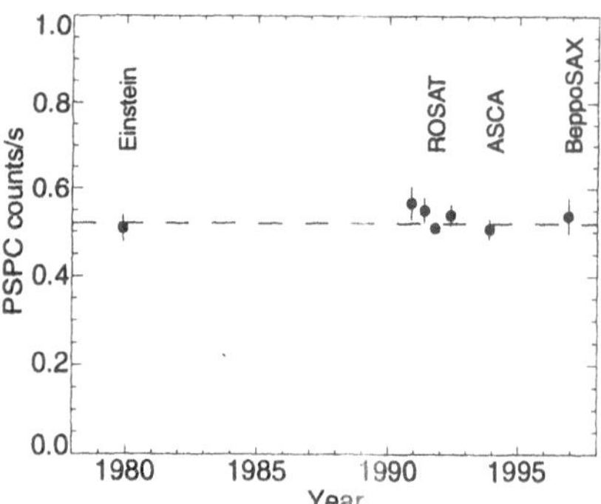

Fig. 1. Long term X-ray light curve of ζ Pup in the energy band 0.9–2.0 keV

A detailed analysis of the ζ Pup data obtained with BeppoSAX will be published elsewhere. Here we provide first results of a timing analysis. Employing the same methods described in Berghöfer et al. (1996) we carried out a search for periodic variations in the BeppoSAX observations of ζ Pup.

Above 0.9 keV both data sets obtained with the LECS and MECS detectors show a period of 15 h. As an example we show in Fig. 2 the BeppoSAX LECS (0.9–3.0 keV) light curve when folded with a 15 h period. The solid curve shows the best fit for a sinusoidal model.

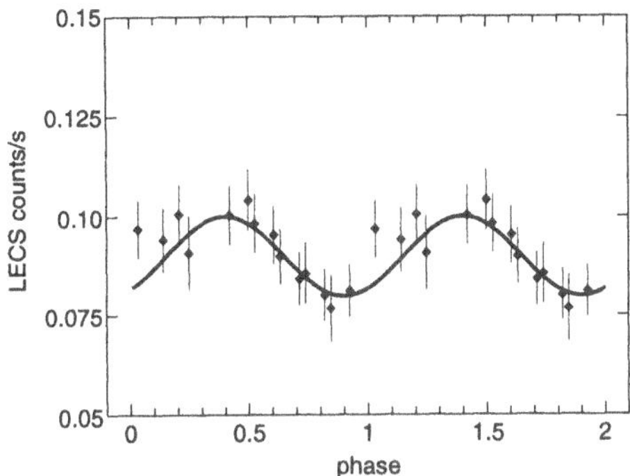

Fig. 2. BeppoSAX light curve of ζ Pup in the energy band 0.9–3.0 keV when folded with a 15 h period. The solid curve shows the best fit sinusoidal model.

The BeppoSAX observations of ζ Pup provide evidence for a 15 h period in the X-ray flux of this star. This period is significantly different to the 16.7 h period detected in the ROSAT observations 5 years before. However, both periods are of the same order as the 19.2 h reoccurrence time of DACs observed in 1995. This strongly supports the idea of a quasi-periodic variability of the wind of ζ Pup triggered by NRPs. The observed modulation of the wind density also provides a perfect trigger mechanisms for the unstable wind to form shocks and hence produce the X-ray emitting hot gas.

T.W.B. acknowledges the support from the Alexander von Humboldt Stiftung (AvH) by a Feodor Lynen Fellowship.

References

Berghöfer, T. W., Schmitt, J. H. M. M. 1994a, A&A 290, 435
Berghöfer, T. W., Schmitt, J. H. M. M. 1994b, ROSAT X-ray light curves of early-type stars, In: Astrophys. Space Sci. 221, 309
Berghöfer, T. W., Schmitt, J. H. M. M. 1994c, *Science* 265, 1689
Berghöfer, T. W., Schmitt, J. H. M. M. 1995, Long term X-ray variability studies of OB-type stars, In: Adv. Space Res. Vol. 16, No. 3, p. 163
Berghöfer, T. W., Schmitt, J. H. M. M., Cassinelli, J. P. 1996, A&AS 118, 481
Berghöfer, T. W., Baade, D., Schmitt, J. H. M. M., et al. 1996, A&A, 306, 899
Chlebowski, T., Harnden, F. R., Jr., Sciortino, S. 1989, ApJ, 341, 427
Hillier, D. J., Kudritzki, R.-P., Pauldrach, A. W. A., et al. 1993, A&A, 276, 117
Lucy, L.B., White, R.L. 1980, ApJ, 241, 300

On the Variable Winds of BA Supergiants

Eugene Chentsov

Special Astrophysical Observatory, 357147 Nighnij Arkhyz, RUSSIA

Abstract. The manifestation of instability and nonhomogeneity of the atmospheres and winds of β Ori, HD 168607, 6 Cas and other highly luminous B7–A3 stars are considered on the basis of visible and near–infrared spectra obtained with CCD–echelle spectrometers of the 1–m and 6–m telescopes SAO RAS. Changes in the profile shapes, radial velocities and differential shifts of lines create an impression that unstable disk– or ring–shaped structures may exist at the wind bases of supergiants such as β Ori. In the process of destruction they occasionally produce more or less radial flows of escaping and infalling gas. It looks that, as the luminosity increased, the geometry and kinematics of the wind simplify. The wind becomes more symmetrical, the sings of compression diappear and even in the lower layers steady (at least over an interval of several years), although unstable expansion is observable.

I shall restrict myself to presenting my observational data on spectroscopic manifestations of variability and nonhomogeneity of winds. The objects include supergiants and hypergiants from B7 to A3. The spectra in the visible and near–infrared regions were obtained with the CCD–echelle spectrometers on the 1–m and 6–m telescopes of SAO RAS.

I am going to start with β Ori B8 Ia. Our set of H_α profiles is noticeably smaller than the unique Heidelberg (Kaufer et al. 1996b) or Toledo (Morrison et al. 1998) sets, but shows the same characteristic of supergiants B7-B9: double-peaked Be-emissions predominate instead of the expected P Cyg profiles. High–velocity absorptions are seen more rarely, and not only blue–shifted but also red–shifted. It seems that discrete absorption components of resonant lines of extra–atmospheric UV can be blue-shifted only.

Is it possible that at the wind base there is some disk– or ring– shaped formation? Being unstable, such a formation could, therefore, give rise to more or less radial structures such as jets, fragments of loops or spirales.

Is it possible that this shows up also on the time scales? While wind UV DACs can be seen for several months, cyclic changes of H_α and H_β profiles, that are formed in relatively compact areas, often last only 1–2 weeks. The H_α profile keeps looking like Be for all this time, although it can also show some weak high–velocity details. At times the radial velocities of photospheric lines vacillate almost in synchrony with the velocities of main H–absorptions.

Occasionally, however, there are events that take more time and, perhaps, space. One of these events was observed in the fall of 1993. H_α and H_β absorptions were clearly bifurcated. At velocities of around -100 and $+70$ km/s, components of H_α can be seen for up to 40 days. H_β,s interval is shorter.

During this time, judging by the absorptions of HeI and FeII, photospheric layers complete several pulse cycles. Fig. 1 displays the evolution of the H_α profiles, obtained through dividing them by the "photospheric" one. With no allowance made for the radial velocity of the star as a whole, the October profile can be described as a direct P Cyg profile. By December it slowly evolves into the shape resembling the inverse P Cyg profile. The spherical–symmetric expansion and compression are absent. It seems that the column of escaping matter on the line of sight is replaced by the column of matter falling towards the star, as a result of its axial rotation (Israelian et al. 1997).

It is more difficult to account for the behaviour of photospheric absorp-

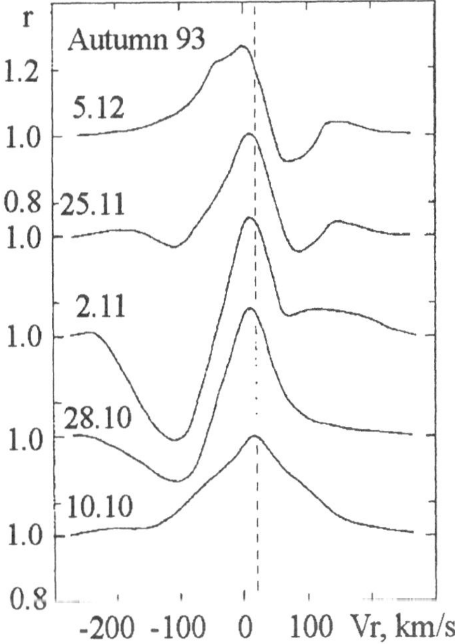

Fig. 1. H_α profiles of β Ori in the fall of 1993 divided by the "photospheric" one. The radial velocity of stellar center of mass (dashed line) was determined by the use of its visual companion.

tions. Their formation area is involved in pulse–like movements and for the strongest lines it even approaches the base of the wind. Fig. 2a demonstrates the relation between radial velocity for the line core and central residual intensity. The differential shifts of the lines are quite real. It would seem that they reflect the radial gradient of the velocity in the atmosphere. The outer layer, where the lines of FeII are formed, is moving outwards with respect to the deeper layer, where the HeI lines are formed. The intermediate layers are represented by the SII, NeI, SiII, MgII lines and show intermediate velocities. In that case why is it so that the strongest FeII lines that are formed higher than the weaker ones show a positive but not negative shift relative to the weaker lines? This unexplained effect persists at any phases of pulsations and wind distortion of the profiles. The profiles show significant changes, while the equivalent width changes only slightly. A system of macroscopic ascending and descending streams would be able to produce this kind of variability.

But perhaps the described effect is a peculiarity inherent in β Ori. Let us shift in temperature and luminosity. HD 183143 B7 Ia is hotter and more luminous. Its H_α emission is quite intense without even dividing it by the photospheric profile and the asymmetry of this emission, and, very likely the shell as well (i.e. wind) is less than in β Ori. The wind of HD 183143 is more stable than that of β Ori. However, the radial velocities are higher along the strong absorptions than along the weaker ones, as is also the case with β Ori (Fig. 2b).

On the other hand, is it possible that the unaccounted errors in the effective wavelengths can account for this? Let us now look at cooler objects with a more developed FeII spectrum. In the case of HD 21389 A0 Ia (Fig. 2c) as compared to β Ori, the only difference is the relative extent of the chains of the HeI and FeII absorptions. The right end of the FeII chain persists in bending upward. For older supergiants, such as ν Cep A2.5 Ia, we can find moments, when shifts disappear but their HeI lines are already rather weak (Fig. 2d). η Leo A0 Ib can completely liberate us from methodical anxiety. It is a supergiant of lower luminosity, in which case all measurable absorptions produce the same velocity (Fig. 2e).

6 Cas is in the same spectral class as ν Cep, but its luminosity is higher. This hypergiant can give us finally confident evidence for the positive gradient of the atmospheric velocity (Fig. 2f). We can even call it, by analogy with the Balmer progression, "iron progression".

It will be recalled that we are talking about the central parts of the profiles only. But the wings, especially the blue wings, of the FeII lines in the spectrum of 6 Cas are more deformed when the line is deeper, regardless of what is going on in the atmosphere. The local depressions of the blue wings of FeII 42) are seen for the corresponding velocities in H_α and H_β. They are displaced along the profiles, increasing the velocity of expansion in synchronism (Fig. 3). In 200 days this velocity grew from 50 to 180 km/s (Chentsov, 1995). This phenomenon is known in O–supergiants, it was revealed by Heidelberg monitoring of hypergiants of early subclasses B (Rivinius et al. 1997) and in HD 92207 A0 Iae (Kaufer et al. 1996a). If I am not mistaken, 6 Cas is so far the coolest hypergiant that exhibits it.

The hypergiant HD 168607 B9.4 Ia0 returns us to β Ori temperature but

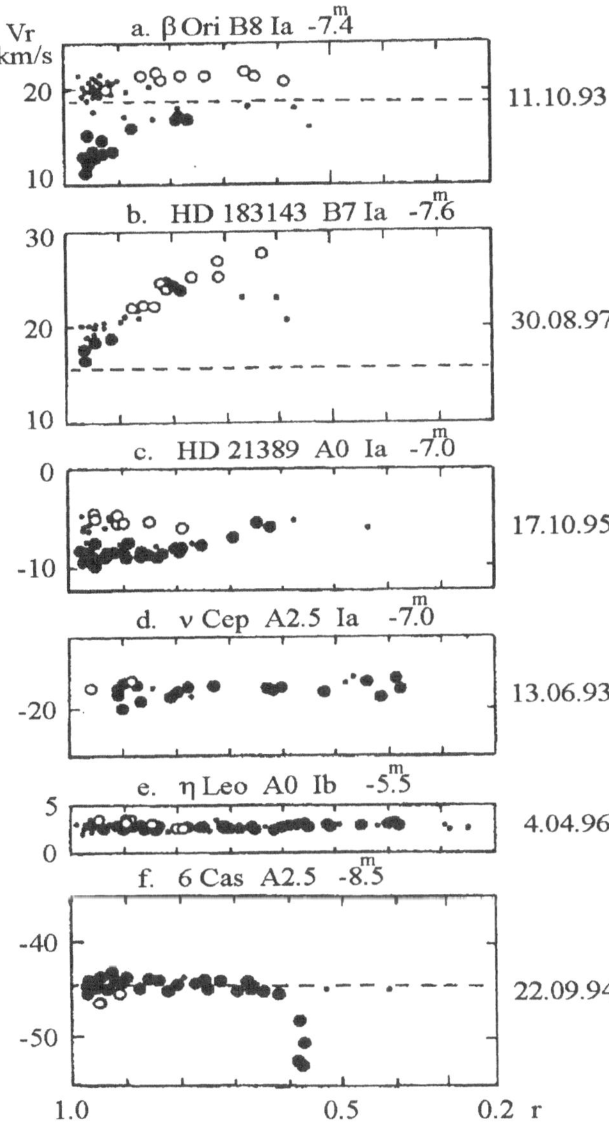

Fig. 2. The relations between radial velocity for the line core and central residual intensity. The weakest lines are on the left, the strongest are on the right. Open circles: HeI; filled circles: FeII; dots: SII, NeI, SiII, MgII.

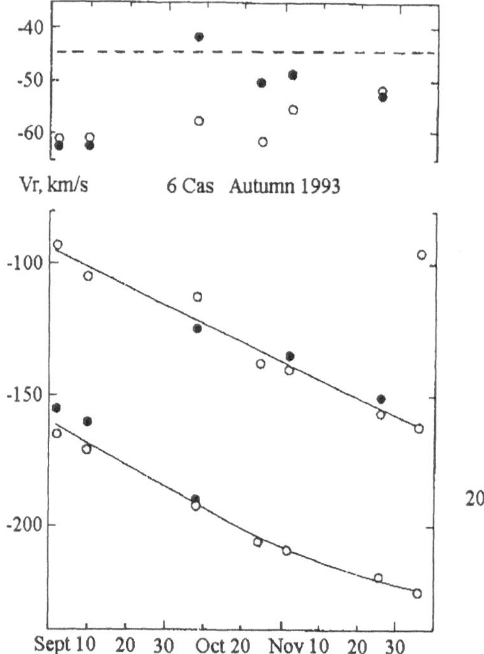

Fig. 3. Changes with time of radial velocities for 6 Cas. Open circles: H_α and H_β, filled circles: FeII.

at higher luminosity. Its photosphere is represented not only by HeI, CII, SII lines, but also by FeII absorptions with excitation potentials over 10 eV. Other lines of FeII are symmetrical and stationary emissions and represent the extended tenuous shell. FeII lines with low excitation potentials have the wind profiles characterized by split absorptions that are clearly seen also in Balmer lines (Fig. 4). In the atmosphere the absorptions reveal gradual increasing of the expansion velocity with height, whereas above it, in the wind, components of any depths show the same fixed velocities (Fig. 5). Do they propagate as in the case of 6 Cas? We are able to observe HD 168607 no more than once or twice a season. Our modest statistics show, however, that radial velocities more or less uniformly fill the interval between -10 and -130 km/s.

Thus, the concept of stable and spherically symmetric wind has the grounds to be rejected for the supergiants. It may still be able to serve us in the case of the hypergiants, even if with some reservation. As the luminosity increases, the kinematics and geometry get more simple. The wind becomes

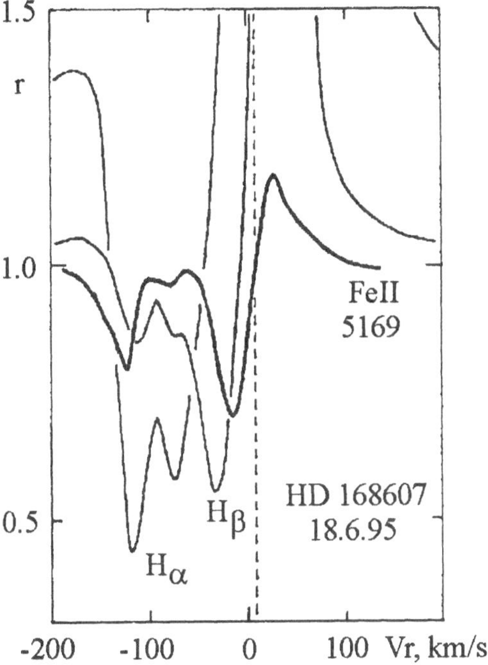

Fig. 4. The coincidence of the positions of three components of H_α, H_β and FeII 5169 for HD 168607.

more symmetrical. The signs of compression disappear, even at the wind base. The continuous (at least for a few years) although unstable expansion is seen. This can serve as support for the methods of "extragalactic stellar astronomy" being developed by Kudritzki (1997).

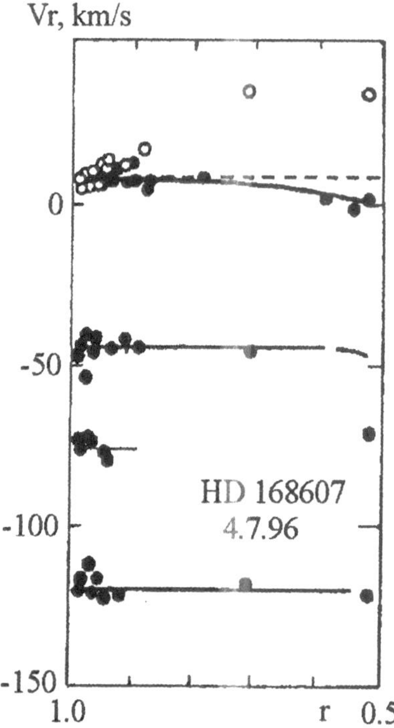

Fig. 5. An exemple of dependencies of radial velocities from residual intensities for HD 168607. Absorption components – filled circles, emissions – open circles.

References

Chentsov E.L., 1995, Ap. Sp. Sci., **232**, 217

Israelian G., Chentsov E., Musaev F., 1997, MNRAS, **290**, 521

Kaufer A., Stahl O., Wolf B., Gang Th., Gummersbach C.A., Kovacs J., Mandel H., Szeifert Th., 1996, A&A, **305**, 887

Kaufer A., Stahl O., Wolf B., Gang Th., Gummersbach C.A., Jankovics I., Kovacs J., Mandel H., Peitz J., Rivinius Th., Szeifert Th., 1996, A&A, **314**, 599

Kudritzki R.P., 1997, Max Planck Inst. Ap., MPA 1023

Morrison N.D., Mullis C.L., Gordon K.D., 1998, *poster at this colloquium*

Rivinius Th., Stahi O., Wolf B., Kaufer A., Gang Th., Gummersbach C.A., Jankovics I., Kovacs J., Mandel H., Peitz J., Szeifert Th., Lamers H.J.G.L.M., 1997, A&A, **318**, 819

Discussion

A. Fullerton: Have you been able to estimate the acceleration of the DACs in the Hα profile of 6 Cas, e.g., in terms of a β-law?
E. Chentsov: The mean acceleration can be estimated easily: $\Delta v_r \sim 100-150$ km/s per 200 d, i.e., $0.5-0.7$ km/s per day. I did not estimate β, but it seems that the nature of these events is the same as in the case of O supergiants and B hypergiants.

A. Moffat: Do you find any periodicities and if so, which ones?
E. Chentsov: I have found no strict periodicities, but cyclical changes in v_r were detected. They are pulsation-like: $1-2$ weeks for β Ori and $1-1.5$ month for 6 Cas. Characteristic times for the HVAs of β Ori are nearly 40 d; for the DACs of 6 Cas, $100-150$ d.
A. Kaufer: Maybe it is worth mentioning that in our extended time series of β Ori in the spring of 1994 we found an HVA event about 110 days after your observations in the autumn of 1993, which revealed a very strong HVA. This time difference is quite consistent with one rotational cycle of this envelope structure.

Nevena Markova and Eugene Chentsov

UV Wind Variability in B Supergiants and its Implications for Wind Structures

Derck Massa[1] and Raman K. Prinja[2]

[1] Raytheon STX, NASA/GSFC, Code 631.0, Greenbelt, MD 20771, USA
[2] Astronomy Department, University College London, Gower Street, London WC1E 6BT, England, U.K.

Abstract. We discuss why B supergiant winds are particularly well suited for wind studies, and present or refer to dynamic spectra which suggest the presence of disks, bifurcated winds, shock formation, rotationally modulated winds and the spontaneous generation of wind enhancements. They underscore the strength and richness of wind variability in B supergiants and the challenges these phenomena present to theoretical studies of stellar winds.

1 Why BIs are well suited for wind studies

B supergiant (BI) winds are ideal for studying wind variability because: 1) their wind lines are often well-developed but *unsaturated*; 2) their wind lines cover a range of ionization, close to the dominant stage of the wind (often C II – N V, and Si III & IV); 3) their winds are less chaotic than O stars, making coherent features easier to identify and track; 4) their winds have lower terminal velocities, so individual structures evolve more slowly and the Si IV $\lambda\lambda1400$ doublets are decoupled; 5) their strong UV photospheric lines are excellent T_{eff} – luminosity diagnostics, and 6) the strongest UV photospheric lines sample the wind-photosphere interface, often displaying wind behavior.

2 The observations

Over its lifetime, *IUE* obtained 15 good time series of 10 BIs, with time coverages of $\sim 1 - 30$ days, and sampling rates from 30 min. to once a day. Four of these have been described in detail by Massa et al. (1995 – evidence for $v = 0$ structure), Prinja et al. (1995), Fullerton et al. (1997 – evidence for Co-Rotating Interaction Regions), Howarth et al. (1998), Prinja et al. 1997 – a 2-component wind and disk structure), Massa et al. (1998, 1999 – evidence for a photospheric – wind connection). Prinja et al. (1999) will provide a complete overview of all these series. Here we demonstrate a few common aspects of them and concentrate on ionization structure.

UV wind line variability in BIs can imply line of sight optical depth variations up to 10 over localized wavelengths and a factor of 2 or more integrated

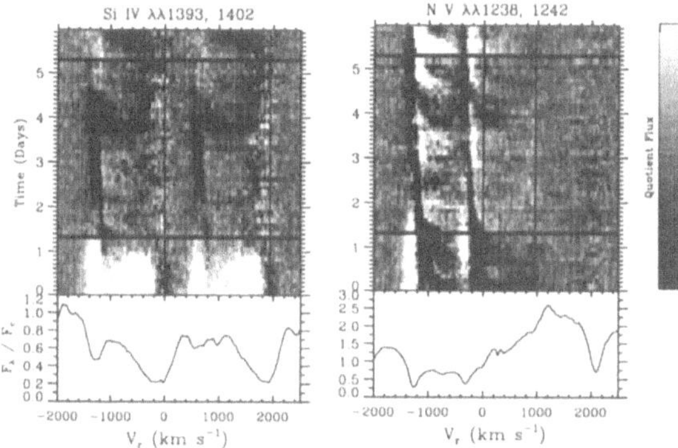

Fig. 1. Dynamic spectra of a 1993 time series of HD 150168. This star is an SB1 binary, and the individual spectra have been aligned on the photospheric velocity of the primary. The spectra are normalized be the series mean (shown at the bottom of each panel), and the extremes of the scale run from 0.75 – 1.25.

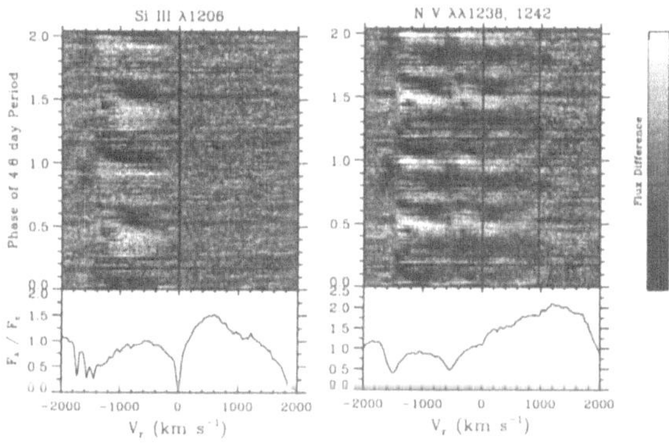

Fig. 2. Dynamic spectrum of 1995 ("MEGA") data for HD 64760. The 16 days of spectra are arranged relative to the 4.8 day period determined by Fullerton et al. (1997). The spectra are differences from the series mean spectrum.

over the entire profile. These variations can occur as **normal DACs**, which appear at intermediate v and narrow as they propagate to higher v; **bowed absorptions**, which appear at intermediate velocity and propagate blueward and redward simultaneously and are thought to be related to CIRs; and **other absorptions**, which often simply appear over a large range of velocity all at once. In addition, the features often extend to $v = 0$ suggesting an origin on stellar surface and, when the extent of the series is adequate, often **recur**, usually on the rotation timescale – also suggesting surface features. Finally, when lines from different ions are compared, we see **ionization stratification**, which means that related structures occur at different velocities in different ions – often just an optical depth effect, and **ionization variability** which implies that the ionization state of a specific feature changes with time.

All of the previous effects can be seen in the accompanying figures. Figure 1 of the B1 Iab-Ib star HD 150168. Near the beginning of the series, there is a maximum in N V which correponds to a minimum in Si IV – *ionization variability*. Near day 4, a bowed structure shows appears with its minimum in N V lagging the Si IV minimum – *ionization stratification*. Figure 2 is for HD 64760, "phased" on a 4.8 day period. Notice that the N V data appear to have nearly twice the frequency of the Si IV data – *ionization variability*.

3 Conclusions

BI wind activity is rich and varied, with similar structures occurring in very different stars and different structures occurring in same star at different times. The variability evolves smoothly, and several lines of evidence point to surface features as the origin of much of the activity, although some features are difficult to reconcile with this interpretation. We have also detected photospheric line variability which is only sometimes associated with variability in the the wind lines.

References

Fullerton, A.W, Massa, D., Prinja, R.K., Owocki, & Cranmer 1997, A&A 327, 699
Howarth, I.D., et al. 1998, MNRAS 296, 949
Massa, D., Fullerton, A.W., & Prinja, R.K. 1995, ApJ 452, 842
Massa, D., Fullerton, A.W., & Prinja, R.K. 1998, in *ESO Workshop on Cyclical Variability in Stellar Winds*, p121
Massa, D., Prinja, R.K., Fullerton, Howarth, & Owocki 1999, in preparation
Prinja, R.K., Massa, D., & Fullerton, A.W. 1995 ApJ 452, L61
Prinja, R.K., Massa, Fullerton, Howarth, & Pontefract 1997, A&A 318, 157
Prinja, R.K., Massa, D., & Fullerton, A.W. 1999, in preparation

Discussion

A. Kaufer: Do you see any relation between the "bananas" and the much more slowly evolving DACs in the CIR picture?

D. Massa: The DACs seem to recur less frequently, if at all; and we just don't have runs long enough to determine whether they are periodic.

H. Henrichs: Have you checked whether simultaneous photometry from HIPPARCOS is available?
D. Massa: No.

S. Shore: For the spontaneous (that is, the non-propagating) features, is there anything like a characteristic velocity at which they appear?
D. Massa: This is somewhat tied up with the definition, in that these features are typically identified at intermediate velocity. However, they can appear over a wide range of velocities reaching to $v = 0$ at some times.

J. Dachs: Why do you suppose γ Arae to show a disk in its wind?
D. Massa: It seems to have two separate winds: one has a terminal velocity of about −750 km/s in Si III and Si IV, and another is seen in the edge of N V with a terminal velocity of about −1200 km/s. Also, the Si IV profiles look like those in HD 93521: flat-bottomed with little emission. These are indicative of a disk. Conversely, the N V profiles have strong emission and weak absorption, which is indicative of a hot wind confined to a cone above the poles.

Derck Massa and Alexander Fullerton

Variability and Evidence of Non-spherical Stellar Winds in A-Type Supergiants

Eva Verdugo[1], Antonio Talavera[2], and Ana I. Gómez de Castro[3]

[1] ISO Data Centre, VILSPA, P.O. Box 50727, 28080 Madrid, Spain
[2] LAEFF/INTA, VILSPA, P.O. Box 50727, 28080 Madrid, Spain
[3] Instituto de Astronomía y Geodesia (CSIC-UCM), Fac. cc. Matemáticas, Universidad Complutense, Av. Complutense s/n, E-28040-Madrid, Spain

Abstract. The profiles of the Hα and uv lines and the observed variability suggest deviations from spherical symmetry for the envelope of A-type supergiants.

1 Introduction

During the last years increasing efforts have been dedicated to study the atmospheres of A-type supergiants. However, compared to the amount of work devoted to OB stars, studies of photospheric and wind properties of A-type supergiants are still scarce.

In this paper we describe several visible and ultraviolet spectral lines of a sample of 41 A supergiants. The spectral ranges observed include the Balmer lines (Hα, Hβ, Hδ and Hγ), the Ca II H and K lines, the Na I D lines and the Mg II line at 4481 Å and the uv spectra obtained with the IUE satellite. These observations represent the best source of data so far for a global view of their spectral variability.

We also present the first results of a quantitatively analysis of the Hα profile by means of a non-LTE model.

2 Line profiles

The most sensitive indicator of winds in A supergiants is the Hα line in the visible range and the Mg II (uv1) lines in the ultraviolet range. Our analysis of these lines shows that these stars can be divided into two groups: the less luminous stars (spectral type Ib) which show symmetric absorption profiles in all the visible lines studied, and only show wind effects in the uv lines of Mg II as a time evolving component. The second group contains the most luminous A supergiants (spectral type Ia and Iab). The Hα line in these stars presents asymmetric, P Cygni and/or double-peaked emission profiles. The uv lines (Mg II, C II, Si II, Al II and Fe II) are also asymmetric with no emission and a sharp blue edge or profiles composed of discret absorption components. These profiles are represented in Fig. 1.

The frontier between these two groups is $M_v \sim -6$.

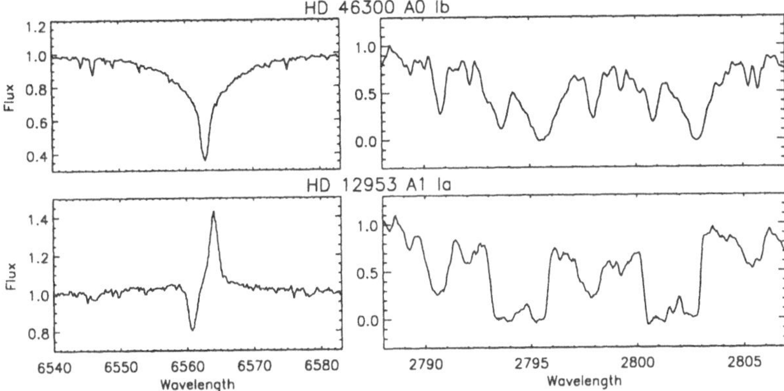

Fig. 1. Hα and Mg II (uv1) profiles for Group I (top) and Group II (bottom).

3 Variability

In the visible range we have only found significant variations in the Hα line. Stars with symmetric absorption Hα profiles do not show any variability. However we have observed variations in most of the stars showing emission profiles. The strongest variations are detected on time-scales of days and most of them are due to a variable double-peaked emission. The observed V/R variability points out deviations from spherical symmetry (Fig. 2).

Contrary to what woud be expected we found that the less luminous stars present the most spectacular variations in the uv spectrum. We have observed the appearance and evolution of a DAC in the Mg II lines (Fig. 2). However such component have not been detected in any other uv line.

The time-scales of variability are consistent with the rotational periods.

Object	$v_{\rm break}$ [km/s]	$P_{\rm break}$ [days]	$v\sin i$ [km/s]	$P_{\rm rot}/\sin i$ [days]
HD 87737	202	10.5	20	106
HD 12953	147	50	49	150
HD 207260	175	26	44	102

4 Synthetic Hα profiles

We have modelling Hα profiles for an expanding, spherically symmetric atmosphere. We have used the equivalent-two-level-atom approach (Mihalas and Kunasz, 1978) to solve the radiative transfer problem and the non-LTE equation of statistical equilibrium in the context of a multilevel hydrogen atom. The transfer equation is solved in the comoving frame of the flow.

We have found several models that match the observed emission strength (Fig. 3). However we have encountered difficulty to math the depth of the absorption feature. Even the most succesful profiles are deeper than the observed profiles suggesting that these models do not include all the physical

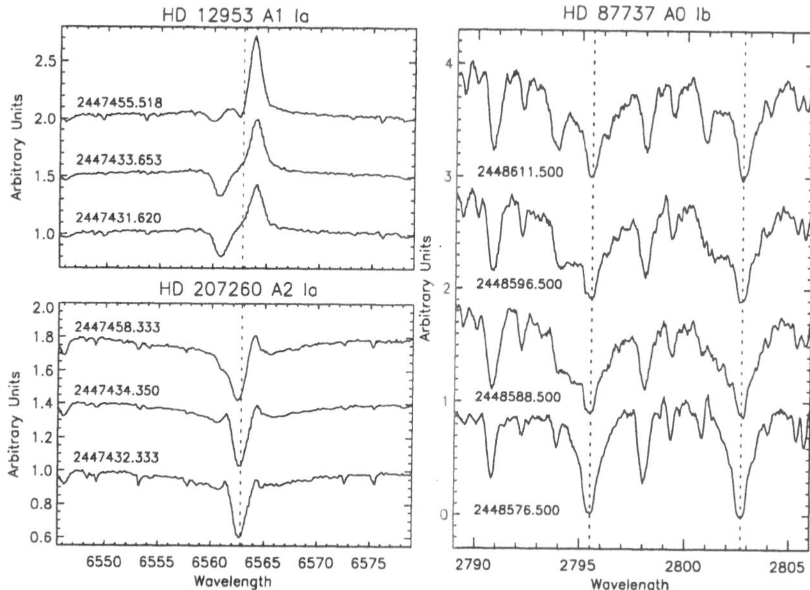

Fig. 2. Variability in the Hα and Mg II (uv1) profiles. The rest position of the lines are marked with a dashed line. Each spectrum is labeled with the Julian Date of the observation.

processes that operate in the envelope of A-type supergiants. Deviations from steady-state flow structure and non-spherical extension are possible.

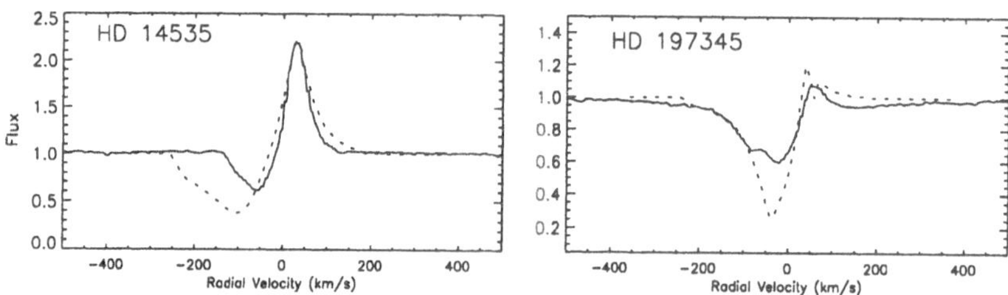

Fig. 3. Synthetic Hα (dashed line) compared with the observations (solid line).

References

Mihalas, D., Kunasz, P.B.: 1978, *Astrophys. J.*, **219**, 635

Discussion

L. Kaper: What fraction of your sample of A supergiants shows double-peaked Hα emission? How do you interpret this?
E. Verdugo: About ten stars of our sample display double-peaked Hα emission. Presently, I cannot explain this. I can only say that these kind of profiles are the main evidence for deviations from spherical symmetry for the envelopes of A supergiants. In fact, we cannot reproduce these profiles with the ETLA code, which assumes an expanding spherically, symmetric atmosphere.

J. Puls: In A super/hypergiants, the Hα profile should have a P Cygni morphology, for theoretical reasons. However, as you have shown, these spectra exhibit pure emission profiles at Hα. Do you have an idea of a reason for the behaviour?
E. Verdugo: No, I have no explanation for this behaviour.
H. Lamers: Maybe clumping.

S. Shore: What is the frequency with which you find the absorption components that look sort of detached? Are they variable? (This could be an indication of clumping in the wind.)
E. Verdugo: I have only detected such a profile once, for two stars of my sample: HD 12953, which has a variable P Cygni profile; and HD 223960, which shows the strongest double-peaked emission profile with clear V/R variability.

Jaques Breysacher and Christiaan Sterken

Variable Winds in Early-B Hypergiants

Bernhard Wolf and Thomas Rivinius[1]

Landessternwarte Königstuhl, D-69117 Heidelberg, Germany

Abstract. Early-B hypergiants belong to the most luminous stars in the Universe. They are characterized by high mass-loss rates ($\dot{M} \approx 10^{-5} M_{\odot} yr^{-1}$) and low terminal wind velocities ($v_{\infty} \approx 400$ km s^{-1}) implying very dense winds. They represent a short-lived evolutionary phase and are of particular interest for evolutionary theories of massive stars with mass loss. Due to their high luminosity they play a key role in connection with the "wind momentum - luminosity relation". Among the main interesting characteristics of early-B hypergiants are the various kinds of photometric and spectroscopic variations. In several recent campaigns our group has performed extensive high dispersion spectroscopy of galactic early-B hypergiants with our fiber-fed echelle spectrograph FLASH/HEROS at the ESO-50 cm telescope. The main outcome was that their dense winds behave hydrodynamically differently to the less luminous supergiants of comparable spectral type. Outwardly accelerated propagating discrete absorption components of the P Cyg-type lines are the typical features rather than rotationally modulated line profile variations. These discrete absorptions could be traced in different spectral lines from photospheric velocities up to 75% of the terminal velocity. The stellar absorption lines show a pulsation-like radial velocity variability pattern lasting up to two weeks as the typical time scale. The radius variations connected with this pulsation-like motions are correlated with the emission height of the P Cyg-type profiles.

1 Main Properties and Variability of Early-B Hypergiants

Early B-hypergiants belong to the most luminous stars in the Universe ($M_{bol} \approx -10 \ldots -11$) with effective temperatures around 20 000 K. They represent short-lived evolutionary phases of massive stars. For the given luminosity the masses are comparatively low. In this connection R 81 (B2.5Ia-O) is of particular interest; it is an eclipsing binary (Wolf et al. 1981, Stahl et al. 1987). A rough estimate from the light curve gives 33 $M_{\odot}$, i.e. R81 has presumably lost more than half of its initial mass in previous evolutionary phases.

IUE observations have shown that early-B hypergiants are characterized by high mass-loss rates ($\dot{M} \approx 10^{-5} M_{\odot} y^{-1}$) with comparatively low velocities ($v_{\infty} \approx 400$ km s^{-1}) implying rather high wind densities. In addition to the usual wind indicators numerous lines originating from metastable states formed in the wind (particularly Fe III lines) were identified (cf. e.g. Wolf & Appenzeller 1979, Appenzeller & Wolf 1979, Prinja et al. 1990, Lamers et al. 1995).

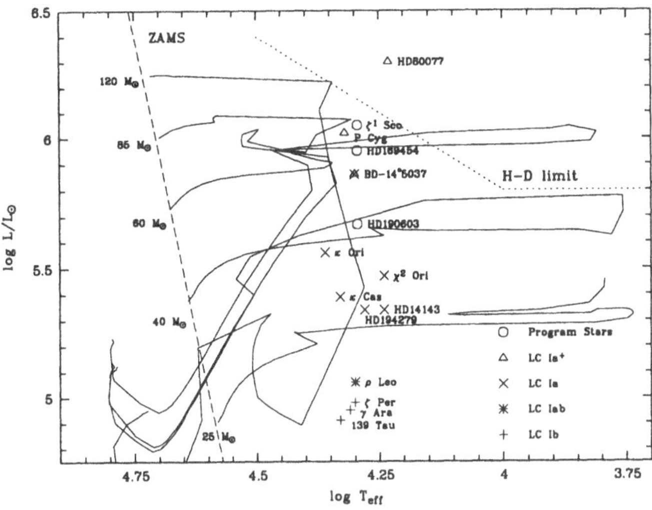

Fig. 1. A sample of early B super- and hypergiants shown in the Hertzsprung Russell diagram

Due to these characteristics early-B hypergiants are important test cases for:
a) evolutionary theories of massive stars with mass loss (cf. e.g. Schaerer et al. 1996)
b) the "wind momentum - luminosity relation" (high luminosity) (cf. e.g. Kudritzki, this Volume)
c) the "bi-stability jump" of v_∞/v_{esc} of 2.6 at $T_{\rm eff} > 21\,000$ to 1.3 at $T_{\rm eff} <$ 21 000 (Lamers et al. 1995).

Photometric variations of the prototype ζ^1 Sco in time scales of 11 to 16 days of about 0.1 mag were reported by Sterken (1977) and Burki et al. (1982). Spectroscopic snap-shot observations of early-B hypergiants in the optical range have shown pronounced line-profile variations of the P Cygni-type lines and radial velocity variations of the photospheric absorption lines with amplitudes of about 20 km s^{-1}. The photospheric absorption lines are asymmetric with blue wings extending to -200 and -300 km s^{-1} (Sterken & Wolf, 1978,1979). From various IUE observations profile variations of wind lines with variable wind absorption components were discovered (Burki et al. 1982). These observations indicate:
a) a pure hydrostatic photosphere does not exist; time and depth dependent velocity fields in the atmosphere are the predominant features.
b) considerable mass-loss variations do occur.

What is the nature of these variations? What kind of hydrodynamic processes, pulsation-like motions etc. are prevailing? Are there obvious corre-

Table 1. Stellar parameters of the observed stars

Object		ζ^1 Sco	HD 169454	HD 190603
$T_{\rm eff}$	[K]	19700	19700	19700
$\log g$		2.30	2.30	2.40
$R_\star$	[R$_\odot$]	91	82	59
$M_{\star,\log g}$	[M$_\odot$]	60	48	32
$M_{\star,\rm evol.\ track}$	[M$_\odot$]	40	40	32
$\log L_\star/{\rm L}_\odot$		6.05	5.95	5.67
$M_{\rm V}$	[mag]	−9.09	−8.85	−8.03
$\dot{M}$	[10^{-6}M$_\odot$ yr^{-1}]	6.2	6.3	2.7
v_∞	[km s^{-1}]	370	483	515
$v_{\rm esc}$	[km s^{-1}]	365	345	360

lations of the wind lines' variability and the variations of the photospheric absorption lines? And how do the B hypergiants compare with the less luminous B supergiants?

To learn more about the atmospheric motions and wind properties of early B hypergiants, snap shot observations are not sufficient - extended spectroscopic campaigns are a prerequisite. Such campaigns were carried out by us with FLASH/HEROS (cf. Rivinius et al. 1997, Kaufer,1998 and this Volume) at the ESO-50cm telescope. The typical stellar parameters of the target stars are given in Table 1 and their location in the HRD is shown by Figure 1. The dates of the observational campaigns are given in Table 2.

2 Line-Profile and Radial Velocity Variations

Our high dispersion spectroscopic campaigns allowed us to follow the wind properties and teir variability down to very deep layers and to diagnose the photosphere-wind interface of early-B hypergiants. We will briefly demonstrate that for the case of the prototype ζ^1 Sco.

Table 2. The observation campaigns

Object	Spectra/Nights						Ave. S/N
	1990	1991	1992	1993	1994	1995	[5350Å-5450Å]
HD 190603	28/87	42/154	–	–	–	–	330
ζ^1 Sco	–	–	37/52	93/114	103/128	57/120	110
HD 169454	–	–	14/45	11/17	25/62	2/2	100

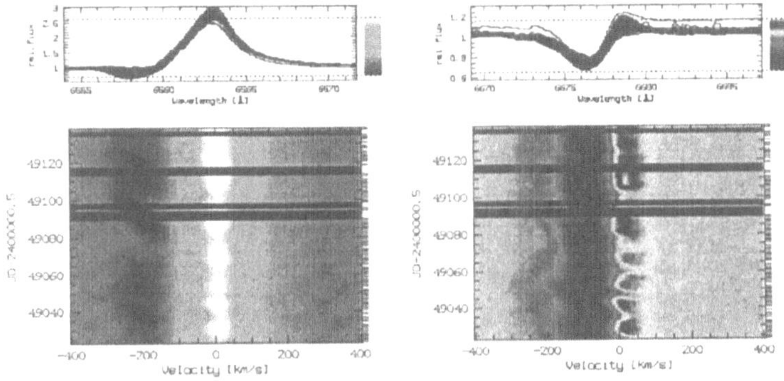

Fig. 2. The dynamical spectra of Hα and He I λ6678 in 1993

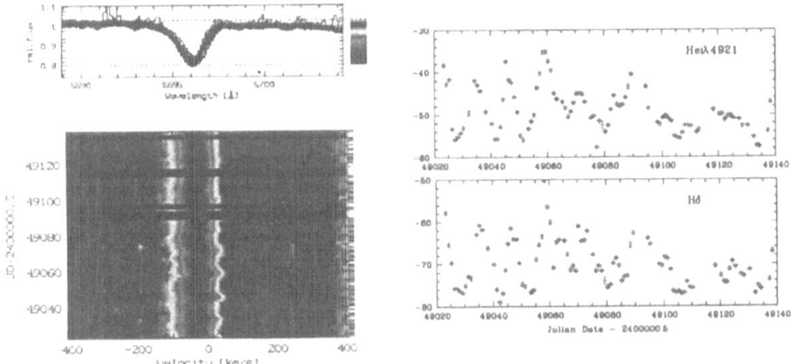

Fig. 3. The dynamical spectrum of Al III λ5696 and the measured radial velocities of He I λ4921 and Hδ in 1993

2.1 Observations

Fig. 2 shows the dynamical spectra of Hα and of He I λ6678 of ζ^1 Sco as observed during the 1993 campaign. As shown by the figure the P Cyg profiles are characterized by accelerated absorption features formed throughout the wind. One has immediately the impression, that this time series is a key finding to investigate the wind variation and that it opens for the first time the chance to derive empirically the velocity law of wind features. There is no obvious connection between emission and absorption variability in the P Cyg profiles; the repetition time of the absorption features is about 24 days. The emission variability is similar in all P Cyg lines; the repetition time is about 15 days and is comparable to the time scale of the photometric variability.

The kind of radial velocity variations of photospheric lines is shown by Fig. 3. It contains the dynamical spectrum of the line Al III λ5696 and the plots of the variations of the photospheric absorption lines He I λ4921 and Hδ. Cyclical variability is present most of the time, the typical cycle times being 10 to 20 days. After several weeks it eventually evolves into icoherent vari-

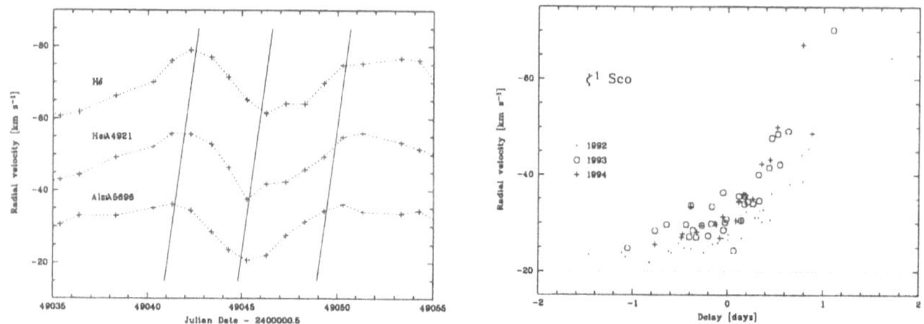

Fig. 4. The temporal shifts of the variation pattern for lines originating in different depths of the photoosphere (left) and the derived pattern delays of different lines plotted versus their mean velocities

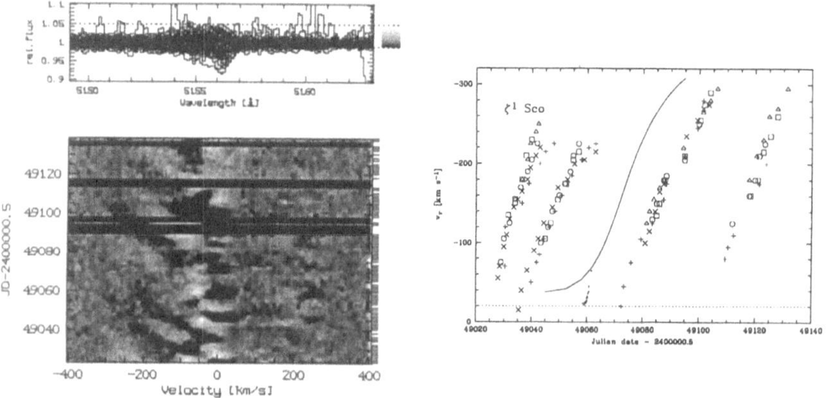

Fig. 5. Dynamical residual spectrum of FeIIIλ5156 in 1993 and the measured velocity increase of propagating disturbances. Also shown is a β type velocity law with $\beta = 2.5$. The dots connecting to the β-law are the values given in Fig. 4 (right)

ability. Pulsation-like motions are evident.(Note that the early-B type hypergiants are located in the HRD in a region where Kiriakidis et al. (1993) have predicted strange mode oscillations). The pulsation-like variability pattern is similar in all lines, but the average velocity depends on the line strength, i.e. the weekest lines show the smallest velocities. Moreover, as shown by Fig.4 the variability of lines of higher velocities (i.e. of lines formed further out in the photosphere) is delayed by up to three days; the delay can be determined by cross correlation. A model comes up that the acceleration of the disturbances has something to do with the pulsation-like motions.

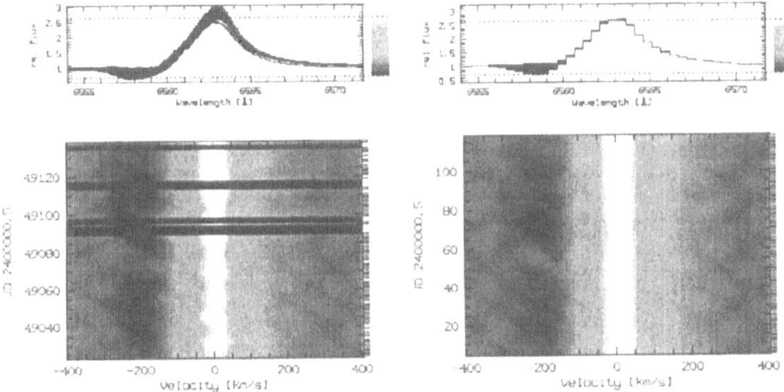

Fig. 6. The dynamical spectra of Hα and the modelled variability in 1993

2.2 Velocity Law of Perturbations

The dynamical spectra of Hα (cf. Fig.2) and Hβ were used to diagnose the propagation of disturbances in the outer part through the wind. In order to trace back the velocity law to photospheric velocities we searched for fainter P Cyg-type profiles in which the absorption features can be traced down to low velocities. The FeIII$\lambda\lambda$5127,5156 lines (cf. Fig.5) have these properties. Four events are visible in these lines in 1993. If we trace these events we get the velocity curves which can be compared with β-type velocity laws (Fig.5). β=2.5 provides a reasonable description, but note that the acceleration of the disturbances is much steeper close to the photosphere than any β law, which are mostly used to describe the winds of hot stars (cf. e.g. Rivinius et. al., 1997) In addition the pulsation-like motions and the delay of the radial-velocity pattern for photospheric lines formed at different depths can be used to diagnose the photosphere-wind interface. As indicated by the diagram in Fig.5 the same steep acceleration is indicated by this delay pattern.

To our knowledge this is for the first time that the velocity law has been directly derived from the propagation of wind features in B supergiants. But

Table 3. Model parameters for the observed stars. The temperature distribution in the wind is given by a square root law

Object		ζ^1 Sco	HD 169454	HD 190603
v_0	$[\mathrm{km\,s^{-1}}]$	18	18	18
β		1.7	1.5	1.7
$v_{\mathrm{turb}}/v_\infty$		0.05	0.037	0.035
T_{inner}	[K]	16000	16000	16000
T_{outer}	[K]	7000	7000	7000

note that this empirically derived law for the propagation of discrete features has to be regarded as an upper limit; the ambient wind is faster. The stationary wind component was calculated with the SEIBALMER4 code to β=1.7 for ζ^1 Sco (in Table 3 the β values of our target stars are listed along with the pertinent parameters of our wind models (cf. Rivinius et al. 1997)). In passing we note that the radial velocity variations observed by the photospheric lines cause radius variations of the order of 10% which in turn lead to gravity variations $\Delta\log g$ of about 0.08 dex. The photometric variations of the Strömgren indices (particularly the amplitutes of c_1) of ζ^1 Sco can be modeled with ATLAS 8 atmospheres for:
$T_{\text{eff}} = 19\,450$ K, $\log g = 2.21$, $R_\star = 98\ R_\odot$ and
$T_{\text{eff}} = 19\,950$ K, $\log g = 2.28$, $R_\star = 90\ R_\odot$, respectively

This causes luminosity variations of about 10%, which are correlated to the emission height variations. This finding is also important in connection with the wind momentum-luminosity relation of luminous hot stars (Kudritzki, this Volume).

For the nature of the disturbances we favour blobs rather than spherical shells. As shown by Lamers (1994) blobs have a large effect on the profiles if in front of the stellar disk and provide a better explanation of our observations. In fact the profile variability of the absorption components in the hydrogen lines could be reasonably fitted by superposing local density enhancements to the stationary wind (Fig.6).

3 Conclusions

The winds of early-B hypergiants are highly variable. The variations are quite different from the wind activity of less luminous BA supergiants (cf. Kaufer, 1998 and this Volume). They are more wind dominated; rotation and corotating interaction regions are not the main effect. Propagating discrete absorption components due to non-spherical density perturbations are typical. The behaviour is similar to the DACs phenomenon in O stars. The photospheric absorption lines show pulsation-like radial velocity variations which presumably trigger the variations in the wind. Our analysis shows for the first time that disturbances which affect the wind are generated very deep in the atmosphere (i.e. in the photospheric or even sub-photospheric region). The velocity law of disturbances could be followed through a wide range of the wind down to photospheric layers by making use of strategic lines distributed over a wide wavelength range.

The large wavelength coverage of modern spectrographs of high resolution and long-term monitoring programs are the prerequisite for this kind of investigation.

References

Appenzeller I., Wolf B., 1979, A&AS **38**, 51
Burki G., Heck A., Bianchi L., Cassatella A., 1982, A&A **107**, 112
Kaufer A., 1998, In: *Reviews in Modern Astronomy* 11, Schielicke R. (ed.), p. 177
Kiriakidis M., Fricke K.J., Glatzel W., 1993, MNRAS **264**, 50
Lamers H.J.G.L.M., 1994, In: *Instability and Variability of Hot-Star Winds*, Moffat A.F.J., Owocki S.P., Fullerton A.W., St.-Louis N. (eds.), Kluwer, p.41
Lamers H.J.G.L.M., Snow T.P., Lindholm D.M., 1995, ApJ **455**, 269
Prinja R.K., Barlow M.J., Howarth I.D., 1990, ApJ **361**, 607
Rivinius Th., Stahl O., Wolf B., et al., 1997, A&A **318**, 819
Schaerer D., de Koter A., Schmutz W., Maeder A., 1996, A&A **310**, 837
Stahl O., Wolf B., Zickgraf F.-J., 1987, A&A **184**, 193
Sterken C., 1977, A&A, **57**, 361
Sterken C., Wolf B., 1978, A&A **70**, 641
Sterken C., Wolf B., 1979, A&AS **35**, 69
Wolf B., Stahl O., de Groot M.J.H., Sterken C., 1981 A&A **99**, 351
Wolf B., Appenzeller I., 1979, A&A **78**, 15

Discussion

J. Puls: Two types of β velocity laws can be found in the literature, namely $v = v_0+(v_\infty-v_0)(1-1/r)^\beta$ and $v = v_\infty(1-b/r)^\beta$, where b has to be calculated from $v_{\rm min} \sim 0.1 \ldots 0.5\, v_{\rm sound}$. If you want to compare observed velocity fields in the lower part with β velocity laws (representing the global wind), you should use the latter type since only this parameterisation is consistent with hydrodynamical models.
B. Wolf: Thank you. We will do that in the future. This second velocity law will probably be in better agreement with our observations at the base of the wind.

A. Moffat: How confident can one be that the β of the ambient wind (~ 1.7) is really different from the β of the disturbances (~ 2.5)?
B. Wolf: We derived $\beta = 1.7$ by fitting the profiles of the Balmer lines (mostly Hγ and Hβ) using a modified version of the SEIBALMER program and the parameters given in Table 3 of my talk. This value provides a substantially better profile fit than $\beta = 2.5$.

L. Kaper: The distance to ζ^1 Sco (and hence the confirmation of its extreme luminosity) is "known" due to its location in an association, which makes ζ^1 Sco a prototype of its class. Is ζ^1 Sco's association confirmed (by HIPPARCOS)?
B. Wolf: Yes, it is. ζ^1 Sco has a large proper motion. Within the uncertainties $\sigma_E = 1$ mas/year, which corresponds to 10 km/s at a distance of 2 kpc. The membership of ζ^1 Sco in Sco OB 1 is confirmed by HIPPARCOS.

Wind Variations of Wolf-Rayet Stars

Anthony F.J. Moffat

Département de physique, Université de Montréal, C.P. 6128, Succ. C-V, Montréal, PQ, H3C 3J7, Canada, and Observatoire du mont Mégantic. Killam Research Fellow of the Canada Council for the Arts.

Abstract. What distinguishes WR stars from most of their progenitor phases is their dense winds, that often completely hide their small hydrostatic cores. Thus, to learn about these cores, which are especially important in an evolutionary context, we are generally limited to studying the winds only. One way to do this is to examine their ubiquitous variability. Variability studies of (single) WR stars using various techniques have revealed two main classes of wind variation: (1) large-scale and periodic, probably related to rotating disturbances (e.g. NRP or magnetic structures) close to the core surface, and (2) small (multi-) scale and stochastic, probably intrinsic to the winds themselves. The former is only seen rarely in WR stars, contrary to other hot stars. The latter appears to be universal in WR and probably (though less obvious) in all hot-star winds.

1 Types of Variability

Wolf-Rayet (WR) stars have the strongest winds among stable massive stars. However, evidence accumulated over the past few years indicates that WR winds are far from steady and smooth, as was once assumed. This is deduced from the fact that (single) WR stars are generally quite variable, mostly in their strong emission lines. Since most WR winds are so dense that we cannot see their hydrostatic cores, this variability must be occuring in the winds themselves in most cases. By quantifying the nature of the variability, we can therefore explore not only properties of the winds, but also the origin of the variability, which may lie hidden from direct view.

Several techniques have been applied to reveal wind variabilty in WR stars (e.g. photometry, polarimetry, spectroscopy, or combinations thereof), but the most effective so far has been the study of *optical* line profile variations (LPVs). Previous work has often concentrated on UV LPVs, obtained using IUE (e.g. St-Louis 1990). However, the UV data are most effective in revealing LPVs in the P Cygni absorption profiles of mainly strong, often resonance, lines that prevail in the UV. This limits their visibility to the part of the wind that lies in the column towards the star. Despite the many interesting UV results, I will concentrate on the more global LPVs seen in emission lines in the *optical*, where it has been easier to get high-quality data. According to a recent study (Lépine & Moffat 1999), optical wind LPVs can be sorted into three categories:

1. Stochastic (S). All hot-star winds observed *appropriately* so far (i.e. monitored *at least* on an hourly basis over several hours, at high signal-to-noise S/N and high spectral resolution) show stochastically varying, narrow emission subpeaks superposed on broad emission lines. This is the case for some ten WR stars (Robert 1992; Lépine 1998; Lépine & Moffat 1999), two [WC]-type central stars of planetary nebulae (Balick et al. 1994; Grosdidier et al. 1997), one Of star (Eversberg et al. 1998) and at least one nova (Garnavitch, priv. comm.). *It is strongly suspected that this is actually a very widespread phenomenon, probably present in all hot-star winds.* The most likely interpretation is that we are seeing the manifestation of a hierarchy of density enhancements, often referred to as "clumps", "blobs" or "shocks", propagating outwardly with the wind. In fact, they probably *are* the wind.

2. Recurrent (R). Broad subpeaks are seen in only a small number of WR winds (Morel 1998; Morel et al. 1998a, 1998b). They tend move across the whole emission line and recur, but only over several cycles. Among the 30 or so WR stars which are apparently bright enough to have been checked for this kind of LPV, only WR6 and WR134 appear to fall in this category, with the possible addition of WR1 (Morel, priv. comm.). These stars also appear to exhibit S-type LPVs, which are often masked by those of R-type. WR11 = γ^2 Vel also shows broad variable subpeaks, but these apparently behave in a stochastic fashion and belong to the S category (Lépine et al. 1998). It is quite noteworthy that WR6 and WR134 are the single WR stars showing the largest amount of line depolarisation (Schulte-Ladbeck 1994; Harries et al. 1998). WR1 is not exceptional at all in this way, however, and another WR star with moderately srong line depolarisation is the WR component in the long-period binary WR137.

The current interpretation of the R scenario in WR winds is that we are witnessing the rotational modulation of a persistant large-scale anisotropic outflow, which is likely related to the propagation of discrete absorption components (DACs) and corotating interaction regions (CIRs) in OB-star winds (e.g. Cranmer & Owocki 1996). While these are seen frequently in OB winds,they are rare in WR winds. Presumably they are generated by magnetic structures in forced rotation or non-radial pulsations close to the stellar (hydrostatic) surface, the former being somewhat more likely. Such features do not last more than several rotations, hence are incoherent on long timescales.

3. Periodic (P). Only binary systems appear to reveal truly periodic LPVs in WR and other hot stars. These are normally interpreted as arising in cyclic modulations due to wind-wind collisions, photospheric and atmospheric eclipses, heating effects and other binary-related distortions (see Marchenko et al. 1997).

2 Evidence for Stochastic Line Profile Variations

In this review, I will limit further discussion to the S type LPVs. As noted above, these small-scale, stochastic variations are probably omnipresent in all WR and other weaker hot-star winds. We now look in more detail at the evidence for these stochastic line profile variations in WR winds. To do this, it is useful to divide up the evidence into classes according to the degree of sophistication of the observations, reflected by their "dimension":

0-D. Here the manifestation of the variations are characterised by a quantity which gives no information about the detailed nature of the actual 3-D clumping: it could be e.g. shells or clumps. In order of $\sim$ decreasing wavelength, there are a large number of 0-D signatures of clumping:

• IR/radio excess, compared to the thermal emission from the underlying star (Runacres & Blomme 1996; Blomme & Runacres 1997). The level of the excess can only be accounted for if clumping occurs, since the f-f emission excess goes as the square of the density.

• Increase of the eclipse width and depth especially at secondary minimum (O-star in front) of the IR compared to the optical light curve of the WR + O eclipsing system V444 Cyg (Cherepashchuk et al. 1984). Again, this must be due to the enhanced f-f emission due to some kind of clumping of the WR wind, although it provides no indication of the true nature of the clumping.

• Dust formation in WC star winds. In such a hostile environment, localised density enhancements are needed both to form the dust and once formed, to shield it from the intense UV radiation field of the central star (Cherchneff et al. 1998). This can occur either in the winds of single, cool WC stars (most WC9-10 and some WC8 stars) or in the wind-wind collision zone in some WC + O binaries (Williams 1996), both leading to IR excess. Additional evidence for dust formation comes from episodic dips in the continuum light curves of some WC9 stars (Veen et al. 1998). Again, no real constraints are provided as to the true nature of the clumping.

• Photometric (e.g. Moffat & Shara 1986) and polarimetric (Robert et al. 1989) continuum variability. The stochastic nature of this variability on timescales of hours to days suggests some kind of clumping in the winds. The fact that the polarisation scatter is normally much less than the photometric scatter implies that the clumps both scatter and emit light (Brown et al. 1995).

• Distortion of line intensity ratios. Nugis & Niedzielski (1995) find an enhancement of the flux ratio of HeI to HeII lines in the IR, that allows one to deduce that clumping must be important.

• Disk-shaped HeI line profiles. These can be reproduced if clumping is important (Antokhin et al. 1992).

• Diffuse electron scattering (Hillier 1991). This creates an enhancement on the red side of emission line profiles, due to scattering in an expanding medium around the star. Since (optically thin) electron scattering is independent of clumping, this enhanced component is correspondingly less than the main direct emission component for the same line when the wind is clumped than when it is smooth.

• Saturation of P Cygni absorbtion profiles (Prinja et al. 1990). This is a consequence of a non-monotonic velocity expansion law of the wind, caused by velocity scatter due to shocks in the wind; a dispersion of $\sim 10^2$ km/s is implied.

• High X-ray fluxes from hot-star winds (e.g. Berghöfer & Schmitt 1994). Although current X-ray satellites are not large enough yet to detect the expected stochastic X-ray variability, the observed X-ray fluxes imply that shocks must be present in the wind, with thermal temperatures of $\sim 10^6$ K (Wesselowski 1996; Feldmeier et al. 1997).

• Erratic X-ray variability in massive X-ray binaries (Kaper et al. 1996). This could be caused either by clumped winds being accreted onto the compact companion, or by instabilities within the accretion disk (if there is one).

1-D. Doppler imaging has now enabled us to separate out various clumps in the wind and at least partly qualify the nature of the clumping. Only one observational signature so far falls in this category for the otherwise unresolved inner WR wind:

• LPV. The information content is greater for emission lines, which arise globally in the wind, compared to P Cygni absorptions, which probe the highly localised column of the wind from the observer to the star. The latter is somewhat ambiguous, unable to clearly distinguish between DACs and clumps. In any case, most WR resonance P Cygni absorptions in WR stars are saturated, making their study virtually impossible except at their sloping edges. Global emission line variability has been detected by various groups now (e.g. Schumann & Seggewiss 1975; Moffat et al. 1988; McCandliss 1988; Robert 1992; Lépine 1998); an update of previous results is given in the next section.

2-D. Resolution of the outer parts of WR winds has now been accomplished via direct imaging in two special cases. Note that even the closest WR star, γ^2 Vel, would require a resolution of ~ 2 mas to distinguish 1 $R_\odot$, a possible clump size in the inner wind, at its (new) distance of ~ 250 pc.

• Radio interferometry. The second closest WR star, WR147 ($d \sim 700$ pc), has been resolved using MERLIN into a thermally clumped wind on a scale of ~ 20 mas (10^3 $R_\star$) (Williams et al. 1997). We (Barlow et al.) hope to do the same soon for γ^2 Vel.

• Optical line imaging of the young WR nebula M1-67 around WR124. Grosdidier et al. (1998) have used HST/WFPC2 to resolve this nebula in Hα, which is filled in right down to the observable limit close to the central star. This remarkable image shows, among other interesting features, a number of strange point-like sources of bright Hα emission surrounded by what appear to be their own local wind blown bubbles. Lack of bow-shock shapes implies that they are travelling out with the ambient wind, suggesting that we might be seeing a few of the largest wind clumps emerging from a previous (slower) LBV stage. The alternative speculation that we are seeing some new kind of instability in the wind lacks a theoretical understanding. Some of these ideas may change, as Grosdidier et al. (1999) are in the process of examining the velocity field across M1-67, based on new high-resolution Fabry-Pérot observations in Hα from CFHT.

3+ -D. Increased resolution might be provided by 2-D imaging combined with velocity (RVs or pm's) or other information. One can only dream for now of Vakili's (this conference) technique of "interfero-spectropolarimetry"!

3 Highlights of New Work on S-type LPVs

3.1 Data Source and Technique

Based on the same spectral data reported in Robert (1992), Lépine et al. (1996) and Lépine & Moffat (1999) have significantly improved the analysis of the 9 WR stars for line variability. These data comprise 3-4 contiguous nights of R $\sim$ 30 000, S/N $\gtrsim$ 200/pixel (continuum), δt $\sim$ 1 h spectrosocpy of the best optical line in a range of WN and WC subtypes.

The key here is twofold: (1) Wavelet analysis has been applied to the line variations relative to a mean profile. This is a better alternative to Fourier techniques or temporal variance spectrum analyses (Fullerton 1990): wavelets allow one to analyse the pattern as a whole, rather than trying to extract individual clumps or looking at one pixel at a time. (2) A phenomenological model has been devised to simulate the observed LPVs. The model is calculated with the same sampling as the observations, in velocity space using a β velocity law to relate projected acceleration to velocity. The model also allows for a scaling power law for the fluxes of the clumps (although not in velocity dispersion) as expected if turbulent cascading is acting. From these simulations, it was deduced that superposition effects are generally very important, leading to a high level of confusion and change in interpretation compared to previous studies (e.g. Moffat et al. 1994).

3.2 Results

One of the WR stars (WR134) in the Robert (1992) sample is dominated by R-type LPVs (cf. Morel et al. 1998b for a more detailed study of this star). The remaining 8 stars all show very similar S-type LPV patterns, which can be summarized as follows:

- All the most obvious emission subpeaks in all stars have velocity widths of FWHM $\sim$ 2-10 Å and amplitudes of $\sim$ 2-8 % of the average line intensity.

• Subpeak motion occurs always away from line centre, both in the blue and red directions. This can be most readily interpreted as clumps propagating radially outward in a spherically symmetric wind, and seen in velocity projection v(r) $\cos\theta$, each with $\theta \sim$ const.

• Each observed subpeak generally is the sum of a large number of independent, discrete wind emission elements (DWEEs). These are assumed to be Gaussian in shape in the simulations.

• A large number of DWEEs ($\gtrsim 10^4$) are needed to account for the low degree of variability observed. This implies that DWEEs must occupy a small volume.

• DWEEs reveal a large velocity dispersion, with dominant scale $\sigma_v \sim$ 100 km/s ($\sim$ 350 km/s in WR134). This is likely a consequence of macro-turbulence and associated shocks.

• A strong anisotropy (radial versus tangential) prevails in velocity dispersion within the DWEEs: $\sigma_{v_r} \sim 4\sigma_{v_t}$. This is compatible with radiative wind driving (e.g. Rybicki et al. 1990). For WR134 $\sigma_{v_r} \sim \sigma_{v_t}$, implying significant angular motion, as expected for R-type LPV.

• There is marginal evidence for optical depth effects within the DWEEs, implying that the escape probablility for photons may be slightly less in the radial than in the transverse direction.

• Relatively low radial acceleration is found in the inner wind and high acceleration in the outer wind, characterised by the parameter product $\beta R_\star/R_\odot$ = 20 ... 80, much larger than obtained from most "standard model" fits, with $\beta \sim 1$ and $R_\star$ of several $R_\odot$.

• The duration of subpeak events agrees with the overall line shape, implying a relatively thin line-emitting region (LER) in velocity space.

• The stochastic behaviour of the DWEEs agrees with model simulations (e.g. Gayley & Owocki 1995), with rapid growth of random fluctuations near the base out to the observed wind.

• The same phenomenon appears to be at work in all WR winds, independent of the spectrum and subclass, e.g. σ_v is independent of the geometry of the LER, v_∞, wind acceleration, ...

However, there still remains one important factor, which turns out to be a rather severe limitation: The number of visible subpeaks in any given line in a given star is proportional to v_∞, as noted by Robert (1992). Does this mean that faster winds have more structure? The answer now appears to be *NO*! Rather, it is spurious, due to superposition effects that are worst in narrow-line (slower-wind) stars. Their is no way around this problem, although observing stars with large v_∞ does help somewhat. Higher spectral resolution or S/N, or shorter δt will not help much.

4 Importance of Clumping

From studies of subpeak behaviour on WR emission lines, the following conclusions can be stated regarding the associated wind clumping:

• Clumping is probably universal in hot-star winds. There is no reason why WR stars should be special, and in any case, evidence exists already for clumping in other types of hot stars.

• All previous smooth-wind based mass-loss rates will have to be reduced by about a factor three. This comes from both observation (polarization modulation and period changes in WR binaries: St-Louis et al. 1993; Moffat & Robert 1994) and theory (e.g. Schmutz 1997; Hillier & Miller 1998; Nugis et al. 1998).

• The revision in mass-loss rates will have an important impact on models of stellar evolution.

• Clumping is probably crucial to explain the formation of dust in WC winds.

• WR winds appear to be excellent laboratories to study turbulence in action.

• Propagating inhomogeneities allow us to trace the wind kinematics, e.g. derive a relation between the observables acceleration and velocity, and thence deduce the velocity law v(r).

• Clumping will tend to "soften" collisions, e.g. with the ISM or other winds.

• Clumpy winds will make a direct contribution to the clumpiness of the ISM.

5 Future Prospects

Several avenues remain to be exploited:

• Resolving clumps interferometrically. This has already started in the radio and will soon (in 10 years from space?) be possible in the optical/IR. Some current "dirty laudry" is likely to take a severe cleaning...

• Spectroscopy: larger samples of stars of different type, longer continuous time-coverage and probing different lines simultaneously.

• Exploring internal structure of clumps, other scaling laws, and establishing a clear relation to turbulence.

• More modeling is necessary: 3-D hydrodynamics with radiative effects included and more quantitative comparisons with observations; calculating CIRs/DACs in WR winds.

References

Antokhin, I.I., Nugis, T., Cherepashchuk, A.M. (1992): Sov. Astron., **36**, 260

Balick, B., et al. (1996): AJ, **111**, 834

Berghöfer, T.W., Schmitt, J.H.M.M. (1994): ApSpSc, **221**, 309
Blomme, R., Runacres, M.C. (1997): A&A, **323**, 886
Brown, J.C., et al. (1995): A&A, **295**, 725
Cherchneff, I., et al. (1998): A&A, submitted
Cherepashchuk, A.M., Khaliullin, K.E., Eaton, J.A. (1984): ApJ, **281**, 774
Cranmer, S.R., Owocki, S.P. (1996): ApJ, **462**, 469
Eversberg, T., Lépine, S., Moffat, A.F.J. (1998): ApJ, **494**, 799
Feldmeier, A., Puls, J., Pauldrach, A.W.A. (1997): A&A, **322**, 878
Fullerton, A.W. (1990): Ph.D. Thesis, Univ. of Toronto
Gayley, K.G., Owocki, S.P. (1995): ApJ, **446**, 296
Grosdidier, Y. et al. (1997): IAU Symp. No. 180, 108
Grosdidier, Y., et al. (1998): ApJ, **506**, L127
Grosdidier, Y., et al. (1999): in preparation
Harries, T.J., Hillier, D.J., Howarth, I.D. (1998): MNRAS, **296**, 1072
Hillier, D.J. (1991): A&A, **247**, 455
Hillier, D.J., Miller, D.L. (1998): ApJ, **496**, 407
Kaper, L., et al. (1996): A&AS, **116**, 257
Lépine, S. (1998): Ph.D. Thesis, Univ. de Montréal
Lépine, S., Moffat, A.F.J., Henriksen, R.N. (1996): ApJ, **466**, 392
Lépine, S., Eversberg, T., Moffat, A.F.J. (1998): ApJ, submitted
Lépine, S., Moffat, A.F.J. (1999): ApJ, **510** (1 Jan)
Marchenko, S.V., et al. (1997): ApJ, **485**, 826
McCandliss, S.R. (1988): Ph.D. Thesis, Univ. of Colorado, Boulder
Moffat, A.F.J., et al. (1988): ApJ, **334**, 1038
Moffat, A.F.J., Shara, M.M. (1986): AJ, **92**, 952
Moffat, A.F.J., Robert, C. (1994): ApJ, **421**, 310
Moffat, A.F.J., et al. (1994): ApSpSc, **216**, 55
Morel, T. (1998): Ph.D. Thesis, Univ. de Montréal
Morel, T., et al. (1998a): ApJ, **498**, 413
Morel, T., et al. (1998b): ApJ, in press
Nugis, T., Niedzielski, A. (1995): A&A, **300**, 237
Nugis, T., Crowther, P.A., Willis, A.J. (1998): A&A, in press
Prinja, R.K., Barlow, M.J., Howarth, I.D. (1990): ApJ, **361**, 607
Robert, C., et al. (1989): ApJ, **347**, 1034
Robert, C. (1992): Ph.D. Thesis, Univ. de Montréal
Runacres, M.C., Blomme, R. (1996): A&A, **309**, 544
Rybicki, G.B., Owocki, S.P., Castor, J.I. (1990): ApJ, **349**, 274
Schmutz, W. (1997): A&A, **321**, 268
Schulte-Ladbeck, R. (1994): ApSpSc, **221**, 347
Schumann, J.D., Seggewiss, W. (1975): IAU Symp. No. 107, 299
St-Louis, N. (1990): Ph.D. Thesis, Univ. College London
St-Louis, N., et al. (1993): ApJ, **410**, 342
Veen, P., et al. (1998): A&A, **329**, 199
Wesselowski, U. (1996): 33rd Liège Ap Coll, 343
Williams, P.M. (1996): 33rd Liège Ap. Coll., 135
Williams, P.M., et al. (1997): MNRAS, **289**, 10

Discussion

G. Koenigsberger: The radio emission associated with the WR 147 wind-wind collision region seems very broad. Could you comment on this?
A. Moffat: According to wind-wind collision models (e.g., Stevens et al. 1992), the relative separation of the shock fronts on either side of the contact surface can indeed be quite large, especially if the collision is more or less adiabatic. This is likely to be the case in long-period systems like WR 147.

A. Feldmeier: Do you think that there is a connection between the X-ray emission from OB stars and the strongest blobs you find? Say, if the blob gas is the cool, dense gas behind the strong X-ray emitting reverse shock, then (since the radiative cooling zone is isobaric) $\rho T = \text{const}$ and one could expect that the largest blob overdensities are given by $\rho_{\text{blob}}/\rho_{\text{CAK}} \sim T_{\text{X-ray}}/T_{\text{wind}} \sim 100$. Can you derive the largest overdensities from your observations?
A. Moffat: Unfortunately, with the extremely large number of estimated clumps ($> 10^4$) combined with the ambiguities (superposition) of spectroscopy, we cannot provide any serious constraints on the density contrast. If compressible turbulence is at play, the small (less massive) clumps are expected to be densest, so that the X-rays could arise in a very large number of clumps.

A. Maeder: Does the revision of the mass-loss rates you are suggesting (factor of ~3) also apply to O-type and WR stars?
A. Moffat: We only have real line-profile variability data for the extreme O star ζ Pup. In that case, the stochastic variability is every bit as high as any WR star in the same line relative to the average line emission. We can only guess that this might lead to similar corrections for ζ Pup. It would not surprise me if all hot-star winds were clumpy, but more work is needed to show what the impact of that might be.
H. Lamers: The mass loss rates from O stars are usually derived from P Cygni profiles.

S. Shore: In the binary WR + O system, is it possible that you would see a change in the stochastic variations of the emission lines with the binary phase? For instance, in V 444 Cyg there is a hole in the wind from the O – WR wind collision, so you might expect to see different wavelet spectra as a function of time.
A. Moffat: In V 444 Cyg, most of the variations are phase dependent, which masks the stochastic component to a large extent. In γ Vel, we find no significant phase dependence of the stochastic line variation component around more or less half the orbit.

Spectral Analyses of Wolf-Rayet Stars: The Impact of Clumping

Wolf-Rainer Hamann and Lars Koesterke

Universität Potsdam, Am Neuen Palais 10, D-14469 Potsdam, Germany

Abstract. Inhomogeneities are accounted for in our non-LTE stellar wind models in a first-order approximation. When applied for spectral analyses, clumpy models yield lower mass-loss rates than homogeneous models, while other parameters are not affected. For representative WR stars, we determine the density contrast from the electron-scattering line wings and obtain mass-loss rate reductions by a factor of two, typically.

1 Model atmospheres with clumping

The so-called standard models for WR atmospheres are based on the assumptions of spherical symmetry, homogeneity and stationarity of the flow. The main features of WR spectra can be reproduced by these calculations, thus validating its basic assumptions as a reasonable approximation. However, there are various evidences that real WR atmospheres are actually inhomogeneous to some degree.

For a first-order approach to clumped stellar winds, we implement the following simplifying treatment of clumping in our non-LTE model atmosphere code. In the clumps the density is enhanced by the factor D with respect to the homogeneous model with the same mass-loss rate. The interclump space is void. The clumps are assumed to have small size, compared to the photon free path.

In the described formulation, the density enhancement D would cancel out in the emissivities and opacities, if they were linear in density. However, Wolf-Rayet spectra are known to be dominated by processes which scale with the *square* of the density. This can be concluded from the scaling property of Wolf-Rayet models, which has been discovered first by Schmutz et al. (1989). They defined a so-called transformed radius R_t as a combination of $\dot{M}$, R_* and v_∞ and found that models with same R_t exhibit the same emission line equivalent widths, irrespective of different combinations of the involved parameters (while, of course, T_*, composition etc. are fixed). This invariance was validated by various numerical experiments with reasonable accuracy.

The concept of the "transformed radius" now is generalized for clumped models. In order to cancel out in quadratic processes, the clump density enhancement D must be compensated by diminishing the mass-loss rate by a factor $\sqrt{D}$. Thus we define

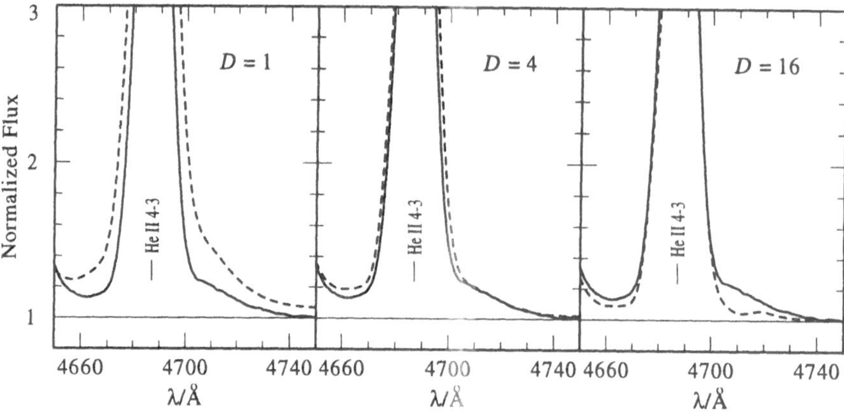

Fig. 1. The line wings of the He II line at 4686 Å, as observed in the WN7-type LMC star Br 24 (solid line). The synthetic profiles (dashed lines) were calculated for different density contrast D as indicated. The mass-loss rate is scaled with $D^{-0.5}$ for compensation, while the other parameters (see caption of Fig. 2) are maintained.

$$R_t = R_* \left[\frac{v_\infty}{2500\,\mathrm{km\,s^{-1}}} \Big/ \frac{\sqrt{D}\dot{M}}{10^{-4}\,M_\odot\,\mathrm{yr}^{-1}} \right]^{2/3} \tag{1}$$

and expect that models with same R_t exhibit the same line equivalent widths, irrespective of different combinations of D, $\dot{M}$, R_* and v_∞. For constant v_∞, the absolute spectra should only differ by a scaling with R_*^2.

Note that this scaling invariance holds only for the ρ^2 processes, i.e. for the main spectral features. The electron scattering opacity, however, scales linearly with density. Thus, in inhomogeneous models the enhanced clump density is already fully compensated by the volume filling factor; scaling down the mass-loss rate in order to keep the same R_t decreases the effective Thomson opacity. Now, Thomson scattering causes a wide frequency redistribution of photons, due to the high thermal speed of the electrons (cf. Hillier 1984). Hence, for a series of models with same R_t the extended line wings caused by this frequency redistribution of electron-scattered line photons become weaker with increasing D, while the main spectral feature remain unchanged.

2 Spectral fits and conclusions

The described dependence of the electron-scattering line wings can be used to determine the adequate value of D from observed spectra. This is demonstrated here by the example of Br 24. The comparison with the observation (Fig. 1) reveals that the homogeneous model ($D = 1$) gives too strong electron-scattering wings, while the model with $D = 4$ fits best.

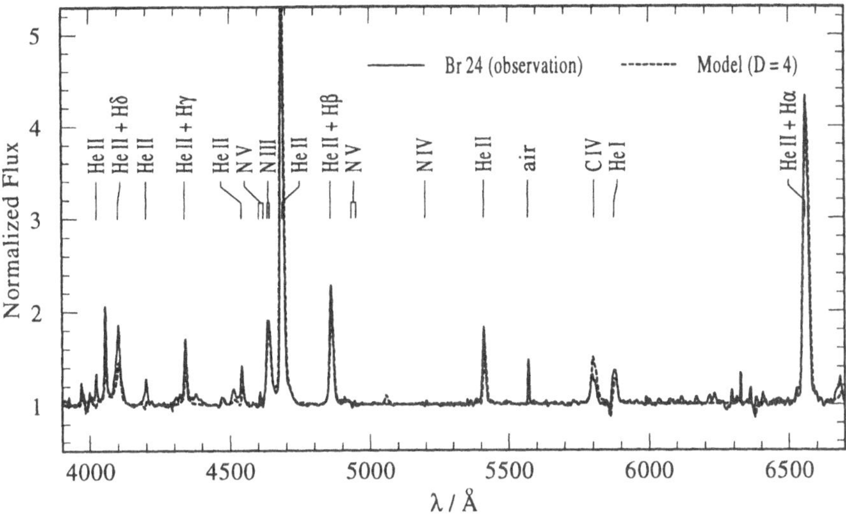

Fig. 2. The observed spectrum of the LMC star Br 24 in the visual, compared to a model calculation with clumping ($D = 4$). Accounting for the clumping effect has nicely improved the detailed agreement between observed WR spectra and our models. Model parameters: $T_* = 40\,\mathrm{kK}$, $\log L/L_\odot = 5.35$, $\log \dot{M}/(M_\odot \mathrm{yr}^{-1}) = -5.0$, $v_\infty = 950\,\mathrm{km/s}$, composition H/He/C/N = 40/59/0.01/0.8 (by mass)

To the first order, the mass-loss rate is the only parameter which is affected by the application of clumped models for spectral analyses. In a recent paper (Hamann & Koesterke 1998b) we selected representative Galactic WR stars of different spectral subclass and found that D=4 is a typical value, which implies that the empirical mass-loss rates become smaller by factor $\sqrt{D} = 2$ than obtained from the use of homogeneous models. In Hamann & Koesterke (1998a) we analyzed the nitrogen spectra of the Galactic WN stars. If their mass-loss rates are now scaled down by a factor of two, the average values of the "momentum ratio" $\dot{M} v_\infty c/L$ become 4.5, 4.3 and 15 for the WNL, WNE-w and WNE-s spectral subclass, respectively. These values are still above the single-scattering limit (i.e. unity), but no longer implausible for radiative acceleration with multiple-scattering effects. The consequences of smaller WR mass-loss rates for the stellar evolution must be considered.

References

Hamann W.-R., Koesterke L., 1998a, A&A 333, 251
Hamann W.-R., Koesterke L., 1998b, A&A 335, 1003
Hillier D.J., 1984, ApJ 280, 744
Schmutz W., Hamann W.-R., Wessolowski U., 1989, A&A 210, 236

Discussion

G. Koenigsberger: Regarding the clumping factor, which appears to be the same for all WR subclasses: could you comment on why "D" should be independent of the properties that determine the wind structure, which in turn lead to the observed spectra?

W.-R. Hamann: Clumping in WR winds is most likely caused by the instability of radiation driving. Existing calculations (Gayley & Owocki 1995, ApJ 446, 801) show that the perturbation growth rate scales with the inverse of the wind momentum ratio, η. Although η is larger for WR stars than for OB stars, the instability is predicted to produce strong clumping. However, fully self-consistent hydrodynamical WR models do not yet exist. Thus, we have no prediction about how the density contrast may depend on the stellar parameters.

R. Ignace: Recent work by Nugis et al. suggests that the free-free continuum slopes of WR stars cannot be explained by constant clumping and filling factors. Can you comment on the consequences of their inferences for your profile modelling?

W.-R. Hamann: Free-free radio emission is a ρ^2-process, as is the line emission. Hence radio mass loss rates are affected by clumping in exactly the same way as shown in my talk. The mass-loss rates derived from radio emission agree fairly well with those from optical/UV lines. This implies that the clumping contrast must be similar in the radio-emitting region (far out in the stellar wind) and the line-forming region, respectively. The slope of the radio free-free continuum is not affected by clumping, unless the density contrast strongly varies with the radial coordinate. The corresponding evidence given by Nugis et al. (1998, A&A 333, 956) is rather weak in my opinion.

Anthony Moffat, Wolf-Rainer Hamann, Knut Ødegaard and Joyce Guzik

The Long-Term Variability of Luminous Blue Variables

Roberta M. Humphreys

Astronomy Department, University of Minnesota, 116 Church St. SE, Minneapolis, MN 55455, USA

Abstract. The stars known as Luminous Blue Variables include two very distinctive subgroups - the S Dor-type variables which basically define what we call an LBV and the much rarer 'giant eruption' LBV's which include famous stars like η Car and P Cyg. The distinctive characteristics and long term variability of these two groups is reviewed. The lesser 1890 eruption of η Car is shown to have been much more significant than previously believed and resembles the second peak seen in the historic light curve of P Cyg. Because so many, if not all, stars in certain parts of the HR diagram appear to be luminous, blue, and variable, I suggest returning to our previous designation - S Dor variables and η Car variables for these two important groups of stars.

1 Introductory Remarks

Throughout this meeting every star or class of stars that has been discussed is luminous, blue, and variable!

So the first question is what is a Luminous Blue Variable or LBV?

The most important point to remember about the definition or description of any class of objects is that we must be talking about the same physics. In previous reviews, we (Humphreys 1989, Humphreys and Davidson 1994) have described LBV's in terms of their behavior i.e. 'luminous, hot stars that exhibit a particular type of instability' whose defining characteristic is the irregular eruption or outburst. During this eruption the star brightens 1 to 2 magnitudes in the visual; the duration is measured in years, and the star maintains nearly constant bolometric luminosity with corresponding changes in its spectrum and apparent temperature. This is a description of the normal or classical LBV's typified by stars like S Dor and AG Car.

A second group of stars also called LBV's are the giant eruption LBV's; stars like η Car and P Cyg which actually increased their luminosity during the eruption. There are only four probable members of this group and they are also included in this review.

The second question is what is meant by 'long-term'? A normal LBV eruption typically lasts a few years followed by a period of quiescence or additional outbursts. So for the purpose of this paper, long-term will refer to a photometric record lasting a decade or more.

2 The Normal or Classical LBV's

On timescales of 10 years or more, the light curves of the normal or S Dor-type variables show a wide range of variability, but all vary by 1 to 2 magnitudes during the eruption. Some have an eruption and then are relatively quiescent for many years. Examples are R71 in the LMC (see Sterken et al 1997), M33 Var.C (Hubble and Sandage 1953, Humphreys et al 1988, Szeifert et al 1996) and M33 Var.2 (Hubble and Sandage 1953). While others appear to repeat in an almost semi-regular way such as M33 Var.B (Hubble and Sandage 1953, Szeifert et al 1996), AG Car, and the proto-type S Dor (see de Koter 1993 for recent variability). Humphreys and Davidson (1994), van Genderen et al (1997), and Sterken et al (1997) have pointed out the quasi-periodic variability seen in some LBV's.

2.1 The S Doradus Instability Strip

When we look at the upper HR diagram (see Figure 9 in Humphreys and Davidson 1994) we notice a very important property of the S Dor-type variables; their temperature at quiescence or visual minimum is luminosity dependent. Wolf (1989 a,b) was the first to recognize that the S Dor variables at quiescence all lie along an inclined strip in the HR diagram. This locus combined with the constant luminosity and the cool temperature limit near 8000K at visual maximum shows that the amplitude of the variability depends on the luminosity. The more luminous the S Dor variable, the higher the amplitude (Wolf 1989a,b). This is a very important characteristic of the S Dor-type variables. It must be providing an important clue to the origin of the instability in these evolved stars ranging in initial mass from $\approx$30 to $100M_{\odot}$.

So what do these stars have in common? Most likely, their L/M ratio. All of them are overluminous for their current mass; consequently, they are near the Eddington limit for their mass. The less luminous variable have most likely shed a lot of mass as red supergiants while the more luminous ones may have undergone giant eruptions like P Cyg or η Car plus repetitive S Dor-type outbursts.

3 Giant Eruption LBV's

The giant eruption LBV's or η Car-like variables actually experience a significant increase in their total energy during their eruption. The four historical members of this group are η Car (1840's), P Cyg (1600's), V12 in NGC2403 (SN54J) in 1954, and SN61V in NGC1058 (1961). The properties of these very rare stars are summarized in Table 1. The recently discovered variable, V1 in NGC2363 (Drissen et al 1997), is a possible member of this group.

Eta Car is famous for what is often called the 'great eruption' around 1840 that lasted more than 20 years during which it briefly became the second brightest star in the sky. Although its famous bipolar lobes were created during the great eruption, measurements of the proper motions of the Weigelt knots within 0.2 arcsec of the central star (Weigelt et al 1995) and the velocities of the same knots (Davidson et al 1997) show that they are in the equatorial ejecta and have an age of 100 years, the time of the lesser eruption in 1890. Smith and Gehrz (1998a) have measured motions of several identifiable features in the equatorial ejecta with a 50 year baseline and also derive an 1890 date.

Table 1. Properties of Giant Eruption LBV's

	ΔM_v (mags)	ΔM_{bol} (mags)	M_{bolmax} (mags)	Total energy released (ergs)
η Car	3-5	~ 2	-14	$10^{49.5}$
P Cyg	~ 3	1-2	-11 to -12	$10^{48.4} - 10^{48.8}$
SN61V	~ 5.5	$\sim$3.5	-17	$10^{49.5}$
V12	4	~ 2	$\leq$-11.6	$\geq 10^{47.3}$

The second or lesser eruption in 1890 doesn't look very impressive on the historic light curve (see Figure 1), but Smith et al(1998b), using the measured motions and their estimate of the mass in the lobes and equatorial ejecta from their infrared images, find that the kinetic energy is $10^{48.4}$ergs and $10^{48.1}$ergs for the 1840 and 1890 eruptions, respectively, compared to $10^{49.5}$ergs total energy in the great eruption (Davidson and Humphreys 1997). Given that the kinetic energy in the two eruptions is comparable, it is worth taking a second look at the 1890 one.

We now know that η Car formed dust after the 1840 eruption and its extreme faintness near 8th magnitude, $\approx$ 4 magnitudes below its pre-eruption brightness, was due to circumstellar dust. The 1890 eruption lasted 7 years with a maximum in 1889 at 6.2 apparent visual magnitude. A spectrum from 1893 shows an early F supergiant absorption spectrum with hydrogen emission like an S Dor-type variable at maximum. Assuming that the luminosity remained constant ($M_{bol} \approx$ -12mag) as in a normal S Dor-type eruption, then its m_v would have been $\approx$ 1.5 mag (Figure 1) minus the CS extinction and the total energy was $\approx 10^{48.4}$ergs.

A second and independent estimate can be made from the probable amount of circumstellar extinction in 1890. From its current energy distribution and assuming that η Car is 20,000 to 25,000K, the CS extinction is presently 2 to

3 mags. But Eta has brightened about two magnitudes since the late 1940's. So in 1890, it was probably suffering 4 to 5 magnitudes of extinction and its m_v was actually 1.2 to 2.2 mag during the lesser eruption.

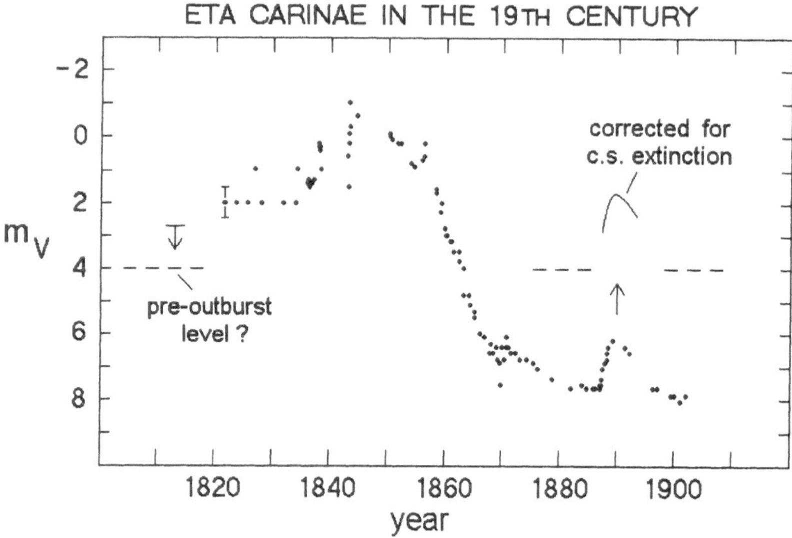

Fig. 1. The light curve of η Car showing the great eruption and the lesser eruption in 1890 corrected for circumstellar extinction.

A similar second peak or eruption has been observed for P Cyg (de Groot 1988 , Lamers and de Groot 1992) and in the light curve for SN61V (Doggett and Branch 1985). It is uncertain whether V12 has had a second peak because of infrequent measurements since 1955. The light curves of P Cyg, SN61V, and V12 in NGC2403 also show a post-eruption plateau or shoulder at or near the pre-outburst magnitude. Eta Car may also have had a brief plateau in 1860 - 1862. Circumstellar dust is also present around η Car, SN61V (Shields and Filipenko 1997) and V12. P Cyg does not have any evidence for CS dust now, but it has been 400 years since its primary eruption. This raises an important question. Why and how do these giant eruptions end? Does the eruption end and then dust forms or does the dust form first, bringing the eruption to an end? The short plateau in the light curves of these stars may be relevant to this question, especially since it occcurs at or near the pre-outburst magnitude.

4 Final Thoughts

Within the general class of stars we now call LBV's, there are two clearly recognizable groups - the S Dor variables and the η Car-like variables. These two groups are undoubtedly related; some of the η Car-like variables have shown S Dor-like variability, but the physics of their distinctive behavior and what we observe is very different.

In addition, there are now numerous candidate LBV's with some shared properties with these two groups. These stars are often called LBV's in the literature. We also have stars like HD5980 (Barba et al 1995, and see paper by Koenigsberger this volume) and the 'pistol star' (Figer et al 1997). Are they LBV's? HD 5980, like many massive stars is luminous, blue and variable, but I do not think it is an S Dor-type variable.

When it was first introduced the term LBV served a useful purpose by helping us recognize the similarities between groups of stars that were variously known as S Dor variables, P Cygni stars, η Car stars, Hubble-Sandage variables etc. But I think it is increasingly clear that use of the term LBV has become confusing. Remember, if are ever going to understand any group of stars we must be confident that its members share the same physics. Therefore I suggest we return to our former terminology:[1]

- *S Dor variables* with their characteristic light and spectral changes at nearly constant luminosity; and
- η *Car-like variables* with a significant increase in luminosity during the eruption; other characteristics may include a brief plateau, obscuration by circumstellar dust, and a second, lesser eruption.

References

Barba, R. H., Niemela, V. S., Baume, G., Vasquez, R. A. (1995): ApJ 446, L23
Davidson, K., Ebbets, D., Johansson, S., Morse, J. A., Hamann, F. W., et al (1997): AJ 113, 335
Davidson, K., Humphreys, R. M. (1997): ARA&A 35, 1
de Groot, M. (1988): Ir. Astron. J. 18, 163
de Koter, A. (1993): PhD Dissertation, (Utrecht, The Netherlands)
Doggett, J. B., Branch, D. (1985): AJ 90, 2303
Drissen, L., Roy, J-R, Robert, C. (1997): ApJ 474, L35
Figer, D. F., Morris, M., McLean, I. S., et al (1997): BAAS 29
Hubble, E., Sandage, A (1953): ApJ 118, 353
Humphreys, R. M. (1989): *The Physics of Luminous Blue Varables*(Kluwer, Dordrecht), 3
Humphreys, R. M., Davidson, K. (1994):PASP 106, 1025
Humphreys, R. M., Leitherer, C., Stahl, O., Wolf. B., Zickgraf, F.-J. (1988): A&A 203, 306

[1] The term P Cygni should be reserved for the line profile.

Lamers, H. J. G. L. M., de Groot, M. (1992): A&A 257, 153
Shields, J. C., Filipenko, A. V. (1997): BAAS 29, 1265
Smith, N., Gehrz, R. D., (1998a) AJ in press
Smith, N., Gehrz, R. D., Krautter, J. (1998b): AJ in press
Sterken, C., van Genderen, A. M., de Groot, M.(1997): *Luminous Blue Variables: Stars in Transition* (Astron. Soc. Pacific, San Francisco) 35
Szeifert, Th., Humphreys, R. M., Davidson, K., Jones, T. J., Stahl, O., Wolf, B., Zickgraf, F-J. (1996): A&A 314, 131
van Genderen, A. M., Sterken, C., de Groot, M. (1997): A&A 318, 81
Weigelt, G. et al (1995): Rev. Mex. Astron. Astrofis. Ser. Conf. 2, 11
Wolf, B. (1989a): *The Physics of Luminous Blue Varables*(Kluwer, Dordrecht), 91
Wolf, B. (1989b): A&A 217, 87

Discussion

R. Ignace: Paczynski has coined the term "hypernova" for gamma-ray burst afterglows, because their light curves closely resemble the light curves of other classes of novae. In a purely phenomenological sense, do the light curves of LBV outbursts fit nicely into this scheme or are they radically different?
S. Shore: In answer to you question, one could say that DQ Her and η Car show analogous behaviour during dust formation. But remember that if you put enough Fe absorption lines in the way and shroud the cental source with dust, all of these objects will look alike. This is unfortunate because you then don't have a unique explanation for many of the observational effects.

I. Appenzeller: Since η Car has been established to be a binary, could the "giant eruptions" be connected to binarity?
R. Humphreys: Tidal interactions with a close companion star are probably not the explanation for the giant eruption (e.g., in 1840). The kinetic energy associated with the expanding debris is a significant fraction of the total gravitational binding energy of the proposed binary system. Therefore, the large mass ejections in η Car are related to the star's instability.

Blitz Model for the Eruptions of Eta Carinae

Nathan Smith, Joyce A. Guzik, Henny J.G.L.M. Lamers,
Joseph P. Cassinelli, and Roberta M. Humphreys

Boston U., Los Alamos, Utrecht U., U. Wisconsin and U. Minnesota

Abstract. Following the "Great Eruption" of 1843, η Carinae underwent a second major eruption around 1890. We suggest a preliminary working model developed during this meeting (in one night, hence the term "Blitz") that attempts to explain the temporal development of the 19th century eruptions of η Car, as well as the formation of the Homunculus nebula (note that we are not offering an explanation for the *cause* of the Great Eruption!). The essence of the model is that after the Great Eruption ends, the star's extended outer envelope re-adjusts itself on a thermal time scale. This re-adjustment allows envelope material to crash back onto the surface of the star, inducing the second eruption in 1890.

We begin with the pre-eruption rotating star near the Eddington limit. We will approximate the geometry of the star by an extended envelope and a dense "core" (not necessarily the nulear burning core; just some region that does not experience significant structural changes during the eruption). The star's rotating envelope is in quasi-hydrostatic equilibrium modified by centrifugal forces, where the density decreases with increasing distance from the equatorial plane (see Fig. 1a). We assume that *something* deep within the star triggers the Great Eruption and causes an increase in energy output from the core, pushing the star beyond the Eddington limit. Humphreys (these proceedings) has shown that giant eruptions show a marked increase in bolometric luminosity. The increased energy output from the core may induce pulsations in the envelope that will help drive material off the surface of the star. These pulsations will travel through the envelope at the local sound speed $c_s^2 = \gamma P \rho^{-1}$. Because there is a stronger density gradient in the poleward direction than in the equatorial direction, the trajectory of radial pulsations will curve poleward, driving mass loss preferentially from the polar regions of the star (Fig 1b). The ejecta expand to form the bipolar lobes of the Homunculus.

After the Great Eruption ends, energy output from the core returns to "normal". At this point, the star's distorted envelope is drastically out of any sort of equilibrium. Most of the mass in the polar regions of the envelope has been "scooped out" by the Great Eruption, leaving something like a toroidal structure around the star. The remaining envelope mass is probably several times greater than the mass in the polar lobes ($2M_{\odot}$), since most of the material in the pre-eruption envelope was concentrated toward the equatorial plane. If we assume that this distorted envelope has 10 - $15M_{\odot}$, it will re-adjust itself on a Kelvin-Helmholtz timescale of about 30 - 45 yrs (for $5 \times 10^6 L_\odot, 100 M_\odot, 200 R_\odot$). As this distorted envelope re-adjusts itself, material from the extended regions in the equatorial plane crashes back onto the core (Fig 1c). The potential energy of this infalling material is transferred into kinetic energy for a small portion of the mass ($0.5 M_\odot$), which is ejected from the star at high velocities in a process analogous to the core bounce in a supernova explosion. The residual angular momentum in the outer regions of the rotating envelope will cause the collapsing envelope to rotate rapidly, which may confine the mass ejection primarily to the equatorial plane and may leave a disk around the post-eruption star (Fig 1d).

This model differs from previous models for the formation of the Homunculus in that the bipolar lobes are *not* wind blown bubbles. The axisymmetric geometry of the Homunculus is the direct result of an explosion on the surface of a rotating star. The clumps and knots in the nebula are linearly expanding shrapnel from the explosion, rather than structures that form from gas dynamic instabilities. Some of the presently observed equatorial ejecta would have originated during the 1890 erutpion. This model is appealing because it requires only a rotating luminous star to reproduce the bipolarity and equatorial spray in the nebula; it does not rely on magnetic fields or a binary scenario. It also gives a possible explanation for the second eruption seen in 1890, which may be a common feature seen in the lightcurves of other giant eruptions (Humphreys, these proceedings).

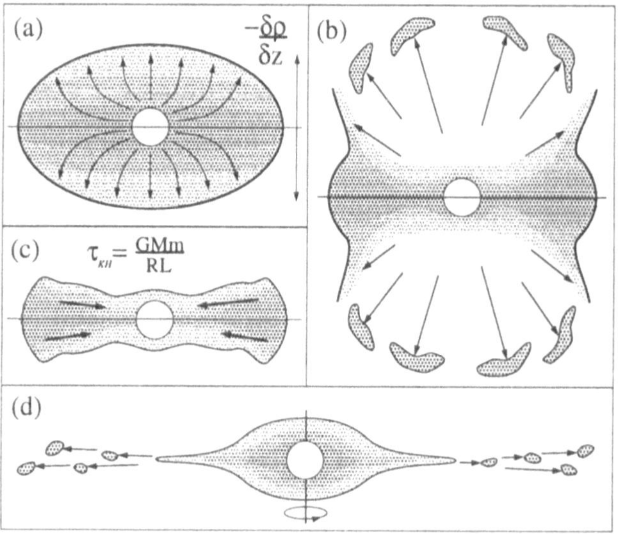

Fig. 1. (a) Pre-eruption rotating star with poleward density gradient. (b) The Great Eruption ejects shrapnel from the poles of the star. (c) The post-eruption stellar envelope re-adjusts itself on a K-H timescale of 30 – 45 years. (d) During the 1890 eruption, the rapidly rotating envelope ejects mass primarily in the equatorial plane.

Short-Term Variations of LBV's

Otmar Stahl

Landessternwarte Königstuhl, D-69117 Heidelberg, Germany

1 Introduction

LBV's are variable on many different time scales from weeks to years or even decades. The typical LBV variations on timescales of years or longer are covered in the paper by Humphreys (this volume). Here I discuss variations of LBV's on timescales from weeks to years.

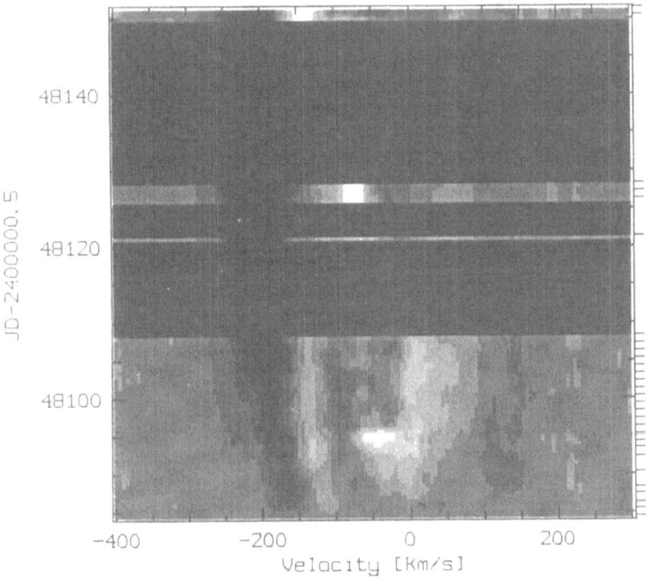

Fig. 1. Dynamical spectrum of P Cygni in the line FeIIIλ5156 obtained at the Landessternwarte Heidelberg in 1990 with FLASH. The mean spectrum has been subtracted to enhance the visibility of variable features. Note the outwards accelerating absorption feature.

Most classes of stars are little studied on these timescales. This also holds true for LBV's. Therefore the hot star group of the Landessternwarte Heidelberg started a monitoring program of a few bright LBV's (and other objects) using the 70cm-telescopes at the Landessternwarte and later the ESO 50cm-telescope at La Silla. For more details about the project see the paper by Kaufer (this volume).

2 P Cygni

The star P Cygni is the brightest LBV. It has been relatively stable in the last centuries and is classified as an LBV only because of its historic outburst in the 17th century. At present it shows a spectrum of a B1.5 supergiant. Most lines show P Cygni profiles.

Our observations show that P Cygni shows variations similar to other early-B supergiants (see paper by Wolf, this volume). The spectroscopic variability seems to be dominated by expanding and accelerating shells.

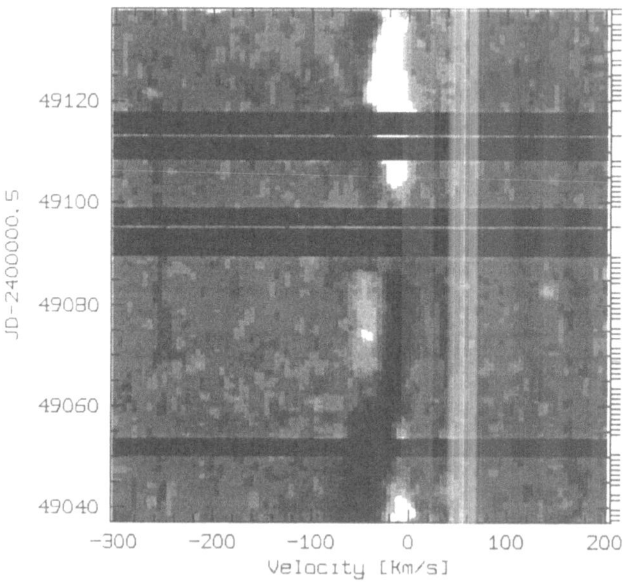

Fig. 2. Dynamical spectrum of HD 160529 around the line FeIIλ6432 obtained with the ESO 50cm-telescope at La Silla in 1993 with FLASH. Note the absortion feature moving towards longer wavelengths.

3 HD 160529

HD 160529 was discovered as an LBV by Sterken et al. (1991). It is a relatively cool LBV (late-B to early-A) with a small photometric amplitude ($\approx$ 0.5 mag) and a comaparatively low luminosity.

The spectroscopic variations of HD 160529 are markedly different from those of P Cygni: Accelerating shells have never been observed in this object. In contrast, absorption features moving from red to blue in the line profiles are often observed. This can be regarded as evidence for rotational modulation, which is common in late-B and early-A supergiants (see paper by Kaufer, this volume).

4 AG Car

AG Car is the brightest of the active LBV's and also known for its nebula. It shows brightness variations from magnitude 8 to 6 on timescales of years. During the outburst, the spectral type varies between Ofpe/WN9 and early-A. Its last outburst started around 1990.

We covered most of this outburst spectroscopically with observations typically spanning four months in each season. The behaviour over the full outburst cycle indicates an asymmetry between the ascending and the descending branch. The emission lines are much stronger after the maximum, possible indicating a higher luminosity.

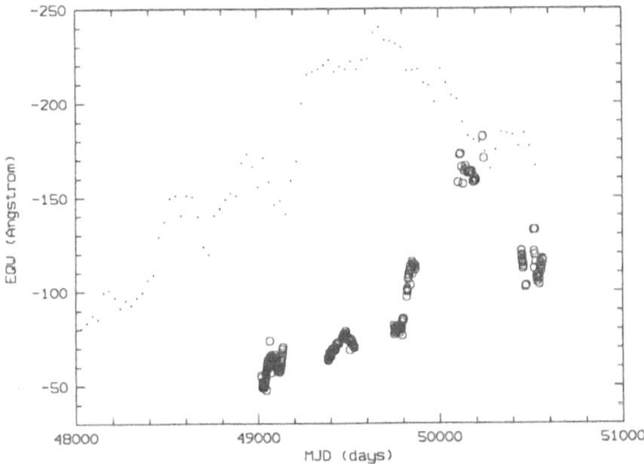

Fig. 3. The development of the equivalent width of Hα of AG Car during the recent outburst (large symbols). The data have been obtained with FLASH, HEROS at the ESO 50cm-telescope. The visual lightcurve is overplotted in small symbols. Note that the maximum of the equivalent widths occurs much later than the visual maximum.

5 Conclusions

The short-term spectroscopic variations of LBV's in relative quiescence are qualitatively similar to variations shown by normal supergiants of similar spectral type. In particular, in LBV's of later spectral type and lower luminosity rotational modulation is more pronounced than in LBV's of higher temperature and luminosity. The timescales in LBV's are probably longer tahn in normal supergiants. This could be due to their higher luminosity.

The short-term variations of LBV's during outburst are very complex and indicate that the details of LBV outbursts are still not understood. The behaviour of AG Car during rise to visual maximum is markedly different from the decline to minimum. This may indicate that a simple change of radius and temperature does not explain LBV outbursts.

Discussion

M. Magalhães: Polarisation data of P Cygni show night-to-night variation of about 0.1 %. You have described longer-term variations, of weeks to months, in the photometric light curve of P Cygni. Could you comment on night-to-night variations in the photometric data of P Cygni?
O. Stahl: I am not aware of any significant night-to-night variations in photometric data. Polarisation seems to be the only quantity which shows these variations in P Cygni.

H. Lamers: You showed that AG Car displays its maximum Hα emission after the photometric maximum. Maybe the star was too cold during the outburst to keep the wind ionised and only after the outburst was the star hot enough to ionise the wind again?
O. Stahl: It is indeed possible that the star is too cool in outburst to keep the wind ionised. This does not, however, explain the asymmetry in the Hα emission with respect to the optical outburst.

Imaging Polarimetry of Eta Carinae with the Hubble Space Telescope

Regina E. Schulte-Ladbeck[1], Anna Pasquali[2], Mark Clampin[3], Antonella Nota[3], John Hillier[1], and O.L. Lupie[3]

[1] University of Pittsburgh, Pittsburgh, PA 15260
[2] ST-ECF, ESO, D-85748 Garching bei München, Germany
[3] Space Telescope Science Institute, Baltimore, MD 21218

Abstract. We have taken advantage of the high spatial resolution attainable with the HST to map the linear polarization in the V band across the nebulosity surrounding Eta Car. There are several new results related to polarization variations on different size scales. First, we present a two-dimensional map of the amount and position angle of the polarization across the Homunculus. Second, we provide measurements of the polarization within prominent features such as the "jet", the "paddle", the "skirt", and the "spot" in the south-eastern lobe. Third, we comment on polarization variations associated with the small-scale structure that can be seen in HST images (and which gives the lobes their cauliflower-like appearance). The new data provide insight into the three-dimensional distribution of dust about Eta Car.

1 The HST observations

The three-dimensional structure of the Homunculus has been a matter of much debate over the last few years. Polarization observations have contributed significantly to our understanding of the nature of the Homunculus. Using the HST's WFPC2, we have recently obtained a high-resolution map of the linear polarization in the V-band (F555W filter) of the Homunculus and other resolved features in the vicinity of Eta Car. The data reduction has been extremely difficult, and is discussed at length in our forthcoming AJ paper, where we also review and give reference to previous polarimetric studies of Eta Car.

2 Results

In view of the limited space for this contribution, we will immediately present our main results and conclusions:

- We confirm that the Homunculus is largely a reflection nebula in the optical (Thackeray 1956, Warren-Smith et al. 1979). The images taken in polarized light are smoother than the flux image (see Fig. 1). This is

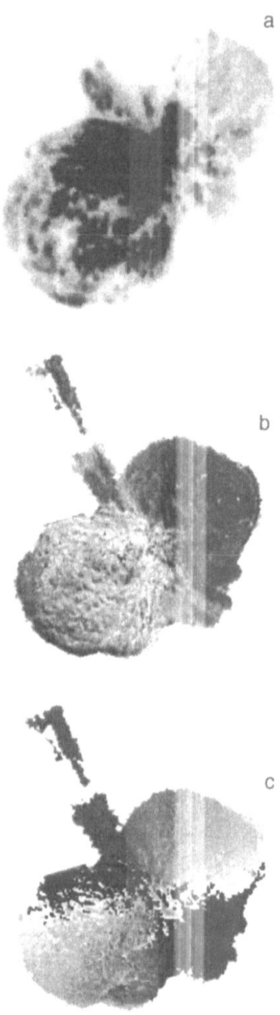

Fig. 1. The V-band image (a), percentage polarization image (b) and position angle image (c) of the Homunculus. Notice how much smoother the lobes are in the polarization images as compared to the flux image, whereas the "jet" and parts of the "skirt" appear as more pronounced features in the polarization image. Orientation of the images: N is up and E is to the left.

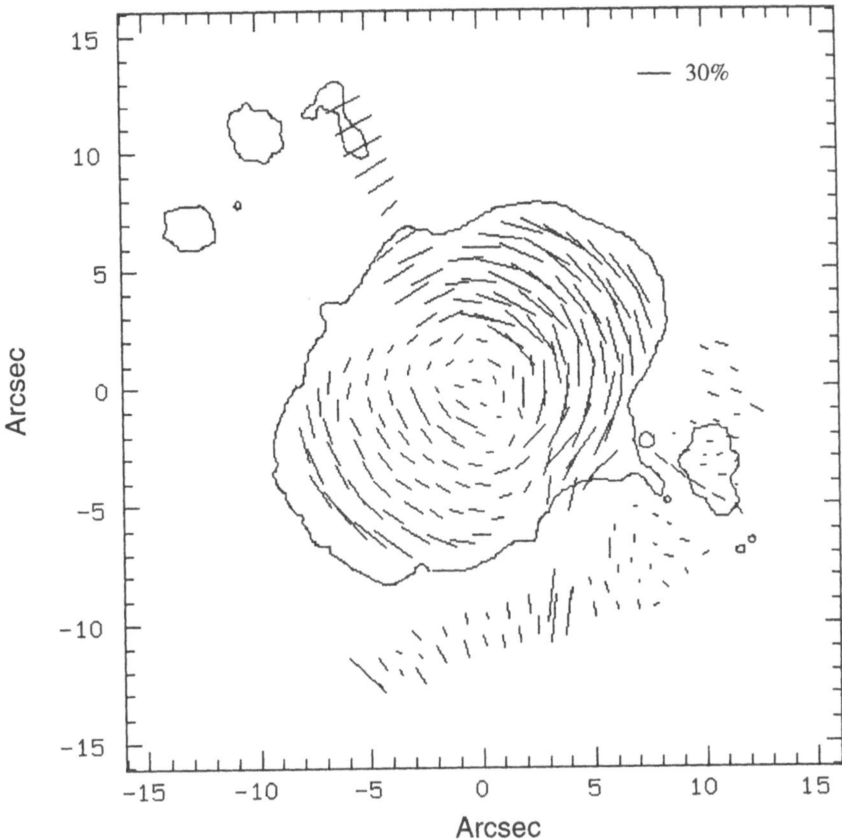

Fig. 2. A map of the polarization vectors. The overall radial distribution of polarization vectors is consistent with scattering of light from a central source by dust in the lobes. The "jet", showing vectors perpendicular to the radius vector from the center as well, clearly is a scattering feature, too. The "S ridge", on the other hand, displays predominantly radial polarization vectors, and is thus not interpreted as an area that is scattering the light from the central source.

a surprise for models which account for the small-scale structure seen in flux images with optically thick patches/thin holes. These models imply that we see the front of the lobes in some places, but the backs of the lobes in other places. Since light scattering off the front of a lobe scatters through different angles than light scattering off the back of a lobe, such "holes" or "spots" might be expected to show a different polarization from their surroundings, which does not appear to be the case.

- The "jet" continuum is largely scattered light whereas the "S-ridge" does not show a scattered continuum (see Fig. 2). The jet is highly polarized. It is seen at a favorable viewing angle that coincides with a very advantageous phase angle of the scattering particles.
- The polarization in the "paddle" and in the streamers of the "skirt" appears to be different from its surroundings (Fig. 1). However, in quantitative measurements through circular apertures the polarization in the paddle does not stand out significantly over its sourroundings in the NW lobe. A radial cut along the minor axis does reveal elevated polarization in one of the three streamers is significant. This suggests that either the line-of-sight geometry of parts of the skirt or the geometry of the particles it contains are favorable for a high polarization.
- Qantitative modeling of the polarization map with a Monte Carlo code indicates the double-flask model is preferred over the double-bubble or bipolar-caps model for the 3-D structure of the Homunculus (see Currie et al. 1996 for the geometry of these competing models).
- Whereas the polarization map and the emission-line profiles observed across the Homunculus can be explained with the basic double-flask model, the shape of the polarized line profiles in the NW lobe remains a mystery.

Acknowledgement: The results presented here are based on observations made with the NASA/ESA Hubble Space Telescope, obtained at the Space Telescope Science Institute, which is operated by the Association of Universities for Research in Astronomy, Inc., under NASA contract NAS 5-26555. We acknowledge financial support through GO program 5840.

References

Currie, D.G., Dowling, D.M., Shaya, E.J., Hester, J.J., Scowen, P., Groth, E.J., Lynds, R. O'Neil, Jr., E.J., The WF/PC IDT, The WFPC2 IDT 1996: Proc. "The Role of Dust in the Formation of Stars", ESO workshop edited by H.U. Käufl & R. Siebenmorgen, p.89

Thackeray, A.D. 1956: Observatory **76**, 164

Warren-Smith, R.F., Scarrott, S.M., Murdin, P., Bingham R.G. 1979: MNRAS **187**, 761

Discussion

R. Humphreys: How do your observations compare with ground-based speckle-polarimetry of Falcke et al. (1996)?

R. Schulte-Ladbeck: Unfortunately, Falcke et al. observed in Hα. The line emission comes from many different places due to direct emission and scattering; I commented on that in the Kona proceedings. One goal of our HST observations was to check on the presence of a "polarisation disk"; however, the exposure time splits/core saturation of the data has prevented us from doing so in the V band.

F. Vakili: With regard to the "Weigelt knots" from the speckle imaging of the central region of η Car, have you zoomed in your polarised map to see if any correlation exists between these knots and local polarisation?
R. Schulte-Ladbeck: Unfortunately, the splits of our exposures (owing to the dynamical range of the WF2 chip) are not suitable to address this question in the V band. Maybe the observations in other filters will have high S/N without overexposing the core.

N. Langer: You did not include the "skirt" in your polarisation model. Could taking it into account change your conclusions with respect to the Homunculus geometry?
R. Schulte-Ladbeck: Maybe I was too brief in explaining the conclusions. Indeed, the "skirt" is not yet included, but you can see what the implications are from the cartoon model. It appears that the polarisation in several areas belonging to the "skirt" are slightly elevated with respect to surrounding areas where the polarisation originates from the front of the NW lobes. Now, in the double-flask model, we "need" to have the front of the NW lobe close to the plane of the sky to get the high and constant polarisation. Alternatively, if we assume that the "skirt" is perpendicular to the long axis of symmetry and thus filled towards the observer, a grain population with properties different from those in the NW lobe (e.g., more forward throwing polarisation) could possibly explain the higher polarisation.

Regina Schulte-Ladbeck and Karel van der Hucht

Non-spherical Outflows in Massive Binary Systems: Circumbinary Disks?

Gloria Koenigsberger, Edmundo Moreno, and Jorge Cantó

Instituto de Astronomía, UNAM

Abstract. The trajectories of wind particles emitted from the surface of a massive star in a close binary system are analyzed. Within a radius 1000 times the separation of the two stars, a significant fraction of the particle trajectories are found to cross the orbital plane, leading to the formation of a large-scale, outflowing circumbinary disk-like structure. The shocks which arise due to the collision of particles which are within crossing streamlines are expected to produce selected regions within the wind, particularly in the orbital plane, in which a higher degree of ionization prevails than in the wind in general. X-ray emission might also be expected from these regions. Such a model is suggested to be applicable to the erupting WR/LBV binary system HD 5980 and other binary LBV systems, during phases when wind velocities are slow.

1 Introduction

The interaction between stars in close binary systems propitiates the presence of non-spherically symmetric mass outflows. Wind-wind collisions and stellar rotation are mechanisms capable of producing departures from spherically symmetric mass loss, as well as the mechanisms assumed to be active in single stars particularly if magnetic fields are present. Many of these mechanisms are described in other papers presented in this Colloquium and, in Drissen et al. (1992). In this paper we turn our attention to the effects that the gravitational and centrifugal forces produce on the emerging wind of one of the stars in the system, and propose that under certain conditions in massive binary systems, a significant amount of the ejected material accumulates in the orbital plane, producing a circumbinary disk-like structure. The conditions most suited to produce this outflow morphology occur in the LBV phase of the star's evolution, when the wind velocity is slow.

Concentrations to the orbital plane of outflowing mass in binary systems has been predicted for low-mass systems in which a common envelope phase occurs (see review by Iben and Livio 1993). Two- and three-dimensional calculations performed by Bodenheimer and Tamm (1984), Livio and Soker (1988), Taam and Bodenheimer (1989, 1991) and Terman et al. (1994) demonstrate that the ejection of envelope material takes place preferentially in the orbital plane in these systems. Although similar to a stellar wind in that it is not corrotating with the binary system, a common envelope, however, differs significantly in its radial velocity and density structure from the stellar winds in massive stars. Hence, the results for the mass-outflow distribution in common envelope phases are not applicable to the massive star case. Thus, calculations need to be performed for conditions prevailing in massive stars.

Mechanisms for producing a concentration of gas in the orbital plane are of interest, because, among other reasons, such a density contrast is believed to be responsible for the bipolar morphologies observed in H II regions surrounding LBV systems such as Eta Carinae, AG Car and other LBV's (Frank 1997; Garcia-Segura et al.1997). Eta Car is now strongly suspected to be a binary system (Damineli et al. 1997; Lamers et al. 1998), and its surrounding H II region was formed at the time of the 1850 Great Eruption. Furthermore, the presence of a wind confined to a disk is suggested for other massive systems, in particular WR 140 (White and Becker, 1995).

2 Formation of a Circumbinary Disk-Like structure in Massive Binaries

In order to analyze the morphology of the outflow from a massive binary system, the trajectories of individual particles within the system were calculated by solving the equation of motion in the frame of reference which is rotating with the binary. We performed a numerical experiment for the case of a $85M_{\odot} + 30M_{\odot}$ (= star A + star B) system, with a circular orbit and separation of $70R_{\odot}$. The assumed stellar radii are $48R_{\odot}$ and $15R_{\odot}$, respectively, and the system is assumed to be in corotation. The wind of star A is assumed to have a velocity of 1000 km s^{-1}. The wind of star B is ignored for the present calculation. Particles were released radially from the surface of star A at spatial intervals of 1 degree in both the θ and ϕ coordinates. The polar angle θ is measured from the north pole of star A and the azimuthal angle ϕ is measured from the line joining the two stars in the direction of the orbital motion. Particles which reached a sphere with radius 1000 times the orbital separation whose origin is the center of mass, were assumed to escape the system and their motion was not followed beyond this point. Particles which collide with Star B are assumed to be trapped and disappear from the calculation.

In Figure 1 we illustrate the velocity field of the particles which cross the orbital plane (x-y plane), measured with respect to the inertial reference frame whose origin is the center of mass of the system. This is the velocity field in the observer's frame of reference. The length of the arrows indicates the magnitude of the velocity component in the x-y plane. These velocities range from close to zero (i.e., the motion is mainly perpendicular to the x-y plane) to more than 500 km s^{-1}. Star A is located at $(-\mu, 0)$ while Star B is at $(1-\mu, 0)$, where $\mu = M_B/(M_A+M_B)$. The values of x and y are in units of the orbital separation. Thus, one finds that a large number of particles intersect the orbital plane at distances of 280 $R_{\odot}$ or more.

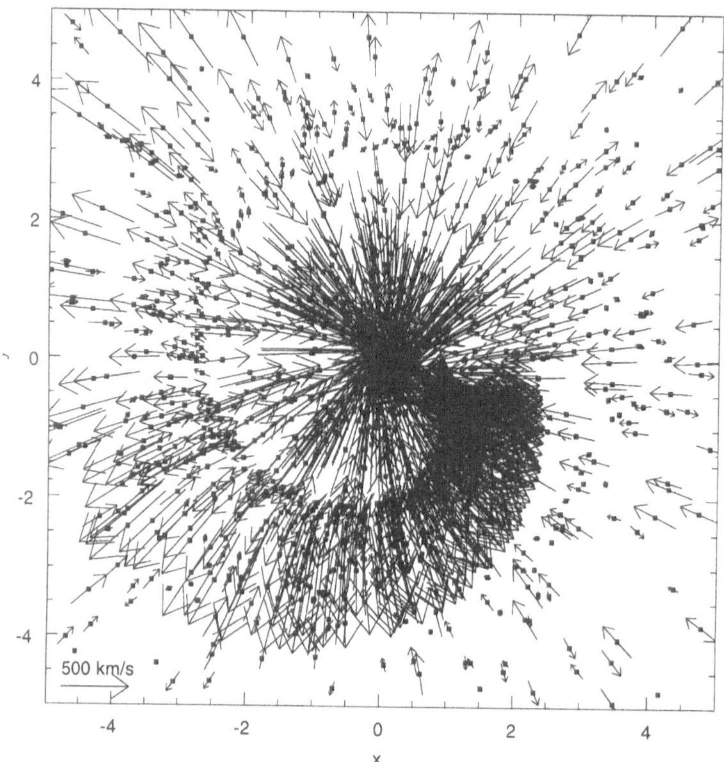

Fig. 1. Distribution of particles which cross the orbital plane. The length of the arrows corresponds to the velocity component in the x-y plane. Orbital motion is counterclockwise. Star A lies nearly at the center of the figure. The x and y axes are in units of the orbital separation.

The particles which are deflected towards the orbital plane are found to emerge from specific regions of the stellar surface. For the parameters listed above and the stated assumptions, the particles which tend to have trajectories which take them towards the orbital plane are those emerging from polar angles $\theta > 53$ degrees. For angles smaller than this, the particles escape the system. Furthermore, the majority of the particles whose trajectories take them across the orbital plane emerged from azimuthal angles between -20 and $+70$ degrees; that is, from the portion of the star facing the companion. For a given set of binary star parameters, the slower the initial particle velocity, the larger is the number of particles which are deflected towards the orbital plane.

It is important to note that the results presented in Figure 1 correspond to particles released only in one of the hemispheres of star A. Since the forces acting in the system are the same in the northern and southern hemispheres and in both cases particles converge towards the orbital plane, a region of shock formation is expected to arise. The strength of the shocks will be proportional to the velocity components perpendicular to the x-y plane. In addition, shocks are expected to be formed above and below the orbital plane, as particle trajectories cross. Depending on the velocities with which these shocks occur, emission lines from a variety of ionization species are expected, most of which with higher degrees of ionization than the general wind. One would also expect soft X-ray emission produced in this process. Before quantifying these effects, however, the collision of particles has to be incorporated into the calculation. Three additional effects need to be considered as well: radiation pressure; the eccentricity of the orbit; and the possibility that the stellar surface from which the wind is assumed to be emerging is not spherically symmetric to begin with (Fliegner and Langer 1995; Cassinelli, Ignace and Bjorkman 1995).

3 A Circumbinary Disk in HD 5980?

The close binary system ($P = 19.3$ days) HD 5980 in the Small Magellanic Cloud, has an eccentric ($e = 0.3$) orbit and a significant orbital inclination ($i > 0.87$). Classified as a WN4 + O7I: Wolf-Rayet system in the late 1970's, its wind velocities were near 3000 km s^{-1}. Over the next few years, the wind velocity systematically decreased and the WN spectral type became cooler. In 1994, it underwent an LBV-type eruption with a more than 2 magnitude visual brightness increase, a significant decrease in wind velocity, and a spectral type of WN11 or B1.5Ia$^+$ Within a few months after light curve maximum, the fast wind resumed at the same time that the spectrum reverted to a hotter WN type (WN6 or 7). A summary of the characteristics of the system and additional references can be found in Barbá et al. (1997), Koenigsberger et al. (1998a; 1998b), Moffat et al. (1998), and Niemela et al. (1997).

The following is a working scenario for HD 5980: There are (at least) two very massive stars (perhaps $85M_\odot + 30M_\odot$) in a close binary system, one of which (star A) systematically increased its radius and mass-loss rate until its dimensions approached its Roche Lobe. The gradual increase in its radius appears to have occurred over the first few years between 1978 and 1991. Starting in 1991, the radius increase appears to be intermittent, with a first important outburst (implying a large radius increase) ocurring in 1993 (see Breysacher (1997) for the long term light curve behavior). It has not yet been determined whether the mechanism producing the outburst is related to evolutionary changes, as in more typical LBV's, or, whether, as proposed by Moreno et al. (1997), the forced oscillations induced by the eccentric orbit have led to the instability. However, the fact remains that star A's radius and mass loss rate grew significantly, and the wind velocity diminished throughout this process. At the time of the eruption, the mass loss rate of star A is $\sim 5 \times 10^{-4}\ M_{\odot}\ \mathrm{yr}^{-1}$, and its spectrum presents lines arising from a wide range of ionization potentials, with a large variety of wind velocities deduced from the P Cygni profiles in the UV lines. Their velocity components go from near 200 km s^{-1} for the low ionization lines up to 1700 km s^{-1} for Si IV and C IV.

In Figure 2 we illustrate the Si IV 1393 and 1402 Å line profiles in IUE spectra obtained at two different epochs: near maximum in the eruption (SWP52888; obtained in 1994 November) and 250 days later (SWP 55394). Both spectra are at the same orbital phase (*phase* = 0.50), corresponding to an orbital position very near apoastron. The Si IV line profiles in SWP 52888 have the following peculiar features: 1) Despite being resonance lines, they do not saturate; i.e., the minimum intensity of the P Cygni absorption component does not go down to zero flux; 2) there are several sharp emission features superimposed upon the absorption component, the most prominent of which are at −900, −1200, and −1480 km s^{-1} (Koenigsberger et al. 1995); 3) the absorption component of the Si IV 1393 line extends out to −1700 km s^{-1} (the same as C IV 1550), while most of the other lines display maximum velocities slower than 800 km s^{-1}. Although orbital phase-dependent line profile changes are generally present, there are many features in the spectrum at the time of the 1994 outburst which remain stable with orbital phase. This implies that the line forming region extends way beyond the orbital separation.

The fact that the Si IV absorption components do not reach zero flux intensity implies that either the region containing these ions does not cover entirely the stellar continuum emission and/or these absorptions are "filled in" by the superimposed narrow emission components. This suggests that Si IV ions could be concentrated to the equatorial plane of the binary, with the morphology of a disk which is thinner than the radius of the erupting star. It is interesting to note that the signature of this possible disk is still present 250 days later, as can be seen in the narrow absorption components at −850 km s^{-1} in SWP 55394.

4 Conclusions

The analysis of particle trajectories in a massive binary system in which one of the two stars has a slow wind indicates that a significant fraction of the wind particles are deflected towards the orbital plane. This suggests that in the LBV phase, a circumbinary disk-like structure can be formed as a consequence of the diminution in the stellar wind velocity, the gravitational force

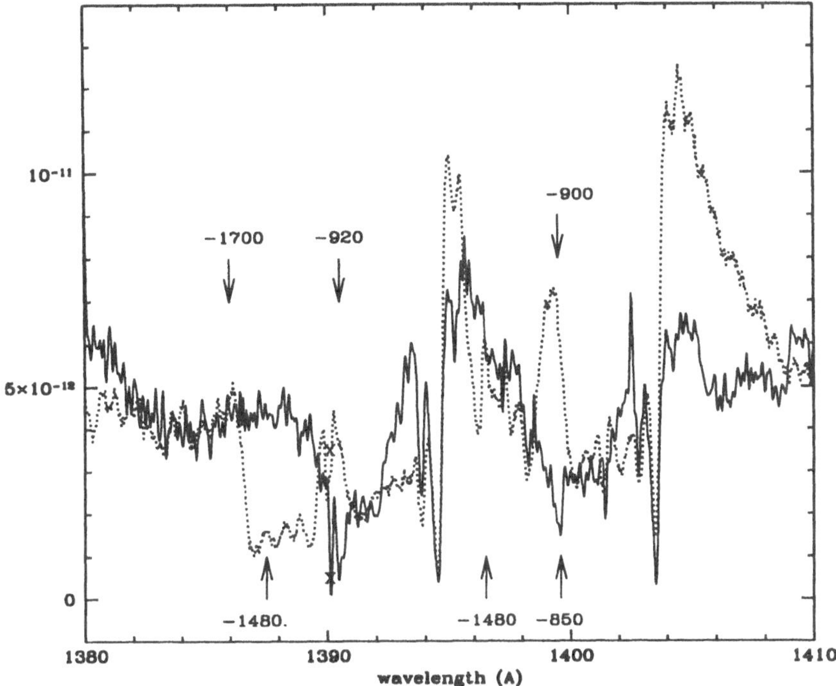

Fig. 2. IUE line profiles of the Si IV 1393, 1402 Å doublet in HD 5980: SWP 52888 (dotted tracing) and SWP 55394. The ordinate is un-dereddened flux in units of ergs cm^{-2} s^{-1} $Å^{-1}$. Velocities of various features, in km s^{-1} and corrected for the SMC motion of +150 km s^{-1}, are indicated."X" indicates a Reseau fiducial point.

of the companion, and the orbital motion. A more detailed analysis, including the effects of radiation pressure and the collision of particles, is required in order to describe more precisely the characteristics of such a circumbinary structure. However, it is important to point out that the preliminary results of this model suggest that the ionization and velocity structures in the wind at large distances from the star in a massive binary system are not smooth, and that multiple shock regions are expected. Some of the UV spectral characteristics observed in the LBV/WR system HD 5980 are consistent with the idea of a circumbinary structure, although the effects of the companion's wind and radiation field need to be incorporated. Finally, it should be noted that the morphology of the wind-wind collision region between the two stars may change significantly from that predicted by stationary-flow models, a circumstance which should be analyzed in the future.

Acknowledgements. We thank Jon Bjorkman for discussing with us the expected effects on the UV line profiles due to the presence of a disk in HD 5980. This work was supported in part by UNAM/DGAPA grants. Juana Orta is gratefully acknowledged for her help in preparing the text.

References

Barbá, R., Niemela, V., Morel, N. (1997): in Luminous Blue Variables: Massive Stars in Transition, ASP Conf. Ser. **120**, 238.

Bodenheimer, P., Tamm, R.E. (1984): ApJ, **280**, 771

Breysacher, J. (1997): in Luminous Blue Variables: Massive Stars in Transition, ASP Conf. Ser. 120, 227.

Cassinelli, J., Ignace, R., Bjorkman, J. (1995) in IAU Symposium 163, WR Stars: Binaries, Colliding Winds, Evolution, eds. K.A. van der Hucht and P.M. Williams, (Dordrecht:Kluwer), p.191.

Damineli, A., Conti, P., Lopes, D.F., 1997, NewA **2**, 107.

Drissen, L., Leitherer, C., Nota, A. (1992): Nonisotropic and Variable Outflows from Stars, ASP Conf. Ser. 22.

Fliegner, J., Langer, N. (1995), in IAU Symposium 163, WR Stars: Binaries, Colliding Winds, Evolution, eds. K.A. van der Hucht and P.M. Williams, (Dordrecht:Kluwer), p.326.

Frank, A. (1997): in Luminous blue Variables: massive stars in Transition, ASP Conf. Ser. 120, 338.

Garcia-Segura, G., Langer, N., MacLow, M.M. (1997): in Luminous blue Variables: massive stars in Transition, ASP Conf. Ser. 120, 332.

Iben, I., Livio, M. (1993): ASP, **105**, 1373.

Koenigsberger, G., Guinan, E., Auer, L. H., Georgiev, G. , (1995): ApJ, **452**, L.107.

Koenigsberger, G., Auer, L. H., Georgiev, L., Guinan, E. (1998a):ApJ, **496**, 934.

Koenigsberger, G., Peña, M., Schmutz, W., Ayala, S. (1998b): ApJ, **499**, 889.

Lamers, H.J.G.L.M., Livio, M., Panagia, N., Walborn, N., (1998), ApJ in press.

Livio, M., Soker, N. (1988): ApJ, **329**, 764.

Moffat, A.F.J., Marchenko, S., Bartzakos, P., Niemela, V., Cerruti, M.A., Magalhaes, A.M., Balona, L., St.-Louis, N., Seggewiss, W., Lamontagne, R. (1998) ApJ in press.

Moreno, E., Georgiev, L., Koenigsberger, G. (1997): in Luminous Blue Variables: Massive Stars in Transition, ASP Conf. Ser. **120**, 152.

Niemela, Barbá, R., Morrell, N., Corti, M., (1997): in Luminous Blue Variables: Massive Stars in Transition, ASP Conf. Ser. 120,

Taam, R.E., Bodenheimer, P. (1989): ApJ, **337**, 849

Taam, R.E., Bodenheimer, P. (1991): ApJ, **373**, 246.

Terman, J.L., Taam, R.E., Hernquist, L. (1994), ApJ, **422**, 729.

White, R.L., Becker, R.H. (1995), ApJ, **451**, 352.

Discussion

J. Echevarría: Do you see any spectroscopic evidence for a disk during or after the LBV eruption event of 1994?
G. Koenigsberger: The Si IV λ1393 line profile is suggestive of an outflowing disk, but modelling is required before a conclusion can be drawn.

H. Henrichs: Have you also analysed the Si III λ1206 line behaviour? Your spectra looked promising.
G. Koenigsberger: I have not, but will do so as soon as possible.

K. van der Hucht: Do you have a series of consecutive IUE observations, let's say 10 or 20 within one 13-day binary period, to separate the variations within one orbital period from long-term variations?
G. Koenigsberger: No, unfortunately at most only a few observing shifts within one orbital period were possible.

M. Friedjung: The mass-loss rate during the eruption is very large and I would expect the wind to be optically thick in the continuum. I presume this mass loss must only be in the orbital plane to avoid occultations.

R. Schulte-Ladbeck: If you are interested in the presence of a circumbinary disk, you might be able to detect it with broad-band polarisation monitoring: analyse the Fourier components in the time series (cf. Brown et al.).

Anatoly Miroshnichenko and Ruslan Yudin

Long-Term Behaviour of the Variable Wind of P Cygni

Indrek Kolka

Tartu Observatory, 61602 Tõravere, Estonia

Introduction. P Cygni has been for a long time an often–used target of observations, and a test object to prove several modelling approaches (cf. e.g. Najarro et al. 1997, and references therein). However, the origin of it's stable variability pattern with moderate amplitudes is poorly known yet. In this contribution I will concentrate on the spectroscopic variability. A more elaborated version of this paper considering many kinds of variability data on P Cygni will be published elsewhere (Kolka 1998).

The Main Spectroscopic Variability Cycle. The striking phenomenon in the spectra of P Cygni is the cyclic drift of subsequent individual components through the absorption part of the P Cyg–type profile. This is demonstrated in Fig.1a in the case of three Balmer lines in 1982 (the data adopted from Markova, Kolka 1989). Every new drift–cycle starts when the component with smaller Doppler–shift becomes deeper than it's neighbour.

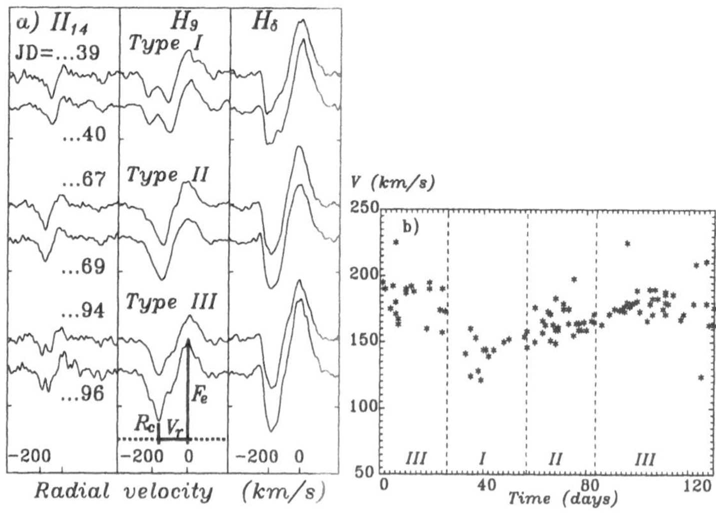

Fig. 1. The drift of components through selected profiles in 1982 (a), and the composite velocity curve of the absorption core in H_9 (b)

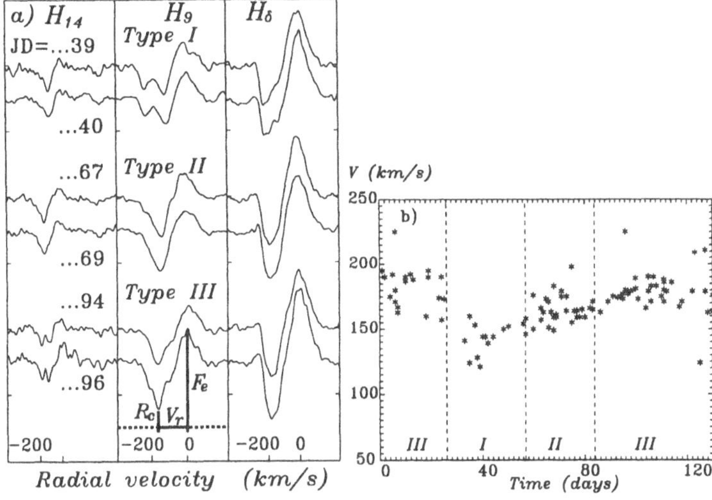

Fig. 1. The drift of components through selected profiles in 1982 (a), and the composite velocity curve of the absorption core in H_9 (b)

The discussed behaviour is indicative to higher Balmer lines which are sensitive to the line-splitting. Other lines in the optical spectral region are reported to have variable Doppler velocities, too (Markova, Kolka 1989 and Stahl et al. 1995). These lines usually do not exhibit *clearly* any separable features in the profile but show rather smooth transition between deeper and shallower periods of their shape. To relate the variability in lines of different origin the curves of velocity variations in H_9 were completed with similar data on a sample of additional lines which cover different intervals on the velocity scale (Fig.2). The cyclical drift is shown for the period in 1990 when an overlap in photographic (Kolka 1994) and CCD data (Stahl et al. 1995) gave the possibility to use for weak lines more reliable CCD-profiles with high S/N – ratio. In Fig.2 one can follow the drift of the velocity perturbation through the profiles from the low to high velocity values. The lower panel in Fig.2 supports the interpretation of the velocity drift: the equivalent width of the purely absorption line (OII 4649) in three specified Doppler-velocity intervals reacts accordingly when the opacity enhancement enters or drifts out of the selected region.

Discussion. We have demonstrated that the main spectroscopic cycle operating in P Cygni has the characteristic length around 100^d. Similar results were obtained by N. Markova (1998). However, other timescales are not ruled out. The line intensity parametres (absorption depths, emission peaks) show

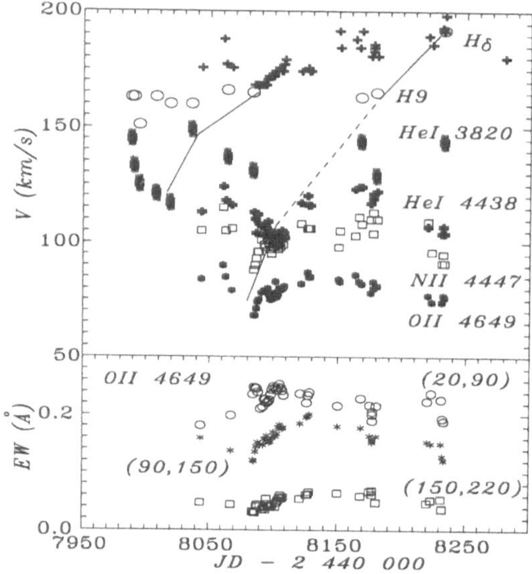

Fig. 2. The drift of the perturbation through different lines, and through different velocity intervals in OII 4649

oscillations both on much longer scales up to 600^d (Kolka 1994, Markova 1998), and on shorter scales down to $15^d \ldots 20^d$ (Stahl et al. 1995).

The natural explanation to moving opacity enhancements (to the main spectroscopic cycle) in the line profiles is an expanding enhanced density shell. But the variability on short timescales which is even better exhibited in polarimetric and photometric data (Taylor et al. 1991, Percy et al. 1996) must be interpreted in the limits of the same scenario. The localized inhomogeneities in moving shells are the obvious possibility but their influence on profiles (localized in the velocity!) is not observed.

Another explanation to the 100^d–cycle could be the corotating spiral density–wave – a new popular approach to interpret the cyclical variability. The shape of the wave which is always far from spherical symmetry provides, perhaps, better possibilities to describe rapid variations, too. We notice that the contemporary "best" stellar parametres of P Cygni (see Najarro et al. 1997) give to the photospheric rotational period a value excitingly near 100^d!

References

Kolka, I. (1994): IAU Symp. 162: Pulsation, Rotation and Mass Loss in Early–Type Stars, eds. L. A. Balona, H. F. Henrichs, J. M. Le Contel, Kluwer, Dordrecht, 536

Kolka, I. (1998): Baltic Astronomy (submitted)
Markova, N. (1998): poster at this Colloquium
Markova, N., Kolka, I. (1989): Tartu Teated No. 103, 3–15
Najarro, F., Hillier, D. J., Stahl, O. (1997): A&A,**326**, 1117
Percy, J. R., Attard, A., Szczesny, M. (1996): A&AS,**117**, 255
Stahl, O., Kaufer, A., et al. (1995): The Journ. of Astron. Data, **1**, (CD–ROM)
Taylor, M., Nordsieck, K. H., et al. (1991): AJ, **102**, 1197

Discussion

H. Lamers: In your figure showing the velocity as a function of time, I see as many cases where the velocity decreases with time as those where it increases. Can you explain this?

I. Kolka: This is the natural pattern of the variability cycle. The velocity of the absorption core decreases when the perturbation enters the representative velocity interval, increases thereafter to the highest value and decreases again when the perturbation drifts out of the specified interval.

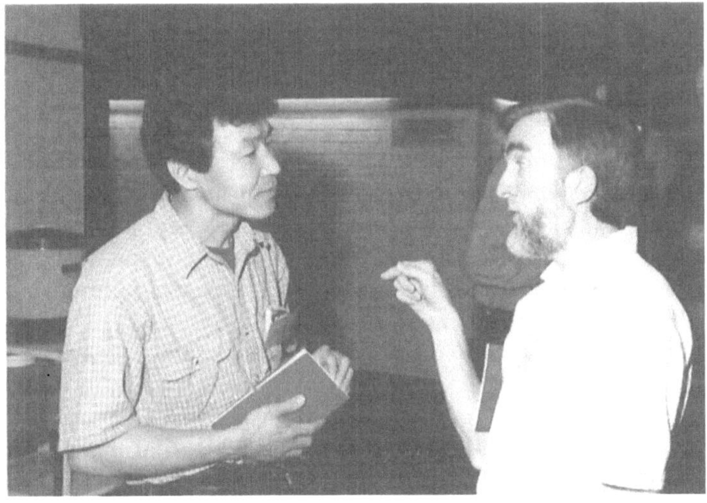

Atsuo Okazaki and Indrek Kolka

High-Resolution Spectroscopy of Stellar Winds in Recently Recognized LBV Candidates

Anatoly S. Miroshnichenko[1], Eugene L. Chentsov[2], and Valentina G. Klochkova[2]

[1] Dept. of Physics & Astronomy, University of Toledo, Toledo, OH 43606, USA
[2] Special Astrophysical Observatory of the Russian Academy of Sciences, Nizhnyi Arkhyz, 357147, Russia

Abstract. We present the results of high-resolution spectroscopic observations of two high-luminosity stars, MWC 314 and AS 314, obtained at the 6-meter telescope of the Russian Academy of Sciences. Both stars are suspected to be candidate LBVs in quiescence.

1 Introduction

MWC 314 = BD +14°3887 was discovered by Merrill (1927), who found hydrogen and Fe II emissions in its spectrum. Photometric observations by Bergner et al. (1995) showed that it is variable with an amplitude of $0^m.3$. Recently it was assigned the name V1492 Aql (Kazarovets and Samus 1997). Miroshnichenko (1996) concluded that MWC 314 is a heavily reddened supergiant ($A_V = 5^m.7$, $\log L/L_\odot = 6.2$, $T_{\rm eff} = 30\,000$ K) with a strong wind ($v_\infty = 500$ km s^{-1} and $\dot{M} = 3\,10^{-5}$ M$_\odot$ yr^{-1}) and suggested it to be a candidate LBV. He also noted that a higher-resolution spectroscopy was needed to obtain more detailed emission line profiles and to detect photospheric lines. The star's temperature estimate was based mainly on a noisy UV spectrum. Moreover, He II lines have not been detected in its spectrum indicating that $T_{eff} <$ (26–27) $10^3 K$ (Schmutz et al. 1991).

AS 314 = LS 5017 = V452 Sct is a poorly-studied heavily reddened ($E_{B-V} \sim 0^m.9$) star. Its spectral type was reported as A3: Ia (Hiltner and Iriarte 1955) or B9 Ia (Stephenson and Sanduleak 1971). Dong and Hu (1991) identified the star with an IRAS source 18365−1353, that made AS 314 potentially interesting object.

2 Observations

The spectroscopic observations were obtained at the 6-meter telescope of the Special Astrophysical Observatory (SAO) of the Russian Academy of Sciences on 1997 July 23 (MWC 314 and AS 314) and on 1997 November 22 (MWC 314). The July spectrum was taken in the range 5370–6670 Å (resolution 0.4 Å) with the echelle-spectrometer LYNX (Klochkova 1995) mounted at the Nasmyth focus and equipped with a 1140×1170 pixels CCD. The November spectrum was obtained in the prime focus with the echelle-spectrometer PFES (Panchuk et al. 1998) in the range 4700–8590 Å (resolution ~ 0.8 Å).

3 Results

In our spectrum of MWC 314 we found 408 emission lines, 63 photospheric lines (not observed previously), and 60 diffuse interstellar bands. Nearly 100 mostly double–peaked emission lines of Fe II, 37 weak single-peaked [Fe II] lines, 8 Fe III lines, and 6 [Fe III] lines were identified. The Balmer lines do not show any noticeable changes in comparison with the data obtained by Miroshnichenko (1996) in 1991. Even at the high resolution they display no P Cyg-type absorption components, that implies that the stellar wind is non-spherical and is viewed not edge-on. The detection of He II lines is doubtful. We found no O II photospheric lines in the spectrum of MWC

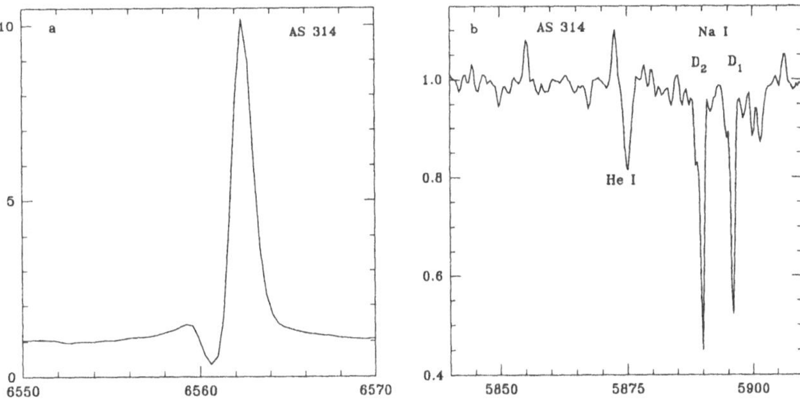

Fig. 1. Parts of the spectrum of AS 314. a. The Hα line. b. The region of the He I 5876 Å and Na I $D_{1,2}$ lines. The wavelengths are given in Angströms, the intensity is normalized to the continuum level.

314, which was also reported for the LBVs AG Car and HR Car (Hutsémekers and van Drom 1991). The heliocentric velocities of the photospheric lines are $\sim +81$ km s^{-1}, while those of the lines of ionized metals and of the Balmer lines are $\sim +41$ km s^{-1}. The latters were used to estimate the distance (D) toward MWC 314 employing differential rotation of the Galaxy. It turned out to be $D = 3.0\pm$ 0.2 kpc, which is in good agreement with the estimate of Miroshinchenko (1996). Thus, we confirmed that MWC 314 is one of the most luminous stars in the Galaxy. The above results are described in more detail by Miroshnichenko et al. (1998).

The spectrum of AS 314 (shown in part in Fig. 1) contains a rather strong Hα (EW = 14 Å) in emission with a narrow P Cyg profile ($\Delta v \sim 100$ km s^{-1}). Many weak Fe II emission lines as well as a few Fe I and forbidden lines were also found. The He I lines at 5876 and 6678 Å are seen in absorption. The Na I $D_{1,2}$ lines are purely interstellar. The strengths of the photospheric lines (S II, N II, C II, Si II, Ne I, and Al III) are consistent with a spectral type of

B9 ± 1. The mean radial velocity of the most spectral lines is ~ -50 km s^{-1} which is significantly smaller than that of stars around AS 314. This might imply that the object has a large peculiar velocity and may be a runaway star and a binary system.

While in the optical and near-IR region AS 314 looks like a reddened B-type star, it shows an excess of longward radiation, which implies the presence of circumstellar dust and is similar to that of AG and HR Car. This may indicate that the star experienced a matter ejection event in the past rather than a steady-state mass loss. Its luminosity was estimated using the strengths of the Si II lines at 6347 and 6371 Å (Rosendhal 1974), which give log $L_{bol}/L\odot \sim 4.9$ and $D \sim 8$ kpc. This brings the star close to the line of LBVs in the Hertzsprung-Russell diagram (Stothers and Chin 1994).

References

Allen, D.A. (1973), MNRAS, **161**, 145
Bergner, Yu.K., Miroshnichenko, A.S., Yudin, R.V., et al. (1995), A&AS, **112**, 221
Dong, Y.S. and Hu, J.Y. (1991), Chin. A&A, **15**, 275
Hiltner, W.A., Iriarte, B. (1955), ApJ, **122**, 185
Hutsémekers, D., van Drom, E. (1991), A&A, **248**, 141
Kazarovets, E.V., Samus, N.N. (1997), IBVS No. 4471
Klochkova, V.G. (1995), Echelle-spectrometer LYNX. User Manual. Spec. Astrophys. Obs. Technical Report N 243
Merrill, P.W. (1927), ApJ, **65**, 286
Miroshnichenko, A.S. (1996), A&A, **312**, 941
Miroshnichenko, A.S., Frémat, I., Houziaux, L., et al. (1996), A&AS, **131**, 469
Panchuk, V.E., et al. (1998), Bull. Spec. Astrophys. Obs., **44**, *in press*
Rosendhal, J.D. (1974), ApJ, **187**, 261
Schmutz, W., Leitherer, C., Hubeny, I., et al. (1991), ApJ, **372**, 664
Stephenson, C.B., Sanduleak, N. (1971), Publ. Warner & Swassey Obs., **1**, 1
Stothers, R.B., Chin, C.W. (1994), ApJ, **426**, L43

Discussions

Discussion

T. Szeifert: Could the secondary maximum in AS 314 be caused by the electron scattering wings around Hγ?
A. Miroshnichenko: This cannot be excluded. However, our first attempts to model the line profile under the assumption of a disk-like geometry show that the secondary peak can be fit quite well.

Evidence for Wind Anisotropies from Dust Formation by Wolf-Rayet Stars

Peredur M. Williams

Royal Observatory, Edinburgh

Abstract. The formation and survival of dust around stars requires a physical environment very different from that believed to hold anywhere in a Wolf-Rayet stellar wind. The observed facts of dust formation by Wolf-Rayet stars force the conclusion that their winds are not homogeneous. They also allow us to deduce the types of inhomogeneity, including clumps and large-scale high-density wakes produced in colliding-wind binaries, that allow the formation of dust.

1 Introduction: The Problem

We have long known from IR photometry that some Wolf-Rayet (WR) stars make dust in their winds. Because these winds are fast and dense enough to disperse the dust, causing it to cool and its emission to fade, persistent, strong IR emission from a WR star indicates persistent formation of new dust (Williams et al. 1987). A few WR stars show IR outbursts at intervals of $\sim$ a decade, indicative of episodic dust formation (e.g. Williams 1997). The significance of these phenomena lies in the great difficulty of forming dust in WR winds: heating of dust grains in the strong UV radiation fields of WR stars restricts dust formation to regions which, in a homogeneous WR wind of known mass-loss rate, are too rarefied by 3–4 orders of magnitude for dust to form (Cherchneff & Tielens 1995). Dust can only form in high-density structures of some sort. Here we consider the evidence for these.

2 Colliding-wind structures

As is often the case in astronomy, the best laboratories are provided by variable objects — such as the archetypal episodic dust-maker WR 140. This is a binary system comprising WC7 and O4-5 stars in a 2900-d orbit (Williams et al. 1990). Dust-formation episodes lasting a few months each recur with the same period, coinciding with periastron passage in the orbit. This phasing of dust formation to the binary orbit provides a crucial clue: the changes in physical conditions in the wind which determine when dust formation occurs must be related to changes with the orbital motion of some long-lived structure in the system. One which could provide the density enhancements required for dust formation is material compressed in shocks formed where the fast winds of the WC7 and O4-5 stars collide. Compression of the wind by

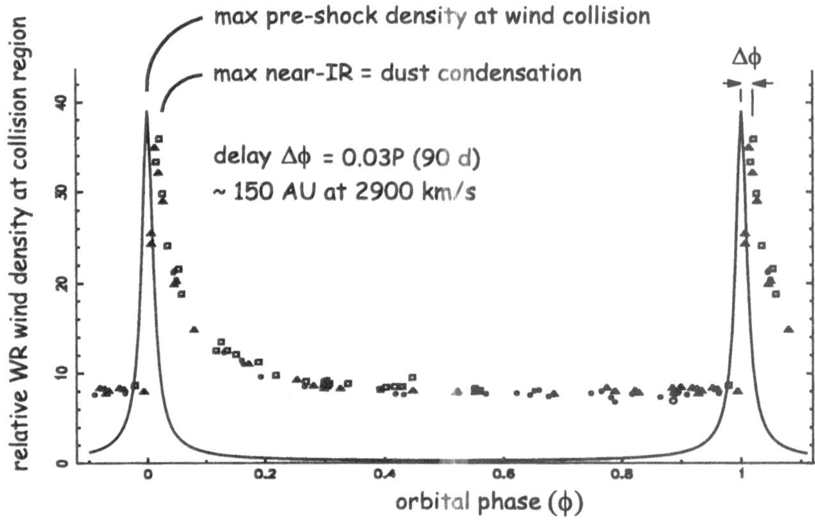

Fig. 1. Comparison of the orbital variation of Wolf-Rayet wind density at the wind-collision region (continuous line) with the observed K-band magnitudes (symbols) phased to the orbital elements.

a factor of $\sim 10^3$ can occur within the shock if the wind cools sufficiently by radiation (Usov 1991). However, the WC7 and O4-5 stellar winds in WR 140 collide and compress wind material all the time, so we then have to ask: what varies round the orbit so as to trigger dust formation for only $\sim 0.02P$ during periastron passage? Consider the systematic variations in the *pre-shock* wind density near the interaction region. This region lies where the momenta of the WC7 and O4-5 winds balance and is much closer to the O4-5 star, whose mass-loss rate is $\sim$ 60 times less than that of the WC7 star. Because the orbit is very eccentric ($e = 0.84$), the separation of the stars and the distance of the interaction region from the WC7 star vary strongly around the orbit. This is especially so around the time of periastron passage: for a very short time, the density of the WC7 stellar wind going into the shock (and being compressed by it) is $\sim$ 50 times greater than that during most of the orbit (Fig. 1). The consequent "spikes" in the pre-shock density appear to be the clock that triggers the dust condensation.

The processes of compression and cooling in the shocks sufficient to allow dust formation by WR 140 have been modelled by Usov (1991). However, dust cannot condense until the compressed wind material has been carried far enough away from the stars so that the grains are not heated to sublimation by the stellar radiation field: a distance $\sim$ 150 AU. Assuming the compressed material moves with the wind terminal velocity $\sim 2900\ \mathrm{km\,s^{-1}}$, this introduces a delay of $\sim$ 90 days between the times of maximum pre-shock wind density (at periastron passage) and maximum dust formation — consistent with the observed phase difference ($\Delta\phi \sim 0.03P$) between maximum pre-shock density and infrared (K band) maximum (Fig. 1).

Extension of the WR 140 paradigm to other dust-makers requires demonstration that they are colliding-wind binaries with appropriate stellar and orbital properties. Spectroscopic companions to some episodic and persistent dust-makers have been found (Williams 1997, Williams & van der Hucht 1996) but determination of orbits will be difficult given the broad emission lines of WR stars and the apparently long periods of episodic dust makers indicated by their IR light curves.

3 Clumps small and large

Some dust-making WR stars (e.g. WR 121) show brief optical occultations by $\sim 10^{-14}$ $M_\odot$ dust clumps forming in the line of sight (Veen et al. 1998). Also, the fading light curves of two of the episodic dust-makers, WR 48a and WR 137, show "mini" infrared outbursts and fadings indicative of minor ($\sim 10^{-9}$ $M_\odot$) episodes of dust formation $\sim$ years after the major outbursts. Whether these phenomena represent part of a continuum of condensing clump masses or come from different processes is an open question.

References

Cherchneff, I., Tielens, A.G.G.M. (1995): Dust formation in hot stellar winds. in: *Wolf-Rayet Stars: Binaries, Colliding Winds, Evolution, IAU Symposium 163* eds K.A. van der Hucht, P.M. Williams, (Kluwer, Dordrecht), 346–354

Usov, V.V. (1991): Stellar wind collision and dust formation in long-period, heavily interacting Wolf-Rayet binaries. MNRAS **252**, 49–52

Veen, P.M., van Genderen, A.M., van der Hucht, K.A., Li, A., Sterken, C., Dominik, C. (1998): WR121 obscured by a dust cloud: the key to understanding occasional "eclipses" of "dusty" Wolf-Rayet WC stars ? A&A **329**, 199–213

Williams, P.M., van der Hucht, K.A. (1996): A search for companions to dust-making Wolf-Rayet stars. in: *Wolf-Rayet Stars in the Framework of Stellar Evolution, Proc. 33rd Liège International Astrophysical Colloquium*, eds J-M. Vreux et al., (Université de Liège, Liège), 353–359

Williams, P.M., van der Hucht, K.A., Thé, P.S. (1987): Infrared photometry of late-type Wolf-Rayet stars. A&A **182**, 91–106

Williams, P.M., van der Hucht, K.A., Pollock, A.M.T., Florkowski, D.R., van der Woerd, H., Wamsteker, W.M. (1990): Multi-frequency variations of the Wolf-Rayet system HD 193793 - I. Infrared, X-ray and radio observations. MNRAS, **243**, 662–684

Williams, P.M. (1997): Formation of dust in hostile environments — what we learn from observing Wolf-Rayet stars. Ap&SS **251**, 321–331

Discussion

G. Koenigsberger: Is it enough to have an increased WR wind density in the wind-wind shock region during periastron or do you need an increased mass-loss rate to form dust?
P. Williams: I have not modelled this aspect; it would help a little, but since the WR 140 system is so wide, even at periastron, any enhanced mass loss is likely to be smaller than in most other systems considered.

J. Bjorkman: Can you use the colour information in the photometry to estimate the maximum dust temperature as a function of time? Is this temperature consistent with constant velocity expansion of the dust?
P. Williams: Yes. The dust temperature falls as the emission fades when dust formation ceases. The cooling and fading of the emission from WR 140 are slower than expected from simple dispersion by the stellar wind, perhaps due to continued grain growth after condensation.

D. Massa: Do you see any evidence for IR spectroscopic features that might tell you what sort of dust is being formed?
P. Williams: No, the spectral energy distribution is smooth; we see no dust features, only interstellar features.

Peredur Williams, Stephen Hulbert, Linda Smith and Regina Schulte-Ladbeck

ISO-SWS Spectroscopy of B[e] Stars

Robert H.M. Voors[1,2], Laurens B.F.M. Waters[3,4], and Patrick W. Morris[5,2]

[1] Astronomical Institute, Utrecht University, The Netherlands
[2] SRON Laboratory for Space Research, Utrecht, The Netherlands
[3] Astronomical Institute, University of Amsterdam, The Netherlands
[4] SRON Laboratory for Space Research, Groningen, The Netherlands
[5] ISO Science Operations Center, Villafranca, Madrid, Spain

Abstract. We present ISO-SWS spectra of B[e] stars. We find a wide diversity of spectral characteristics of B[e] stars, suggesting different origins for the circumstellar matter. Most B[e] supergiants show hot dust with weak amorphous silicate emission. MWC 300 has a warm dust shell with strong crystalline silicate emission; its evolutionary status is unclear.

1 Introduction

B[e] stars are defined by the following characteristics: (**1**) Spectral type B (**2**) Optical emission lines; hydrogen recombination and low ionisation metal lines, both permitted and forbidden (**3**) IR excess due to (hot) dust. The main problem with this definition is that it is purely *phenomenological* and does not constrain the *evolutionary state* of the object. This is illustrated by the fact that the above defined group of B[e] stars contains such diverse objects as B[e] supergiants, HAeBe stars, PPNe, symbiotic stars and also a group of stars of which the evolutionary state is unknown (Lamers et *al.* 1998). We will use spectra taken with the ShortWave Spectrometer (SWS) on board the Infrared Space Observatory (ISO) of a number of B[e] stars, with a focus on the B[e] supergiants, to study the region between 2.4 and 45 μm.

2 B[e] Supergiants

The widely accepted model for B[e] supergiants was first proposed by Zickgraf et *al.* (1985). In this model, the star has a two-component wind. In the polar region, a fast (v_{exp} $\approx$1000 kms^{-1}), hot (C IV, Si IV), low density wind is present, similar to normal B supergiants. In the equatorial region, both the expansion velocity ($\approx$ 100 kms^{-1}) and the temperature (Si II, Fe II) are much lower; assuming a constant mass flux, the density is much higher. The wind in the equatorial region is effectively shielded from UV radiation, and at some point has cooled enough and still has a sufficient density, for dust to form.

This model predicts an energy distribution which shows the central star plus some free-free excess and, superposed on that, the thermal continuum from the dusty disk. Dust condenses typically between 1000 and 1600 K, which corresponds to a few 10^3 R_*. A temperature of 1300 K corresponds to a peak in the energy distribution (F_ν) at 4 μm. If we assume that the formation of the dust does not influence the large scale density distribution of the wind, then the dust density distribution will follow the same r^{-2} distribution as the gas. So we expect a more or less normal B supergiant continuum, and superposed a thermal dust continuum that peaks roughly between 3 and 5 μm, with a steep decline (because of the steep density gradient) towards longer wavelengths.

This is exactly what is observed in several sources, e.g. in CPD−57°2874. The optical spectrum of this star is dominated by H I and He I emission lines (Carlson & Henize 1979). McGregor et *al.* (1988) discovered CO first overtone emission at 2.3 μm, which indicates the presence of a high density region ($n > 10^{10}$ cm^{-3}) around the star. Using an optically thin, spherically symmetric model to fit the SWS spectrum and assuming an outflow velocity in the dust forming region of 100 kms^{-1}, we derive a total dust mass-loss rate for CPD−57°2874 of a few 10^{-9} $M_\odot$ yr^{-1}. McGregor et *al.* derive a total mass-loss rate, based on the strength of Brγ of a few 10^{-6} $M_\odot$ yr^{-1}. Assuming a gas/dust ratio of 100, which is the canonical value found in the interstellar medium, this suggests an opening angle of the disk of the order of 20 degrees. These are very rough numbers but it confirms the idea that only in the equatorial region dust is formed around B[e] stars. It also indicates that probably this region is not extremely thin; less than 1 degree seems unlikely. Since it will be difficult to get a good constraint on the gas/dust ratio, a very tight constraint on exactly how much of the wind contains dust – and how much of it takes part in the formation process – will be difficult to determine.

3 MWC 300: is it a supergiant?

One of the objects with an unclear evolutionary status is MWC 300. Wolf & Stahl (1985) conclude it must be a supergiant, primarily based on the presence of multiplets 115 and 117 of Fe III in emission in the optical spectrum. However, it is also included in a number of pre-main-sequence studies (e.g. Thé et *al.* 1994), even though it does not clearly show any pre-main-sequence characteristics. Henning et *al.* (1994) derive a mass of 300 $M_\odot$ of circumstellar dust and gas, based on their 1.3 mm flux (54±15 mJy) and on the distance of 15.5 kpc, as derived by Wolf & Stahl (1985). Even though there is a very large uncertainty in this mass (they estimate at least a factor of 5), it seems too large to be explained as a post-main-sequence shell.

The SWS spectrum of MWC 300 is shown in figure 1a. The dust in MWC 300 shows a very wide range in temperatures and is much cooler than that of more "typical" B[e] supergiants, such as CPD−57°2874. This suggests a flat density gradient and/or significant optical depth effects. The 9.7 μm amorphous silicate feature is seen in absorption, and this may very well be of interstellar origin. The SWS spectrum of MWC 300 shows prominent narrow emission bands that can be attributed to Mg-rich, Fe-poor crystalline olivines (19.5, 23.5, 27.5 and 33.5 μm) and pyroxenes (most of the narrower peaks). These dust components are often observed in cool stars with very dense, dusty outflows (e.g. red supergiants, OH/IR stars), and also in objects with circumstellar or circum-binary disks (e.g. HAeBe stars).

The origin of the dust is not clear. It is unlikely that it is formed in a present day outflow, because (**1**) the slope of the SED does not agree with an r^{-2} density distribution, and (**2**) the dust-forming layers are at leare distance from the (hot) star, where the density is low; crystalline silicates are only observed in objects with very high density in the dust-forming layers. Another possibility is a non-outflow disk. This alleviates the above mentioned difficulties, but the origin of such a disk in unclear. From low mass stars it is observed that disks containing crystalline dust are often associated with binaries.

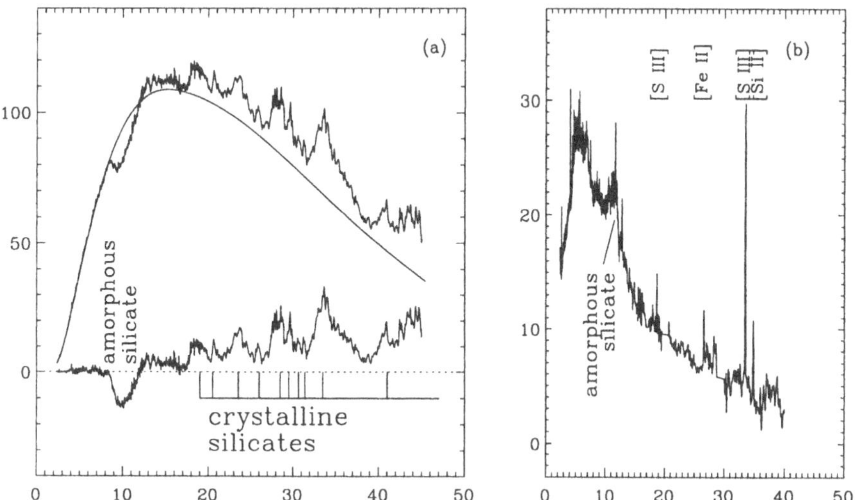

Fig. 1. SWS spectra of MWC 300 (a) and CPD$-57°2874$ (b). The lower spectrum of MWC 300 is continuum (drawn line) subtracted. Abscissa: wavelength in μm; Ordinate: flux density in Jansky

References

Carlson E.D., Henize K.G., 1979, Vistas Ast. **23**, 213

Henning Th., Launhardt R., Steinacker J., Thaumm E., 1994, A&A **291**, 546

Jäger C., Molster F.J., Dorschner J., et *al.*, 1998, *submitted to* A&A

Lamers H.J.G.L.M., Zickgraf F.-J., de Winter D., et *al.*, 1998, *submitted to* A&A

McGregor P.J., Hyland A.R., Hillier D.J., 1988, ApJ **324**, 1071

Thé P.S., de Winter D., Perez M.R., 1994, A&AS **104**, 315

Wolf B., Stahl B., 1985, A&A **148**, 412

Zickgraf F.-J., Wolf B., Stahl O., et *al.*, 1985, A&A **143**, 421

Discussion

S. Shore: A question and a comment: Do the B supergiants (or B[e] stars) that show CO 2μ emission show the PAH features? Since HD 87643 is also known as He 3–365, I should mention that this star shows variations in the UV of $\sim 30\,\%$ on timescales of about a decade. It has one of the very strongest UV iron curtain absorption spectra of any LBV.
R. Voors: Stars in our sample that are known to show CO emission in the K-band do not show PAH features.

M. Magalhães: Was ISO sensitive enough to observe B[e] supergiants in the Magellanic Clouds?
R. Voors: Some B[e] supergiants in the LMC/SMC were observed with the PHOT–S instrument onboard ISO ($\lambda/\Delta\lambda \sim 100$), but they are too faint to be observed with the Short Wave Spectrometer.

P. Williams: Considering the unusual density distribution in the disk, have you looked at the infrared photometric history? I have ground-based $11\,\mu$m observations from 20 years ago.
R. Voors: The ISO SWS fluxes are consistent with the IRAS fluxes. But I would certainly be interested to see your even older data of the silicate emission bump.

F.-J. Zickgraf: Could you see any $10\,\mu$m silicate feature in Magellanic Cloud B[e] supergiants?
R. Voors: Maybe, I do not know.

Roberta Humphreys, Norman Trams and Rens Waters

Session IV

Theories of Wind Variations

chair: D. Massa

The Line-Driven Instability

Achim Feldmeier

University of Kentucky, Lexington, KY 40503, USA

Abstract. The line-driven instability may cause pronounced structure in winds of hot, luminous stars, e.g., fragments of dense shells, strong reverse shocks, and fast cloudlets. We discuss the linear stability theory, including the line-drag effect, phase reversal due to the diffuse radiation field, and the relevance of so-called Abbott waves. Recent hydrodynamic simulations focuss on the influence of a time-dependent source function on the flow structure, and on the X-ray emission from wind shocks and cloud collisions.

1 Introduction

Lucy & Solomon (1970) described a new, *line-driven* instability for OB star winds, which may be connected to (some of) the following observational facts: (i) the appearence of discrete absorption components, periodic absorption modulations, black troughs, and variable blue edges in P Cygni line profiles (see reviews by Fullerton, Henrichs, Kaper, Kaufer, or Massa in this volume); (ii) the X-ray emission from hot star winds, and their superionization; and (iii) cloud formation in O star and Wolf-Rayet star winds (Moffat 1994).

In the following we shall discuss some aspects of the instability from a mostly hydrodynamical viewpoint.

2 Linear theory

2.1 Mechanism of the instability

MacGregor et al. (1979) and Carlberg (1980) calculated growth rates for the line-driven instability assuming optically thin flow perturbations. Contrary, Abbott (1980) found zero growth rates when he applied the Sobolev approximation to the perturbations. He found then a new type of marginally stable, radiative-acoustic waves, which we shall term 'Abbott waves' in the following. Owocki & Rybicki (1984) unified these contradictory results by showing that they refer to different wavelength regimes λ of the perturbations, namely $\lambda < L$ (with L being the Sobolev length) in the work of MacGregor et al. and Carlberg, while $\lambda \to \infty$ in Abbott's analysis.

The regime $\lambda > L$ or $\lambda \gg L$ in between these extremes is especially interesting: while the growth rate drops there as $\Omega \propto \lambda^{-2}$, the instability is very strong ($\Omega_{\rm max}\, t_{\rm flow} \approx 50$; Owocki & Rybicki 1984), so that even rather long-scale perturbations can grow into saturation, and become the most pronounced flow structures in terms of velocity, density, and temperature jumps. Having wavelengths larger than the Sobolev length, these perturbations can be viewed as *unstable* Abbott waves: the propagation speed follows from the usual, first order Sobolev treatment, whereas the small growth rate is of second order (Feldmeier 1998).

The physical basis of the instability for different wavelength regimes is illustrated in Fig. 1: region (a) shows an unstable short-scale perturbation, $\lambda = O(L)$ (where L corresponds to the 'thickness' of the thermal band, indicated in the plot by double lines): an arbitrary, positive velocity fluctuation shifts the gas parcel out of the absorption shadow of gas lying closer to the star, and the enhanced flux accelerates the parcel to even larger speeds, deshadowing it further. With the line force scaling as $g_l \propto \exp(-\tau)$, the general instability cycle can be written $\delta v \to -\delta\tau \to \delta g_l \to \delta v$. Region (b) around the node of a long-scale, sinusoidal perturbation shows the occurence of inward propagating Abbott waves from first order Sobolev approximation: the steepening of the thermal band at the node raises the Sobolev line force, $g_l \propto v'^{\alpha}$ (where $v' = dv/dr$, and $0 < \alpha < 1$), and the gas is accelerated to larger speeds. This corresponds to an inward shift of the node, i.e., an inward phase propagation of a wave. The wave cycle can be written $\delta v' \to \delta g_l \to i\delta v \to -\delta v'$. Finally, region (c) around the velocity maximum of the long-scale perturbation shows that this Abbott wave is unstable from a second order treatment: due to the negative curvature of the thermal band the optical depth is *smaller* there than for the unperturbed flow. Again, a larger line force results which accelerates the gas; this now makes the maximum more pronounced, wherefore $-v''$ increases *further* (we assumed here that the node separation or wavelength is essentially unaffected). Thereby, τ drops further, g_l grows further, and one has unstable growth. The general instability cycle can then be further specified to become $-\delta v'' \to -\delta\tau \to \delta g_l \to \delta v \to -\delta v''$. Notice also the kinematical steepening and finally braking of the wave into a strong reverse shock.

2.2 Information propagation

Yet, this unified picture of Abbott waves and the line-driven instability is oversimplified. In a remarkable paper, Owocki & Rybicki (1986) show from a Green's function analysis that *information propagation* in an unstable, pure *absorption* line flow is limited to the sound speed; contrary, radiative-acoustic waves propagate inward at a phase or group speed equal to the much larger (negative) wind speed (Abbott 1980).

This is an example of the non-equivalence of signal or information speed and group speed in unstable media (e.g., Bers 1983).

To demonstrate this physically, Owocki & Rybicki (1986) consider a Gaussian pulse which is broader than the Sobolev length, and therefore should propagate upstream at the wind (or Abbott) speed. This is indeed confirmed,

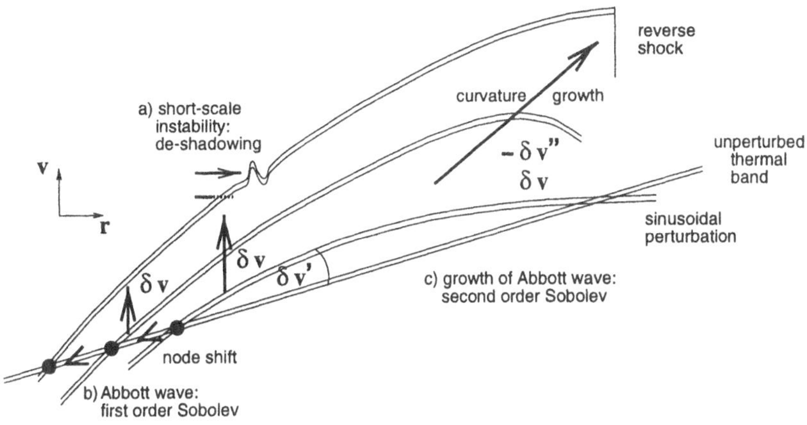

Fig. 1. Line-driven instability and Abbott waves.

even when the Green's function for zero sound or signal speed is used to propagate the pulse! However, Owocki & Rybicki (1986) claim that no *signal* is propagated in this case. Namely, due to its smoothness, information is not localized in the pulse, and properties from any small neigborhood can be used to infer distant properties via a Taylor series expansion, without need for information propagation.

That this is the case in the above example (i.e., that the folding of the Green's function with the signal is equivalent to a Taylor series extrapolation of a smaller to a larger space-time area) is seen if a truly localized information is introduced into the pulse, here by setting its amplitude to zero for all $x > x_0$, with arbitrary x_0. In accord with $a = 0$, the discontinuity at x_0 does not propagate, but remains there. For $x < x_0$ then, the *full*, smooth Gaussian *without* any discontinuity is reconstructed in course of time, and propagates upstream to smaller x! This awkward fact is due to the one-sidedness of the pure absorption line force, i.e., that a perturbation at $x > x_0$ cannot affect the upstream flow at $x < x_0$. Since for $x < x_0$ all derivatives, curvatures, etc. are those of the *full* Gaussian, the latter is reconstructed for $x < x_0$, and propagates upstream as a 'false' signal. We leave here out a discussion of the region $x > x_0$.

The key point in this discussion is the one-sidedness of the absorption line force, and the situation could be fundamentally different for a non-zero source function. Corresponding numerical simulations show then indeed the inward propagation of a front at Abbott speed following a delta-function perturbation (Owocki & Puls 1998).

One reason it is important to decide whether Abbott waves are real are recent claims on the role of *kinks* in the wind velocity law, which propagate upstream at Abbott speed (for corotating intercation regions: Cranmer &

Owocki 1996; for wind clouds: Feldmeier et al. 1997b). Future analytical work will hopefully bring further clarification.

2.3 The line-drag effect

Besides for these matters of wave propagation, which we shall take up again in the next section, line scattering is also important for instability growth rates. Lucy (1984) questioned the occurence of the line-driven instability in hot star winds altogether, by noting that the winds are driven essentially by scattering lines (as opposed to absorption lines), and that the *diffuse* radiation field should cancel any extra line force gained by Doppler-shifting gas into the *direct* radiation field.

However, this exact cancellation occurs only near the star, at the wind base. Due to sphericity effects and the decreasing angular size of the stellar disk with distance, the growth rate is back to 50% of its pure absorption line value within a stellar radius of the stellar surface, and approaches 80% of this value at large radii (Owocki & Rybicki 1985).

The line-drag is therefore most relevant in deep wind layers, and may be important with regard to the photosphere-wind connection, i.e., whether the formation of wind structure is externally triggered or self-excited.

3 Numerical simulations

3.1 SSF and EISF

While line scattering is of central importance for the formation of wind structure, the exact solution to the radiative transfer equation in instability simulations is prohibitively cpu-time consuming. The 'smooth source function' approximation (SSF; Owocki 1991) uses instead a formal integral approach, assuming a prespecified source function from Sobolev approximation. Via this averaged or mean diffuse radiation field, the line-drag effect is incorporated in SSF calculations.

One-dimensional numerical simulations for a spherically symmetric O star wind (Owocki et al. 1988; Owocki 1992) perturbed by a harmonic, photospheric sound wave show that the continuous flow breaks up into a sequence of strong reverse shocks, each decelerating inner, thin, fast gas and compressing it into narrow, dense shells. The shells propagate roughly according to a stationary wind velocity law. For a discussion of self-excited wind structure and issues of periodic vs. chaotic wind structure, we refer to Owocki (1994).

Recently, Owocki & Puls (1996, 1998) proposed a new, 'escape integral source function' approximation (EISF) which accounts for the first time for the *perturbed* diffuse radiation field. The idea is here to replace the photon escape probabilities, β_s, from the smooth, Sobolev source function, $S = \langle \beta_s I_* \rangle / \langle \beta_s \rangle$ (brackets indicate angle averaging) at each time step with the

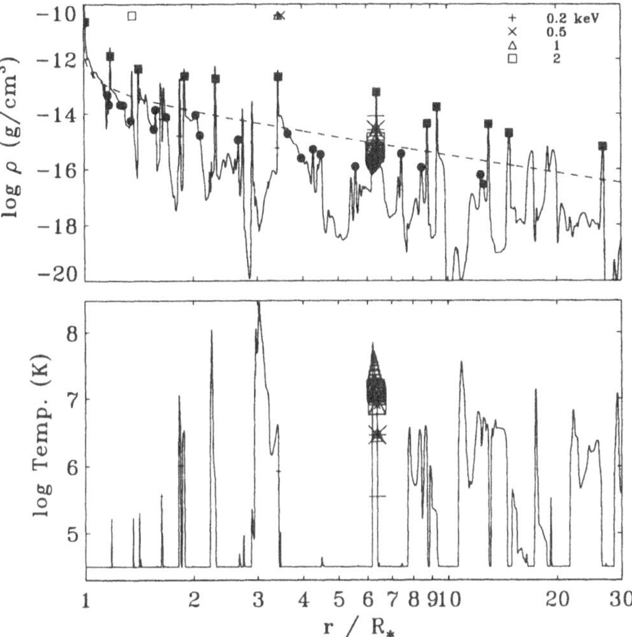

Fig. 2. Density and temperature snapshot for a wind model of ζ Ori. Filled bullets mark fast cloudlets, filled squares mark dense shells. Symbols (+, $\times$, etc.) indicate strong X-ray emission at the given energies.

escape probabilities from the time-dependent wind simulation, β_i. The β_i are the central quantities which distinguish instability simulations from stationary wind models applying the Sobolev approximation, since they include both (de-)shadowing effects of neighboring and widely separated gas parcels.

As Puls et al. (1994) noticed, the inclusion of the correct diffuse radiation field is important since, as was shown by Owocki & Rybicki (1985) from an exact, linear analysis, the perturbed diffuse radiation field can turn anti-correlated, inward propagating density and velocity fluctuations into correlated, outward propagating fluctuations. In the nonlinear, wave braking phase, the former steepen into reverse shocks, the latter into forward shocks. SSF calculations show the dominance of reverse shocks. The question is whether EISF simulations are instead dominated by forward shocks.

The answer is essentially 'no'. A phase reversal occurs only for short-scale fluctuations below the Sobolev length (Owocki & Rybicki 1985). Steepening the thermal band over short lengthscales until it becomes optically thin – and hence the instability ceases since no further de-shadowing is possible – leads only to small velocity jumps of order the thermal speed. The EISF structure appears therefore as short-scale, low-amplitude noise superimposed on the long-scale, large-amplitude sequence of reverse shocks. Still, these results indicate that the Sobolev length as an intrinsic lengthscale of line-driven flows separates two different regimes of the (inverse) turbulent cascade. Caution is therefore required in applying results from, e.g., supersonic Burgers turbulence to hot star winds. Furthermore, we add here that cloudlets which are important for the X-ray emission from O stars (cf. the next section) have lengthscales not too different from the EISF noise. Since the cloudlets are anti-correlated perturbations, future simulations have to show whether they are affected by the inclusion of the perturbed diffuse radiation field. Finally, we refer to Owocki & Puls (1998) for a discussion of the modifications of the *stationary* solution for thin winds due to the inclusion of fore-aft asymmetric (e.g.: EISF) escape probabilities around the sonic point, which are not present in usual Sobolev approximation.

3.2 X-ray emission

One main interest in the line-driven instability is that it may create shocks which are responsible for the observed X-ray emission from hot star winds, and partially (Pauldrach 1987) also for their superionization.

After overcoming numerical problems which lead to a collapse of cooling zones (Cooper & Owocki 1992; Feldmeier 1995), the temperature structure behind strong reverse shocks can be calculated (Fig. 2), and their X-ray emission synthesized. For the self-absorption of X-rays in the dense wind shells, NLTE opacities from stationary wind models are presently used (Feldmeier et al. 1997a).

In agreement with estimates by Hillier et al. (1993) from properties of reverse shocks as deduced from isothermal wind simulations, we find that these shocks can only account for 1 to 10% of the observed X-ray emission during their quasi-steady appearance, i.e., when thin, fast gas is being fed through the front. However, by applying chaotic perturbations at the wind base, we find that short, strong X-ray flashes in the wind can account for the observed X-ray emission, even after time-averaging. The flashes originate from collisions of small, fast cloudlets with the pronounced, dense wind shells (Feldmeier et al. 1997a). Both the continuous stream of thin gas and the discrete cloudlets are ablated from gas which moves ahead (i.e., at somewhat larger radii) of the next inner, pronounced shell. The shells and cloudlets are indicated in Fig. 2.

So far the modeling assumes a spherically symmetric, radial wind (for first 2-D instability simulations, see Owocki, this volume), and leads to major variability in X-ray fluxes. However, cloudlets which form due to photospheric turbulence should have a rather small lateral scale. With independent cloud-shell collisions taking place along neighboring wind cones, (near) constancy of X-ray fluxes should then be achieved by angle averaging (Cassinelli & Swank

1983). The present, 1-D wind models suggest that a few thousand wind cones should be sufficient to achieve the observed flux constancy.

On the other hand, the pronounced shells possibly form due to long-periodic, coherent photospheric perturbations, wherefore their lateral scale may be large, and they may extend over many such neighboring wind cones. Possibly, larger shell segments fragmentize due to the Rayleigh-Taylor instability. Future 2-D simulations have to bring clarification.

Acknowledgements. I thank J. Cassinelli, A. Fullerton, R.P. Kudritzki, C. Norman, S. Owocki, A. Pauldrach, J. Puls, and I. Shlosman for many interesting discussions. I thank the organizers of this meeting for a generous travel grant. Work in this project was funded by DFG projects Pa 477/1-1 and 1-2, and by NASA grant NAG 5-3841.

References

Abbott D.C., 1980, ApJ 242, 1183
Bers A., 1983, in Galeev A.A., Sudan R.N. (eds.) Handbook of Plasma Physics, Vol. 1. North Holland, Amsterdam, 451
Carlberg R.G., 1980, ApJ 241, 1131
Cassinelli J.P., Swank J.H., 1983, ApJ 271, 681
Cooper R.G., Owocki S.P., 1992, PASPC 22, 281
Cranmer S.R., Owocki S.P., 1996, ApJ 462, 469
Feldmeier A., 1995, A&A 299, 523
Feldmeier A., 1998, A&A 332, 245
Feldmeier A., Puls J., Pauldrach A.W., 1997a, A&A 322, 878
Feldmeier A., Norman C., Pauldrach A.W., et al., 1997b, PASPC 128, 258
Hillier D.J., Kudritzki R.P., Pauldrach A.W., et al., 1993, A&A 276, 117
Lucy L.B., 1984, ApJ 284, 351
Lucy L.B., Solomon P.M., 1970, ApJ 159, 879
MacGregor K.B., Hartmann L., Raymond J.C., 1979, ApJ 231, 514
Moffat A.F., 1994, Rev. Mod. Astron. 7, 51
Owocki S.P., 1991, in Crivellari L., et al. (eds.) Stellar atmospheres: beyond classical models. Kluwer, Dordrecht, 235
Owocki S.P., 1992, in Heber U., Jeffery S. (eds.) The atmospheres of early-type stars. Springer, Heidelberg, 393
Owocki S.P., 1994, Ap&SS 221, 3
Owocki S.P., Puls J., 1996, ApJ 462, 894
Owocki S.P., Puls J., 1998, ApJ, in press
Owocki S.P., Rybicki G.B., 1984, ApJ 284, 337
Owocki S.P., Rybicki G.B., 1985, ApJ 299, 265
Owocki S.P., Rybicki G.B., 1986, ApJ 309, 127
Owocki S.P., Castor J.I., Rybicki G.B., 1988, ApJ 335, 914
Pauldrach A.W., 1987, A&A 183, 295
Puls J., Feldmeier A., Springmann U., Owocki S.P., Fullerton A.W., 1994, Ap&SS 221, 409

Discussion

H. Lamers: Do I understand from your modelling that the clumping disappears at $r \gtrsim 10\,R_\star$? If this is the case the radio mass-loss rates are not affected by clumping.
A. Moffat: But then how do you explain the clumpy structure in the MERLIN radio image of the thermal wind around the nearby WN8 star WR 147 at $r \sim 10^3\,R_\star$? Also: P Cygni has a clumpy, resolved wind.
A. Feldmeier: We have one simulation which extends out to $100\,R_\star$ and which shows that the shells disappear at $\sim 50\,R_\star$. This is in agreement with estimates from re-expansion due to internal thermal pressure. Maybe clumps seen at very large distances have a different origin than the line-driven instability.

G. Mellema: Do your numerical models include explicit thermal conduction?
A. Feldmeier: It is coded but usually switched off, since heat conduction should only be important at temperatures significantly higher ($> 10^7$ K) than those deduced from X-ray observations.

R. Ignace: A popular model for explaining WR winds is that of multiple scattering (e.g., Lucy & Abbott 1993, ApJ 405, 738). How do you expect your results for O stars to change for WR stars?
A. Feldmeier: Gayley & Owocki (1995, ApJ 446, 801) have calculated the linear growth rates of the line-driven instability for WR stars, using a diffusion treatment of multiline scattering. They find that the growth rates are reduced relative to O stars by a factor of ~10. However, these growth rates are still large enough that blobs or clumps should also develop in WR winds.

J. Bjorkman: Cohen et al. (1997, ApJ 487, 867) found that the observed ROSAT X-ray fluxes from B stars require an X-ray emission measure larger than that available in a smooth wind. Do you think that the clumping and colliding clouds in your models can explain the observed X-ray levels in B stars?
A. Feldmeier: First, it may be that the mass-loss rates of these winds are higher than is presently assumed, and therefore the dilemma with large emission measures could be avoided. Otherwise, since the cloudlet density is fixed, namely at roughly the stationary wind density, it seems that a larger *number* of cloud collisions per unit time is needed to enhance the X-ray emission. An alternative possibility is that a few adiabatic shocks could heat large volumes of thin wind gas in B stars. Future simulations have to clarify this.

J. de Jong: How large is the density fluctuation from clumping at small radii? Would you expect significant variations in, e.g., the Hα line?

A. Feldmeier: The dense shells occur above $\sim 1.5\,R_\star$ and have densities which are factors of 10 to 100 higher than those of stationary wind gas. According to Puls et al. (1996, A&A 305, 171) the Hα line, which forms between 1 and $1.5\,R_\star$ for O stars should not be significantly affected.

D. Massa: Do you have an idea of the size of the lateral spatial coherence of the instabilities you are modelling?
A. Feldmeier: Presently not, since 2D simulations are still missing. My favourite idea is that the fast cloudlets are caused by photospheric turbulence, and have a similarly short lateral length scale. The dense shells, which move at roughly the stationary wind speed, may be connected to long-period, coherent perturbations, and have a much larger lateral scale; maybe they fragment due to the Rayleigh-Taylor instability.

S. Shore: What happens when you that the photospheric turbulence and organize it with photospheric pulsation? Is it possible to place some limits on the ratio of the energy in the turbulent vs. organized velocity field on the basis of your models?
A. Feldmeier: I also favour such a picture. The pulsations could trigger the formation of dense shells (or shell segments), and the turbulence could trigger the formation of the tiny, fast cloudlets. But we have no quantitative limits so far.

Linda Smith and Achim Feldmeier

Co-Rotating Interaction Regions in 2D Hot-Star Wind Models with Line-Driven Instability

Stanley P. Owocki

Bartol Research Institute of the University of Delaware, Newark, DE 19716, USA

Abstract. I review simulations of Co-rotating Interaction Regions (CIRs) in line-driven stellar winds. Previous CIR models have been based on a local, Sobolev treatment of the line-force, which effectively suppresses the strong, small-scale instability intrinsic to line-driving. Here I describe a new "3-ray-aligned-grid" method for computing the nonlocal, smooth-source-function line-force in 2D models that do include this line-driven instability. Preliminary results indicate that key overall features of large-scale CIRs can be quite similar in both Sobolev and non-Sobolev treatments, *if* the level of instability-generated wind structure is not too great. However, in certain models wherein the unstable self-excitation of wind variability penetrates back to the wind base, the stochastic, small-scale structure can become so dominant that it effectively disrupts any large-scale, CIR pattern.

1 Introduction

Co-rotating Interaction Regions (CIRs) are the sprial-shaped density compressions that form from the interaction of higher and lower speed streams in the wind from a rotating star. CIRs have long been observed and studied in the solar wind, but Mullan (1984a,b; 1986) was the first to suggest them as a possible mechanism for producing variable absorption features in UV lines formed in the line-driven stellar winds from hot-stars. To be visible as direct variations in line-profiles formed from globally integrated radiative flux, the associated flow structure must be on a relatively large scale, on order the stellar radius; in these highly supersonic winds, this is of order $v/v_{th} \gg 1$ larger than the Sobolev length $L \equiv v_{th}/(dv/dr)$, over which the mean flow speed v increases by a ion thermal speed v_{th}. This suggests that the dynamical evolution of large-scale structure might be adequately simulated using the computationally efficient, *local*, CAK/Sobolev expression for the line force (Castor, Abbott, and Klein 1975; Sobolev 1960), and indeed this has been a key simplification in previous simulations of CIRs in line-driven winds (Cranmer and Owocki 1995, hereafter CO96). A recent review (Owocki 1998) has summarized efforts to apply such Sobolev-theory CIR simulations towards modelling various types of line-profile-variability in hot-star winds, including both the classical, slowly evolving Discrete Absorption Components (DACs), and more recently identified Periodic Absorption Modulations (PAMs).

The central question I wish to address here is: To what degree is the derived CIR structure sensitive to this approximate Sobolev treatment of the line-force? One particularly important question regards the formation of pre-CIR velocity plateaus; CO96 have been suggested these as a possible origin of DACs, but it is still not clear whether they might in fact be just an artifact of the (artificially?) fast inward wave propagation obtained in a Sobolev treatment (cf. Owocki and Rybicki 1986). Another issue regards the role of the strong, intrinsic, small-scale instability of line-driving; this is effectively suppressed in a Sobolev treatment (Owocki and Rybicki 1984), but, if included, it might substantially alter, or even entirely disrupt, any larger-scale CIR structure.

In the next section (§2) I outline a newly developed, "3-ray-aligned-grid" approach for extending the nonlocal, "Smooth Source Functiorn" (SSF) line-force method, used in previous 1D instability simulations (Owocki 1991; Owocki and Puls 1996, 1998), into 2D models that can include both large-scale CIRs and small-scale instability-generated structure. I then describe (§3) some initial, still preliminary results indicating that key Sobolev-model features (like velocity plateaus) can indeed still form, but only if the overall level of intrinsic variability initiated near the wind base is not too strong. I conclude (§4) with a brief summary discussion.

2 3-Ray SSF Approach for 2D Instability Simulations

A key aspect of line-driven-instability simulations regards the computation of the line force. For the instability-generated flow structure at scales near and below the Sobolev length, a local Sobolev approach fails completely. (See, however, Feldmeier 1998.) Instead, one must apply much more computationally expensive, *integral* forms that take approximate account of the inherently *nonlocal* scattering character of the radiative transfer for most important driving lines. Owocki and Puls (1996, 1998) discuss various levels for approximating this nonlocal force, including the *Smooth-Source-Function* (SSF) approach that efficiently accounts for key effects (e.g. line-drag; see Lucy 1984) that control the level of instability. The line-force components are obtained from flux-weighted moments of nonlocal escape functions, each of which requires spatial integrations to obtain the optical depth over a wide range of line frequencies, nested within a frequency averaging over the line-profile. The computational requirements of evaluating such nested, nonlocal force integrations at each time-step of a hydrodynamical model have till now limited simulations of wind structure to just a 1D temporal evolution in radius, effectively suppressing, quite artificially, any lateral variations.

In developing multidimensional instability models that include lateral structure, a central challenge is thus to develop an approach for efficient evaluation of these escape integrals over a suitable collection of directional rays. As an initial approach, I have been experimenting with an approximate

3-ray SSF method for computing the nonlocal line-force in 2D wind models in radius r and azimuth ϕ. At any given grid point, the nonlocal escape probabilities are evaluated along one radial ray, plus two nonradial rays on opposite sides of the radial direction, set always to have a fixed impact parameter $p < R_*$ toward the stellar core. A key trick is to *choose* the radial spacing so that each nonradial ray intersecting a grid point with indices $\{i, j\}$ will also intersect other grid points $\{i \pm n, j \pm n\}$, for integer $n \geq 1$. This avoids the need to carry out a conceptually complex and computationally costly interpolation between a (p, z) ray grid for the radiation transport, and the (r, ϕ) grid for the hydrodynamics (cf. Figures 1a and 1b). For uniform azimuthal spacing $\Delta\phi$, such ray alignment occurs for radial grids satisfying

$$r_i = \frac{p}{\cos\left[i\Delta\phi + \arccos(p/R_*)\right]} . \tag{1}$$

Figure 1 illustrates the grid and ray alignment for $\Delta\phi = 5^\circ$ ($= \pi/36 = 0.089$ rad) and $p/R_* = \sqrt{0.5}$. Through spatial integration along the 3 such rays for each of the N_ϕ azimuthal zones, one obtains a "6-stream" description (i.e. in 2 directions along each of the 3 rays) for the required nonlocal escape probabilities from each of the $N_r \times N_\phi$ grid nodes.

The escape along the two nonradial rays provides a rough treatment of the lateral radiation transport. Because these rays are restricted to always impact within the stellar core radius ($p < R_*$), they are best suited for approximating the *direct* component of the line-force; but in the crucial wind-acceleration region near the star, they have a substantial azimuthal component, and so also provide a rough approximation of the azimuthal part of the *diffuse* line-force, including, for example, the important lateral "line-drag" effect that is predicted to strongly damp small-scale azimuthal velocity variations (Rybicki et al. 1990). As the rays become increasingly radial at larger radii, this capacity to approximate the lateral, diffuse radiation is lost, but the 3 rays still provide a quite accurate representation of the finite-disk form for the direct line-force. A more serious limitation arises from the severe loss of radial resolution at large radii, as demonstrated by the radial/azimuthal grid aspect ratio,

$$\frac{\Delta r_i}{r_i \Delta\phi} \approx \sqrt{(r_i/p)^2 - 1}, \tag{2}$$

which increases as r_i/p at large radii. This means that small-scale radial structure can be relatively well-resolved in the inner wind, but then becomes strongly damped by grid-averaging in the outer wind.

¿From the frequency-dependent optical depth along the 3 rays, one can obtain the corresponding integral escape probabilities in the outward and inward directions, and then from the flux moment of these, compute the direct and diffuse contributions to the line-force. (See eqs. 65 and 67 of Owocki and Puls 1996.) In addition to the radial line-force that drives the wind outflow, there is, in general, a nonzero *azimuthal* force component as well. In a rotating wind, this can arise even when the wind itself is axially symmetric, because

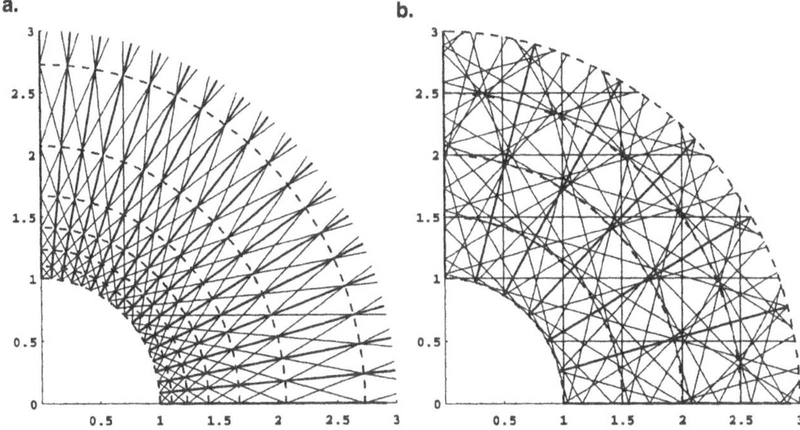

Fig. 1. a.) Illustration of 3-rays with $p/R_* = 0, \pm\sqrt{0.5}$ spaced azimuthally every $\Delta\phi = 5^\circ$, together with the radial grid spacing (dashed arcs) that aligns the rays to intersect multiple (r, ϕ) grid nodes; this allows very efficient evaluation of the ray escape integrals needed for computation of the SSF line-force in 2D wind models. b.) Contrasting the much greater complexity of a nonaligned ray-grid with about the same total number of rays, now with a coarser azimuthal resolution $\Delta\phi = 15^\circ$, but spaced every $\Delta p = \Delta r = R_*/2$ over the full model range $0 < p < 3R_*$; the chaotic, nonaligned ray intersection means that extensive interpolation would be needed both to carry out (p, z) ray integrations, and then to apply these toward computing the line-force on the (r, ϕ) hydro grid.

the different velocity gradients in the prograde and retrograde directions still imply a corresponding asymmetry in the escape probabilities (Grinin 1978). The resulting torque can lead to a moderate ($\sim 20-30\%$) spin-down of the wind rotation. (See, e.g., figure 3 in Owocki et al. 1998.)

3 Preliminary Results of 3-Ray SSF Models of CIRs

Following the general approach introduced by CO96, CIRs and other azimuthal wind variations are induced here by enhancing the wind driving from an isolated, bright spot on the rotating stellar surface. However, whereas CO96 took the line-force to be strictly radial, with a fixed enhancement factor given by the relative proximity to the spot, the simulations here compute both radial and azimuthal components of the line-force directly from the 3-ray quadrature (with the surface brightness contribution to each ray appropriately averaged to account for the larger angle range visible from greater heights). In addition to the overall spindown effect noted above, there is now also an azimuthal line-force contribution directed away from the bright spot. The models here invoke a spot with amplitude and width similar to the

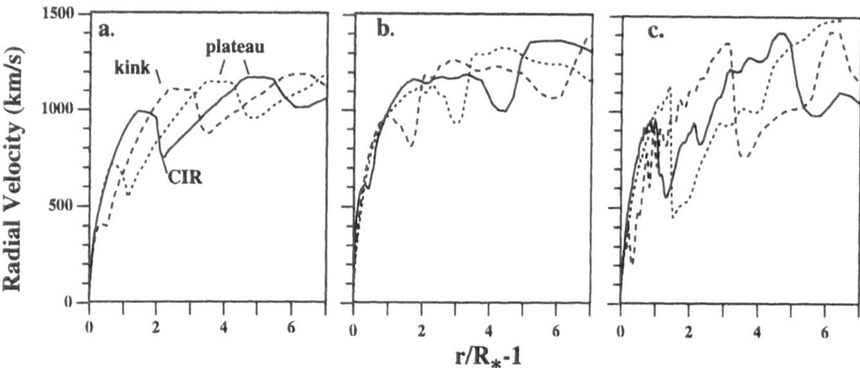

Fig. 2. Radial variation of radial velocity along selected azimuthal angles in spot-induced CIR models, with line-force treatments based on (a.) the local, CAK/Sobolev method, and the 3-ray SSF method with (b.) $p/R_* = \sqrt{0.5}$ and no azimuthal forces, and (c.) $p/R_* = \sqrt{0.8}$, including azimuthal forces.

"standard bright spot model" of CO96, with, however, a horizontal periodicity of just 45° (vs. 180°), simply to reduce computational expense. This azimuthal range is divided into $N_\phi = 157$ azimuthal zones of equal width, $\Delta\phi = 0.005\,\mathrm{rad} = 0.287°$.

Below I compare results of a CAK/Sobolev CIR simulation ("Model A") with two representative SSF models ("B" and "C") distinguished by the relative weighting of the lateral transport. Model B neglects the azimuthal force, and uses a moderate impact parameter $p/R_* = \sqrt{0.5}$. Model C includes the azimuthal force, and uses a larger impact parameter $p/R_* = \sqrt{0.8}$ that gives greater weight to the lateral radiation. The lower boundary $r = R_*$ begins with horizontal/azimuthal grid aspect ratios of 1 and 1/2 for models A and B, which require respectively $N_r = 200$ and $N_r = 141$ radial zones to reach the assumed maximum model radius $R_{max} \approx 8R_*$. As in CO96, the stellar and wind parameters are chosen to represent a standard O-supergiant (e.g. ζ Pup) with rotation speed, $v_{rot} = 200$ km/s.

Figure 2 compares results for the radial variation of radial velocity along selected aziumthal angles in various spot-induced CIR models. Panel (a) shows the characteristic structure of the CAK/Sobolev model A, with a nearly flat velocity plateaus upwind from the velocity minima that signify the dense CIR, extending back to a velocity gradient discontinuity, or "kink", that marks the connection to the unperturbed, outward-accelerating wind. As discussed in CO96, this weak, kink discontinuity propagates inward (relative to the wind outflow) at a characteristic speed $c_- \approx -v$ that is nearly as fast as the local outflow speed v, yielding a quite slow net outward propagation in

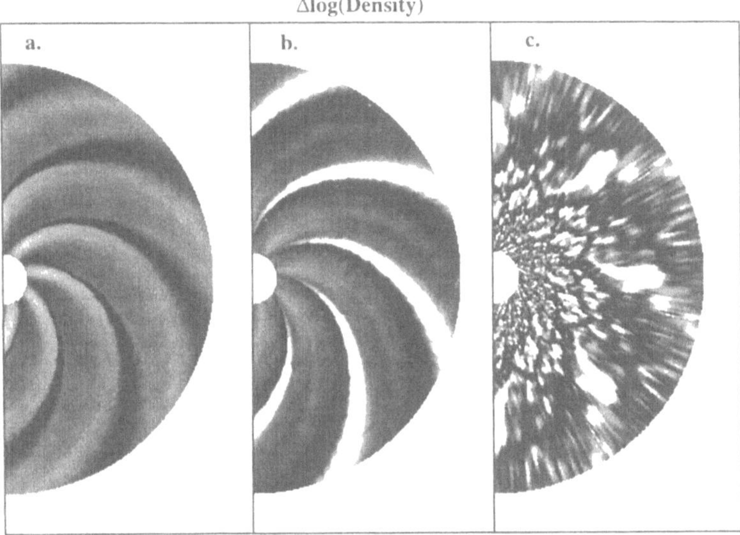

Fig. 3. 2D grey-scale representation of the density variation in the same three CIR models depicted in figure 2.

the fixed stellar frame. The slowly evolving flat velocity plateau that forms between the kink and the CIR moreover gives rise to an overall line-profile variability that has many of the observed characteristics of slowly evolving DACs (CO96).

Figure 2b shows that the corresponding 3-ray SSF model B has a roughly similar overall velocity variation, including extended velocity plateaus. In fact, apart from some moderate waviness in the velocity, there is little of the small-scale structure expected from the line-driven instability. By contrast, analogous 1D models typically show extensive, intrinsic wind structure within about $R_*/2$ from the stellar surface, even without any explicit perturbations. As discussed by Owocki and Puls (1998), this is apparently the consequence of backscattered radiation from the structured outer wind "self-seeding" perturbations at the wind base, which are then amplified by the strong instability growth. The lack of such self-seeded structure here is likely an artifact of the rapidly increasing radial grid spacing, which smooths out small-scale variations in the outer wind, effectively breaking the self-excitation cycle.

Remarkably, figure 2c shows that the 3-ray SSF model C with $g_\phi \neq 0$ and just a slightly different grid parameterization has a dramatically different velocity structure, dominated by small-scale instability variation throughout the wind, including right down to the wind base. This effectively destroys any large-scale features, including the extended velocity plateaus.

Figures 3 a-c show 2D grey-scale plots of the corresponding density variation, measured logarithmically relative to the azimuthal average. The clear spiral CIR pattern of the CAK/Sobolev spot model in panel (a) is still quite apparent in the somewhat more variable 3-ray SSF model in panel (b); but again this large-scale CIR pattern is almost completely disrupted by the extensive small-scale structure arising in the intrinsically unstable case in panel (c).

Other simulation models we have run indicate that both the inclusion of the azimuthal force and the larger lateral ray impact parameter tend to increase the intrinsic variability, with the latter being the stronger effect. The reasons for this are still unclear, and indeed run contrary with the expectation from 3D linear stability analyses (Rybicki et al. 1990) that indicate the diffuse lateral line-force should *dampen* azimuthal velocity variations, and so presumably *stabilize* the flow.

4 Summary Discussion

I caution that the results presented here are still very preliminary. Because the reasons for the marked difference in the level of intrinsic, small-scale variability are still unclear, it is not yet possible to say which of these two extreme scenarios is more likely to represent conditions in actual stellar winds. Nonetheless these results do point to two tentative new conclusions. 1.) The formation of pre-CIR velocity plateaus is not strictly an artifact of using a CAK/Sobolev form for the line-force. 2.) However, such plateaus, and indeed the entire large-scale CIR structure, can be completely disrupted if the level of small-scale, intrinsic instability becomes too strong. In this sense, the 3-ray SSF method introduced here has, despite its still very approximate nature, provided some intriguing first insights into the dynamical processes that control the multidimensional and multiscale structure that likely exists in the highly unstable and highly variable line-driven stellar winds from hot, luminous stars.

Acknowledgements: This work was supported in part by NASA grant NAGW-2624. The computations were carried out using an institutional allocation of supercomputer time from the San Diego Supercomputer Center. I acknowledge numerous helpful discussions with S. Cranmer, A. Feldmeier, A. Fullerton, K. Gayley, and J. Puls.

References

Castor, J. I., Abbott, D. C., & Klein, R. I. 1975, Ap. J., 195, 157 (CAK)
Cranmer, S. R., & Owocki, S. P. 1996, ApJ, 462, 469 (CO96)
Feldmeier, A. 1998, Ap. J., in press
Grinin, A., 1978, Sov. Astr. 14, 113

Lucy, L. B. 1984, Ap. J. 284, 351
Mullan, D. J. 1984a, Ap. J., 283, 303
Mullan, D. J. 1984b, Ap. J., 284, 769
Mullan, D. J. 1986, Astron. & Ap., 165, 157
Owocki, S. P., 1991, Stellar Atmospheres: Beyond Classical Models, L. Crivellari, I. Hubeny, D. G. Hummer, eds. (Kluwer: Dordrecht), 235
Owocki, S.P. 1998, *Cyclical Variability in Stellar Winds*, L. Kaper and A. Fullerton, eds., ESO Astrophysics Symposia Series, (Springer: Heidelberg), 325
Owocki, S. P., Castor, J. I., & Rybicki, G. B., 1988, Ap. J. 335, 914 (OCR)
Owocki, S.P., Cranmer, S.R., Gayley, K.G., 1998, in proceedings of Workshop on *B[e] Stars*, held in Paris, France, June, 1997, A. Hubert, ed., Kluwer, in press
Owocki, S. P., & Puls, J. 1996, Ap. J. 462, 894
Owocki, S. P., & Puls, J. 1998, Ap. J, in press
Owocki, S. P., & Rybicki, G. B. 1985, Ap. J., 199, 365
Owocki, S. P., & Rybicki, G. B. 1986, Ap. J., 209, 127
Rybicki, G. B., Owocki, S. P., & Castor, J. I. 1990, Ap. J., 349, 274
Sobolev, V. V. 1960, Moving Envelopes of Stars (Cambridge: Harvard University Press)

Discussion

T. Rivinius: Can the spin-down in the wind account for detached Be star disks?
S. Owocki: No, the effect only works on outwardly accelerating flows. There is no effect for material in a stable Keplerian disk.

A. Feldmeier: You mentioned that an inward propagating perturbation is only expected when the diffuse force is included in the calculation. Would you expect that the inward propagating velocity plateau in your CIR model disappears when using the pure absorption line force?
S. Owocki: In principle, I would expect this to be the case, because the pure absorption model does not allow inward propagation of information faster than the sound speed. However, because of the very strong instability of the pure-absorption case, including base variability associated with the degeneracy of the overall wind solution (see Poe et al., 1990, ApJ 358, 199), it is difficult in practice to test this notion, because all large-scale features like velocity plateaus simply become totally disrupted by small-scale structure.

N. Langer: Is the spin-down effect for hot star winds you found the end of wind-compressed disks or even wind compression per se?
S. Owocki: No, but the discovery of spin-down is what led me to investigate WCD models with non-radial forces. But it turns out that the latitudinal force is much more important for inhibiting wind compression. In fact, in early phases of P. Petrenz's independent simulation he accidently left out the latitudinal force, while including the spin-down effect, and found such models show little reduction in WCD. With all non-radial forces included,

our independent codes now agree very well on WCD inhibition.

J. Cassinelli: When you included the most sophisticated radiation treatment, you lost the CIRs. Then you went back to the less sophisticated treatment and found the decrease of wind angular momentum. So, we might be left with the impression that when you again do the sophisticated model you will lose the angular momentum decrease and perhaps also the inhibition.
S. Owocki: Sorry, this was mostly a poor choice in the order of presentation. The spin-down effect was discovered using the "most sophisticated" non-local treatment, but its basic cause can be understood in local CAK/Sobolev models. However, non-local effects actually amplify the effect near the sonic point. Furthermore, it is not quite right that CIRs are necessarily lost in the "most sophisticated models". If instability is not initiated too close to the wind base, CIRs are quite distinct even in the non-local force models. Finally, I agree that the effect of instability on WCD inhibition needs to be investigated.

G. Mellema: Would the shape of the PAMs depend on the pitch angle of the spiral pattern? Could we then use this to derive wind parameters?
S. Owocki: That is an interesting possibility that can in principle be investigated with 3-D spot/CIR models. These would be quite doable with modern computers.

F. Vakili: Do you have any idea about the origin of the asymmetry we detect with our interferometric observations of P Cygni in Hα, noting that this asymmetry is found very close to the central star?
S. Owocki: Whenever the mass loss varies with latitude, whether increasing toward the pole or equator, it seems emission from near the star will be asymmetric from any perspective that is not looking directly pole-on.

H. Henrichs: May I confront your wonderful theory with observations of ξ Per (showing Si IV DACs)? How do you define PAMS in this plot?
S. Owocki: I would concentrate on whether there is a clear increase in absorption or a modulation between increased and decreased absorption. The former suggests DACS; the latter PAMs. At first glance, it seems your example might be a case where PAMs in the inner wind evolve outward into DACs.
J. de Jong: ξ Per does show phase bowing in Si IV and the time cross-sections are quite sinusoidal. This seems to point to PAMs rather than DACs.

Session V

Pulsation

chair: A. Maeder

Pulsations in O Stars

Hubertus F. Henrichs

Astronomical Institute "Anton Pannekoek", Univ. of Amsterdam, Kruislaan 403, 1098SJ Amsterdam, Netherlands

Abstract. O stars are located in a domain of the HRD where nonradial pulsations are expected. Photometric surveys did not reveal pulsating O stars, showing that the amplitudes must be very small. Intensive spectroscopic studies yielded pulsation modes for very few O stars only, although many are line-profile variables. Wind contamination of many spectral lines is a major difficulty. Future concentrated spectroscopic efforts will undoubtedly increase the number of pulsating O stars.

The current status of our knowledge of pulsations in these stars is reviewed. From one specific example a critical attitude emerges towards the quantitative results reached so far. We also address the question whether non-radial pulsation can be the cause of the non-spherical time-dependent winds of these stars.

1 Introduction

Pulsating stars are found in nearly every part of the HR diagram. Among O stars, however, only very few pulsators are known. Fullerton et al. (1996) listed 3 confirmed and 6 suspected pulsating O stars and noticed that all of them are located in the instability strip predicted by Kiriadikis et al. (1993).

Besides its asteroseismological potential, the search for nonradial pulsations (NRP) in early-type stars is also motivated by the unknown origin of the widely observed cyclic variability in their winds, notably in the absorption parts of the ultraviolet P Cygni profiles. Most prominent are the migrating discrete absorption components (DACs) with a recurrence time scale that can be interpreted as (an integer fraction of) the stellar rotation period (e.g. Kaper et al. (1998)). The cyclic recurrence of DACs is attributed to corotating wind structures, caused by large-scale inhomogeneities. The unsolved issue is where the modulation comes from. Either magnetic fields or non-radial pulsations could equally provide the required differentiation of the emerging wind (Cranmer & Owocki (1996)). In the first case the number of wind structures matches the number of magnetic footpoints and the modulation comes directly from the stellar rotation, whereas in the case of a single NRP mode the value of the azimuthal number m determines the azimuthal distribution of the wind structures, and the modulation is caused by the traveling speed of the pulsation superposed on the stellar rotation in the observers frame. A third case could also be considered, in which coadding amplitudes of multiple modes may give rise to traveling local perturbations (de Jong et al. (1998b)).

Enhanced equatorial mass loss, to produce disks, as a consequence of sectoral NRP is a third reason for interest in pulsation in these stars.

2 Non-radial pulsations among O stars

All discoveries of NRP in O stars are based on the recognition of systematic changes in the shape of photospheric line profiles (e.g. Fig. 1). A detailed analysis of the moving features yields, in principle, the pulsation properties. In photometric searches for pulsating O stars, for example by Balona (1992), no new pulsators are found. This shows that the amplitudes must be very small, notably for high-order modes, as expected. A study of HIPPARCOS data by Marchenko et al. (1998) revealed a number of short-term periodicities in O stars, but these could not be associated with pulsations.

A few selected key discovery papers of pulsating O stars are Smith (1978) on 10 Lac, Walker et al. (1979) and Vogt & Penrod (1983) on ζ Oph, Baade (1991) on ζ Pup, and Baade et al. (1990) on the companion of γ^2 Vel. The best suitable lines are those formed deep in the photosphere. Typical lines used are Si IV λ4654, He I $\lambda\lambda$4471, 4713, 5875, 6678, C IV $\lambda\lambda$5801, 5812. Most H and He lines are often confused with variable wind contributions with comparable timescales. Typical periods are between 1 and 12 hours, and $\ell = |m|$ values up to 17. Both prograde and retrograde modes have been reported. The reliability of the pulsation parameters is limited, however (see the caveat in section 4). The systematic survey among 31 O stars by Fullerton et al. (1996) yielded line-profile variations in more than 75% of the sample, most of which are likely due to pulsations, but due to lack of coverage, no parameters could be derived in many cases. There is little doubt that a concentrated effort on such stars will reveal the pulsation properties.

The nine O stars for which pulsations are found or suspected are collected in Table 1. Typical mode properties are listed, together with wind periodicities in the last column. Figure 2 gives their positions in the HR diagram. The sample is the same as in Fullerton et al. (1996), but the number of confirmed pulsators has increased by recent work on ξ Per, λ Cep and ζ Pup.

To compare the occupied domain in the HRD with the predicted location of strange-mode occurrence in these massive stars (Kiriadikis et al. (1993), Glatzel & Mehren (1996)) a conversion from [M_V, Sp. type] to [L, $T_{\rm eff}$] is needed. The uncertainties in this conversion inhibit, however, a firm conclusion regarding the evolutionary and pulsational status of the sample stars.

Table 1. O stars with confirmed or suspected pulsations

Name	HD	Sp. Type	$v\sin i$ (km/s)	$P_{\rm NRP}$(h)	Mode($\ell = \|m\|$)	$P_{\rm DAC}$(d)
ζ Pup	66810	O4I(n)f	208	8.4, 4.3	2, 4	0.8
λ Cep	210839	O6I(n)fp	214	12.3	3	1.3:
	34656	O7II(f)	106	8.2	–	1.1
ξ Per	24912	O7.5III(n)((f))	200	3.5	3	2.0
γ^2 Vel	(WR 11)	O9I	200	8.4 - 43	6	
10 Lac	214680	O9V	32	4.9	2	7:
α Cam	30614	O9.5Ia	85	–	–	
	93521	O9.5V	400	1.8, 2.9	9	
ζ Oph	149757	O9.5Vne	400	1.1, 1.3	4 - 17	0.9

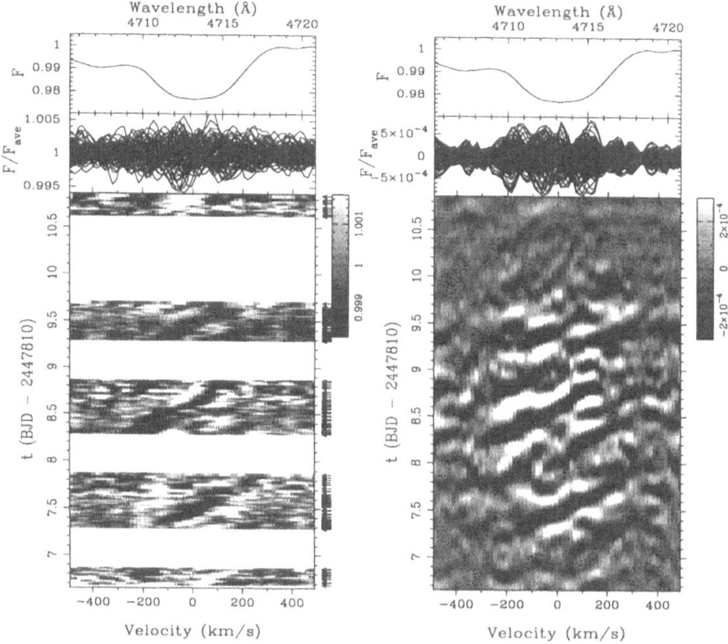

Fig. 1. Dynamic quotient He I spectra of λ Cep, along with the inverse Fourier transform of a selected frequency range, showing moving features attributed to NRP. The top panel shows the average profile. (From de Jong et al. (1998b))

3 Mode identification techniques

Since the early atlas of predicted line-profile variations as a function of pulsation parameters by Kambe & Osaki (1988), a number of powerful methods have been developed to retrieve periodicities and modes from spectroscopic time series. We mention the moment method by Balona (1986) and Aerts et al. (1992), the wavelet analysis by Townsend (1997), cross-correlation techniques by Howarth et al. (1998) and several methods based on Fourier analysis in various forms (Gies & Kullavanijaya (1988), Kambe et al. (1990), Telting & Schrijvers (1997), Kennelly et al. (1998)), each method having their own advantages and specific requirements regarding data quality and coverage. Temperature effects have been considered e.g. by Gies (1991), Lee et al. (1992) and Schrijvers & Telting (1998). Extensive applications to generated data and their success rate for mode retrieval are often included. Most, but not all these methods can be applied to O stars.

An interesting feature emerging of some of these studies is that if $\ell \neq m$, it is the value of ℓ (rather than m) that can be determined, and an additional study of the relative amplitude and phase behavior of the first harmonic frequency is needed to determine m (see Telting & Schrijvers (1997)).

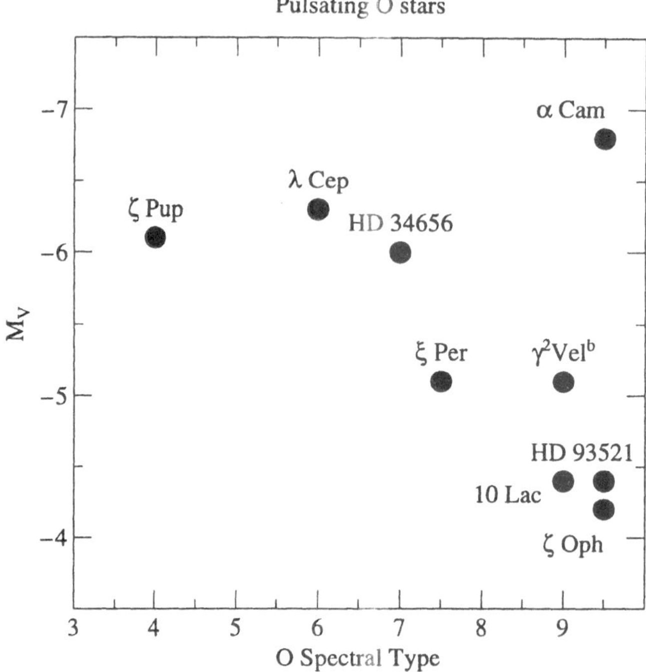

Fig. 2. The nine presently known pulsating O stars in the HR diagram. The stellar parameters are from Puls et al. (1996)

4 Multiplicity of modes: a caveat

The derived pulsation properties are limited by the signal-to-noise ratio and sampling of the dataset. Higher quality data and/or denser coverage will undoubtedly reveal more accurate values, and multiple modes, if present. As an illustration we summarize the evolving knowledge of the well studied O9.5Ve star ζ Oph during the past five years. The two main periodicities in the line profiles at 3.33 h and 2.43 h are considered as $\ell(=-m) = 4$ and 7 respectively by Kambe et al. (1993), but 4 and '5 or 6' by Reid et al. (1993) based on 360 spectra during 10 days, whereas Kambe et al. (1997) find '4 or 5', and '7 or 8', for these modes, respectively. In contrast, a recent study by Jankov et al. (1998) based on 242 spectra during 3 days yields $\ell(=-m)$ $= 5$ for the first mode, and resolves the second mode into 4 other pulsation modes with small differences in period. In the latter study modes up to $\ell \approx$ 17 and periods down to 1 h were identified. This example of the best studied O star so far clearly demonstrates that our present quantitative knowledge of pulsation modes should be considered as fragmentary in all cases of Table 1. This inhibits a fair comparison with theoretical models and shows that any asteroseismological attempt will be premature at present.

5 Wind – NRP connection?

An important question is whether non-radial pulsation can be the origin of the cyclical wind variability in O stars (Henrichs (1984), Abbott et al. (1986)). It appears in several cases that the pattern speed of the waves of a single mode running around the star superposed on the rotation is too high (usually a factor of 5 to 10, see Table 1) to be compatible with the observed wind periods in the DACs. This will be probably true for most short-period single mode pulsations, since for most O stars the rotation rate is much slower than in these examples. As discussed by de Jong et al. (1998b)), the presence of multiple modes could be relevant for the origin of cyclical wind variability. Consider for example a case with two different sectoral modes, traveling around the star with different frequencies. Matching crests with coadding amplitudes above some threshold to generate wind differentiation will appear at different longitudes and epochs. The occurrence of these cyclical surface perturbations is related to the relative traveling speeds and relative m values of the NRP waves. The temporal consequences for the wind behavior will therefore depend on the superposed rotation rate. The simultaneous presence of more than two modes will increase the complexity even more. This means that in order to match the observed periods of wind variability a fine tuning of the pulsation parameters is required in each individual case. This is not obvious because the cyclical wind periods are found to scale with the rotation periods among O (and B) stars, and this is not easily expected for a sample of different stars with a range of different pulsation parameters, but we cannot exclude this on the basis of the presently very small sample. We therefore think that the best candidate for the cause of the cyclical wind variability still remains the presence of weak magnetic fields on the surface, corotating with the star. A proof has to wait for a systematic deep survey of these fields. Preliminary upper limits on the magnetic field strength of ξ Per were presented by Henrichs et al. (1998).

6 Summarizing remarks

The evidence for the presence of pulsations in O stars has substantially increased during the last decade. In this strongly data-limited problem it can be expected that every attempt to detect pulsations in a given O star will be awarded, provided the signal to noise ratio and the time coverage are sufficient. An increase of the quality of the data will undoubtedly reveal many more details of the pulsation properties, in particular the presence of multimodes. From the timescales of the O stars studied so far, it is unlikely that pulsations alone can cause the observed cyclic wind variability. Clearly, much more studies are needed before asteroseismology can be seriously attempted.

Acknowledgments
I am very grateful to Coen Schrijvers, Jeroen de Jong, John Telting and Lex Kaper for their deep insight, continuous support and never-ceasing efforts.

References

Abbott, D.C., Garmany, C.D, Hansen, C.J. et al. 1986, PASP 98, 29
Aerts, C., de Pauw, M., Waelkens, C. 1992, A&A 266, 294
Baade, D., Schmutz, W., van Kerkwijk, M. 1990 A&A 240, 105
Baade, D. 1991, *Proc. ESO Workshop on Rapid variability of OB stars: Nature and diagnostic value* (Ed. D. Baade), p. 21
Balona, L.A., 1986 MNRAS 219, 111
Balona, L.A., 1992 MNRAS 254, 404
Cranmer, S.R., Owocki, S.P. 1996, ApJ 462, 469
de Jong, J.A., Henrichs, H.F., Schrijvers, C. et al., A&A, in press
Fullerton, A.W., Gies, D.R., Bolton, C.T. 1996 ApJS 103, 475
Gies, D.R. 1991, *Proc. ESO Workshop on Rapid variability of OB stars: Nature and diagnostic value* (Ed. D. Baade), p. 299
Gies, D.R., Kullavanijaya, A. 1988, ApJ 326, 813
Glatzel, W., Mehren, S. 1996, MNRAS 282, 1470
Henrichs, H.F. 1984, *Proc. 4th European IUE conf.* (Ed. E. Rolfe, B. Battrick), ESA SP-218, p. 43
Henrichs, H.F., de Jong, J.A., Kaper, L., Nichols, J.S. et al. 1998, *Proc. UV Astrophysics Beyond the IUE Final Archive*, ESA SP-413, 157
Howarth, I.D., Townsend, R.H.D., Clayton, M.J. et al. 1998, MNRAS 296, 949
Jankov, S., Janot-Paccheco, E., Leister, N.V. 1998, preprint
Kambe, E., Ando, H., Hirata R. 1990, PASJ 42, 687
Kambe, E., Ando, H., Hirata R. 1993, A&A 273, 435
Kambe, E., Hirata, R., Ando, H. et al. 1997, ApJ 481, 406
Kambe, E., Osaki, Y. 1988, PASJ 40, 313
Kaper, L. Henrichs, H.F., Nichols, J.S., Telting, J.H. 1998, A&A, in press
Kennelly, E.J, Brown, T.M., Kotak, R. et al. 1998, ApJ 495, 440
Kiriadikis, M., Fricke, K.J, Glatzel, W. 1993, MNRAS 264, 50
Lee, U., Jeffery, C.S., Saio, H. 1992, MNRAS 254, 185
Marchenko, S.V., Moffat, A.F.J., van der Hucht, K.A. et al. 1998, A&A 331, 1022
Puls, J., Kudritzki, R.P., Herrero, A., Pauldrach, A.W.A. et al. 1996, A&A 305, 171
Reid, A.H.N., Bolton, C.T. Crowe, R.A. et al. 1993, ApJ 417, 320
Schrijvers, C., Telting, J.H. 1998, A&A, in press
Smith, M.A. 1978, ApJ 224, 927
Telting, J.H., Schrijvers, C. 1997, A&A 317, 723
Townsend, R.H.D. 1997, MNRAS 284, 839
Vogt, S.S., Penrod, G.D. 1983, ApJ 275, 661
Walker, G.A.H., Yang, S., Fahlman, G.G. 1979, ApJ 223, 199

Discussion

P. Petrenz: Is there any chance to infer constraints on the absolute rotation rate by a comparison of NRPs of rotating and non-rotating stars?
H. Henrichs: If one finds different modes caused by rotational m-splitting, one can directly derive the rotation rate. This has been done for β Cep, for instance, but not for O stars yet, mainly because the data are not good enough to resolve the splitting. In principle, precise values can be derived, since the discrete nature of l and m constrain the allowed range strongly.

S. Shore: What happens in the case of a rotation law that is not a rigid one?
H. Henrichs: A good point: I am not aware of any line profile calculations for pulsating massive stars that take differential rotation into account, but it could be done. It would be interesting to compare such calculations with the standard ones for the purpose of diagnosing non-rigid rotation.

I. Appenzeller: What are the velocity amplitudes found in the O stars of your list and what is the present velocity detection limit?
H. Henrichs: Velocity amplitudes of pulsation are derived using a model and depend rather strongly on the model parameters (in particular the assumed inclination). Nevertheless, most amplitudes are found in the range from 5 to 10 km/s. The detection limit depends on the S/N of the spectra and is in the range of a few km/s.

Kresimir Pavlovski and Otmar Stahl

Non-radial Pulsations of BA Supergiants and Be Stars

Dietrich Baade

European Southern Observatory, Karl-Schwarzschild-Str. 2,
D-85748 Garching b. München, Germany

1 Commonalities of Be Stars and BA Supergiants

If observing time and number of photons are not the limit, it will probably be very difficult to find any Be star or BA supergiant that is not variable. Moreover, there is hardly any major set of observations that is not tempting to explain at least partly in terms of nonradial (g-mode) pulsations. Since a few years ago, such conjectures are also theoretically permissible because improved opacity calculations have established the classical κ-mechanism as a viable source of pulsation driving (cf. Pamyatnykh, these proceedings).

Contrary to Be stars, it can for any given BA supergiant nevertheless be arbitrarily difficult to diagnose nonradial pulsations (NRP's) with certainty because they need to be detected against considerable background 'noise' of other physical processes, most of which are related to mass loss and/or rotation. To make things worse, there is some evidence that NRP's can have some effect on the dynamics of the mass loss. On the other hand, variable and non-spherical winds is the subject of this Colloquium, and this paper is accordingly biased towards the interplay between pulsation and mass loss.

2 BA Supergiants

2.1 Overview

Extensive new ground- (e.g., Spoon et al. (1994)) and space-based (e.g., van Leeuwen et al. (1998)) photometry confirms the long-known universality of variabilities in luminous BA stars but puts it on a more systematic basis.

In optical spectroscopy, the tell-tale signatures of nonradial pulsations are at most occasionally detected against the dominating variations due to rotation and mass loss (Kaufer et al. (1997); see also Kaufer, these proceedings). In normal BA supergiants, Kaufer et al. (1997) do not find a dependency of the radial velocity (RV) variations on the depth of formation of the spectral lines. Therefore, the RV variability is due to some global process, as expected for NRP, and not a consequence of the variable wind. This is different in early B-type hypergiants (Rivinius et al. (1997); see also Wolf and Rivinius, these proceedings), where such a depth dependency does exist. However, when the wind is unperturbed, the Hα emission strength of these stars is well correlated with the time integral of the RV curve, which probably scales with the change in radius. Therefore, if the RV variations are due to pulsations, the latter would contribute significantly to the mass loss.

In the large majority of luminous BA stars, their periods seem to be confined between the maximal acceptable period of the radial fundamental mode and the shortest possible rotation period, which is the domain covered by nonradial *g*-modes (e.g., Kaufer et al. (1997) and Lamers et al. (1998)). Lamers et al. (1998) find, that the pulsation constants of S Dor star are typically twice as large as in normal BA supergiants, and attribute this to the larger fraction of the initial mass previously lost by the S Dor stars. They also conclude that the observed periods are an order of magnitude longer than the ones expected for strange modes. However, since virtually all stars show, often major, season-to-season variations of their periods, the identification of the variability with *g*-mode pulsation is still only a statistical one.

2.2 HD 64760: Nonradial Pulsation and Mass Loss

In a fine example of maximal extraction of information from data with poor signal-to-noise (S/N) ratio, Howarth et al. (1998) simultaneously derive photospheric and wind periods from a 16-d series of high-resolution IUE spectra. They find an 8.9-hr (or its 1-c/d alias, 14.2 hr) and a 1.2-d period in the photospheric line profile variability. Remarkably, the broad absorption troughs of the UV wind lines show the same 1.2-d period, at times also its first subharmonic of 2.4 d. Accordingly, the seed of photospheric perturbations by the nonradial pulsation possibly propagates into the high-velocity wind regime.

Because of the low S/N, the significance of these results is limited. But the agreement with similar variability patterns in O stars (cf. Henrichs, these proceedings) is convincing. Towards later spectral subclasses comparable findings are probably more difficult to obtain because (i) the preponderance of few NRP modes decreases, (ii) the periods are becoming comparable to the length of an observing season, and (iii) other wind instabilities and rotational modulation gain in importance.

3 Be Stars

Until the end of the 1970's the typical time scales sampled by observations of Be stars were weeks to years, and the phenomena covered were the waxing and waning of the circumstellar disks, V/R variations of emission line profiles, and suspected and actual orbital motions. Only in the 1980's was it discovered that Be stars also undergo comparatively more subtle spectroscopic and photometric variations with periods near one day. Henceforth many observational efforts have focussed on properly sampling this newly discovered time scale.

The at present only spectroscopic series of observations at high resolution and S/N, that covers both short (1 day) and medium (months) time scales, is the one obtained by Rivinius et al. (1998a, b, c, d, e) of μ Centauri (HR 5193, B2 IV-Ve). Since by virtue of its temporal structure this data set permits much more far-reaching hypotheses to be formulated than any other one currently does, this section intentionally concentrates on only this one star (but due to space limitations merely a very brief account of its extremely complex variability can be given). In Sect. 3.2, μ Cen is then put into a perspective with numerous observations of other Be stars.

3.1 Pulsation-driven Mass Loss Events in μ Centauri?

Multiperiodic line profile variability
The line profile variability (lpv) of this star can be decomposed into at least 6 periods which could be determined to within nearly one part in 10,000 (Rivinius et al. (1998b)). Two periods are near 0.28 d whereas the other four cluster around 0.5 d. Furthermore, the latter four show a distinct pattern in that the difference between the two lowest and the two highest frequencies is 0.034 c/d each whereas the separation of the two middle frequencies is 0.017 c/d. The lpv is the same within either of the two groups of periods but differs between the groups. A first preliminary analysis (Rivinius 1998, private communication) yields satisfactory fits with NRP velocity fields corresponding to ℓ=2 and m=–1 for the 0.5-d periods; ℓ=3 and m=–1 is still an acceptable solution but ℓ=1 appears excluded. The fitting of the 0.28-d periods is still more tentative. But –2 seems to be the best value for m, and ℓ is probably in the range 3-5 with ℓ=3-4 reproducing the observations best.
Cyclically repeating line-emission outbursts
Mu Cen undergoes frequent outbursts with a common morphology (Rivinius et al. (1998a); Rivinius et al., these proceedings). Matter is ejected above the stellar surface and possibly partly merges with a pre-existing Keplerian disk (for references concerning the Keplerian or non-Keplerian character of disks of Be stars and their discussion see J. Bjorkman, these proceedings).

The most intruiging discovery (Rivinius et al., 1998c, d, e) is that the temporal spacing between outbursts corresponds to the beat periods of the 0.5-d modes having the second and third largest amplitude with the strongest mode, i.e. 55.5 and 29.5 days. There seems to be a threshold for outbursts in the combined amplitude of the participating modes: Only these 2 combinations exceed it whereas the sum of the highest and the lowest amplitude as well as the sum of the second and the third largest amplitude are too small. Moreover, only modes with the same surface velocity field, i.e. the same ℓ and m, and/or very similar periods can contribute to an outburst, because the sum of the respective strongest mode in the 0.28-d and 0.5-d periods exceeds the threshold but outbursts with their beat period of only $\sim$0.6 d are not observed. This situation is fortunate in so far as a larger number of possible combinations would have put the star into a permanent outburst stage in which no temporal regularity would have been detected. The presence of *two* photospheric beat periods in circumstellar processes doubles the significance of the inferred causal connection between NRP and mass loss events and re-confirms the multiplicity of the 0.5-d period.

Osaki (these proceedings) has proposed a model in which the superposition of modes leads to so strong non-linearities that the involved waves break. In this case the matter velocity temporarily equals the phase velocity, which for prograde low-$|m|$ modes is larger than the equatorial rotation velocity, and may, therefore, exceed the critical rotational velocity.

3.2 Supporting Observational Evidence from Other Stars

With not even one other Be star known to exhibit the same processes as μ Cen, their relevance for the Be phenomenon in general cannot even be speculated about. But partial comparisons can be made with other Be stars and entertain the possibility that μ Cen is not the only star of its kind.

Different NRP patterns in Be and Bn stars
In addition to the presence or not of emission lines, i.e. a disk, Be and Bn stars, both of which are rapid rotators, also differ in their NRP properties (Smith and Penrod (1985), Baade (1987)): Be stars generally exhibit both low- and higher-order lpv whereas low-order lpv has not so far been detected in very broad-lined B stars without emission lines. A possible general working hypothesis is, therefore, that low-order NRP and the formation of a disk are somehow related as the case of μ Cen appears to show in more concrete terms.

Outbursts
The number of reports of outbursts in Be stars is considerable (e.g., Balona (1990), Oudmaijer and Drew (1997)) although they are often left unidentified as such. One of the most interesting results may be hidden in the MACHO database (Cook et al. (1995)). The so-called "bumpers" found in the LMC often seem to be Be stars. Of the two types of light curves, the triangular one with a steeper rise than decline is clearly reminiscent of μ Cen. But there is also a rounder and more symmetrical type which does not have its correspondence in μ Cen. This discovery is exciting because of the lower metallicity of the LMC and the metal-dependence of the κ-mechanism.

Recently, Mennickent et al. (1998) reported photometric outbursts with a period of 470 d for the proto-typical rapidly variable Be star λ Eri. However, if the data (of this publically available set) with phases between 0.9 and 1.1, which dominate the power spectrum, are plotted versus time, it becomes apparent that they are mainly due to only two events separated by $\sim$470 d. More observations are therefore needed to consolidate this result.

Sensitivity of observational quantities to outbursts
The above examples show that photometry provides an excellent sensor for outbursts. This may result either from intrinsic variations of the central star that are linked to the outburst physics or from the temporary presence of circumstellar matter or a combination of these two effects.

By contrast, the total emission strength is a poor tracer of minor outbursts also in other Be stars (Hayes and Guinan (1984), Bjorkman (1994)). Obviously, in most cases the amount of additionally ejected matter is negligible in comparison to the steady-state disk. Only at high spectral resolution and signal-to-noise ratio can variations in the wings of hydrogen emission lines be observed which are due to electron scattering close to the star. The equatorial concentration of these electrons also causes a significant polarimetric signal of the outbursts (Hayes and Guinan (1984), Bjorkman (1994)). The two effects have, in fact, also been observed simultaneously (Baade (1986)).

Competing seasonal and short-periodic variations

The strong photometric response to outbursts and similar activities implies a significant handicap of photometric techniques in the search for short-term periods, especially if the latter are very closely spaced. The need to correct for a much larger signal than the rapid variability may even discourage observers to attempt a time series analysis (e.g., Pietrzynski (1997)). Even if such a correction is performed very meticulously (e.g., Štefl and Balona (1996)), there is still the question whether the resulting redistribution of the power in the power spectrum makes the detection of multi-periodicity more difficult.

Mysteriously "variable" periods

One of the enigmas of Be stars is that observers frequently report periods with an accuracy of about 0.1% for single-season data but are forced to adjust the numbers by about 1% from season to season (e.g., Štefl and Balona (1996), Sareyan et al. (1998)). This is in marked contrast with μ Cen in which there is no indication of period variability and the beat periods and phases are reconciled with outbursts several years prior to the lpv observations (Rivinius et al. (1998c), Rivinius et al. (1998d)) and also permit predictions of the future behaviour to be made (Rivinius et al. (1998e)). The recent observations of o And by Sareyan et al. (1998) are of particular interest.

In observing runs from 5 to 11 nights with 6 telescopes at three different longitudes Sareyan et al. (1998) accumulated several thousand photoelectric measurements within 29 days. Typically, the observations spanned 6 hours per night, and with the exception of one observing run all nights seem to have been useful. In spite of this superb sampling, the authors felt compelled to conclude that the primary period of 1.6 d (plus its first harmonic which leads to a pronounced double-wave light curve) differs by up to ±4% with formal intrinsic errors of only 1% or less in different data strings. Moreover, the authors report a correlation between the value of the period and the amplitude of the variation. Over a couple of years, the amplitude can even vary by an order of magnitude and drop to the detection limit.

If this long-term variation is indicative of multi-period beating, the apparent short-term period variations could be due to the same phenomenon: If the 1.6-d period actually consists of 2 closely spaced periods, the position of the corresponding peak in the power spectrum does not match either of the two real frequencies if it is unresolved. In a light curve with this pseudo-period, there will inevitably be a phase drift, leading to the impression of a variable period. On this basis, o And could be the most promising candidate to date for a similar behaviour as found in μ Cen. However, the confusing effect of additional, transient periods during outburst phases of Be stars (Rivinius et al. 1998a, b; Štefl, these proceedings) must serve as a caveat. Since Sareyan et al. (1998) adjusted the mean differential magnitude of o And to 0.0 mag, the actual relevance of this concern cannot be readily estimated.

4 Conclusions and Perspectives

It may require quite some time or a fair amount of luck until nonradial pulsations can be convincingly diagnosed in any given luminous BA star. But the collective evidence that there are NRP's in BA supergiants (α Cygni variables), B hypergiants, S Dor stars, and possibly also η Car stars is persuasive. In some cases, an influence of the pulsation on the structure and dynamics of the radiatively driven wind is seen. The most sensitive tracers of this effect are the Hα emission line profile variations and especially the Periodic Absorption Modulations (PAM's) of UV resonance lines (cf. Owocki, these proceedings). Marlborough (1997) has recently compared major outbursts of Be stars to the eruption mechanisms considered for Luminous Blue Variables. Since there are hints that in some Be stars NRP's may contribute to the driving of outbursts, this raises the reverse question whether the eruptions of very luminous BA stars could also be seen in the light of the role that NRP's may play in Be stars.

A weird detail is that in the PAM power spectra of HD 64760 (Howarth et al. (1998)), HD 91969 (Massa's oral presentation of his paper in these proceedings), and ζ Pup (cf. Berghöfer, these proceedings) strong peaks appear close to one-half of a stellar pulsation frequency. The latitudinal node lines of tesseral modes could lead to such an odd/even effect along the line of sight to the stellar disk. However, the dominance of the first subharmonic of the NRP period also in the X-ray flux of ζ Pup would argue against such a conjecture.

Among the Be stars, by contrast, *bona fide* nonradial pulsators are quite readily identified. But only in μ Cen have so far closely spaced periods been found, that are theoretically expected for high-nonradial order g-modes. In this star, the beating of such modes with different n but the same (ℓ, m) values, i.e. identical structures of their surface velocity fields, seems to be a key ingredient to the strongly episodic mass loss. Rivinius (private communication) has investigated the question whether the two historical periods around 1918 and 1977-1989, during which persistent Hα emission was not detectable, can be understood in the same context. Although the numerical precision of the four half-day periods is not sufficient for a firm conclusion, it appears possible that at those epochs these four NRP modes were maximally out of phase. However, so long as a second star with such a behaviour has not been found, any generalization is anyway extremely premature.

Acknowledgement: I thank Stanislav Štefl for helpful comments on the manuscript and Thomas Rivinius for the permission to quote not yet published results of his PhD thesis work.

References

Baade, D. (1986): Be Star Newsletter, No. 13, p. 5

Baade, D. (1987): *Physics of Be Stars*, eds. A. Slettebak and Th.P. Snow, Cambridge Univ. Press, Cambridge, p. 361

Balona, L.A. (1990): MNRAS, 245, 92

Bjorkman, K.S. (1994): Astrophys. Space Science, 221, 335

Cook et al. (1995): *Astrophys. Applic. of Stell. Puls.*, eds. R.S. Stobie and P.A. Whitelock, ASP Conf. Ser., Vol. 83, p. 221

Hayes, D.P., Guinan, E.F. (1984): ApJ, 279, 721

Howarth, I.D., Townsend, R.H.D., Clayton, M.J., Fullerton, A.W., Gies, D.R., Massa, D., Prinja, R.K., Reid, A.H.N. (1998): MNRAS, 296, 949

Kaufer, A., Stahl, O., Wolf, B., Fullerton, A.W., Gäng, Th., Gummersbach, C.A., Jankovics, I., Kovács, J., Mandel, H., Peitz, J., Rivinius, Th., Szeifert, Th. (1997): A&A, 320, 273

Lamers, H.J.G.L.M., Bastiaanse, M.V., Aerts, C., Spoon, H.W.W. (1998): A&A, 335, 605

Marlborough, J.M. (1997): A&A, 317, L17

Mennickent, R.E., Sterken, C. Vogt, N. (1998): A&A, 330, 631

Oudmaijer, R.D., Drew, J.E. (1997): A&A, 318, 198

Pietrzynski, G. (1997): Acta Astronomica, 47, 211

Rivinius, Th., Stahl, O., Wolf, B., Kaufer, A., Gäng, Th., Gummersbach, C.A., Jankovics, I., Kovács, J., Mandel, H., Peitz, J., Szeifert, Th., Lamers, H.J.G.L.M. (1997): A&A, 318, 819

Rivinius, Th., Baade, D., Štefl S., Stahl O., Wolf B., Kaufer A. (1998a): A&A, 333, 125

Rivinius, Th., Baade, D., Štefl S., Stahl O., Wolf B., Kaufer A. (1998b): A&A, 336, 177

Rivinius, Th., Baade, D., Štefl S., Stahl O., Wolf B., Kaufer A. (1998c): *Proceedings of A Half Century of Stellar Pulsation Interpretations: A Tribute to Arthur N. Cox*, eds. P.A. Bradley and J.A. Guzik, ASP Conf. Ser., Vol. 135, p. 348

Rivinius, Th., Baade, D., Štefl S., Stahl O., Wolf B., Kaufer A. (1998d): *Cyclical Variability in Stellar Winds*, eds. L. Kaper and A.W. Fullerton, ESO Astrophys. Symp., ESO, Garching (in press)

Rivinius, Th., Baade, D., Štefl S., Stahl O., Wolf B., Kaufer A. (1998e): Be Star Newsletter, No. 33, p. 15

Sareyan, J.P., Gonzalez-Bedolla, S., Guerrero, G., Chauville, J., Huang, L., Hao, J.X., Guo, Z.H., Adelman, S.J., Briot, D., Alvarez, M. (1998): A&A, 332, 155

Smith, M.A., Penrod, G.D. (1985): *Relat. betw. Chromosph.-coronal Heat. and Mass Loss in Stars*, eds. R. Stalio and J.B. Zirker, Osserv. Astron. Trieste, p. 394

Spoon, H.W.W, De Koter, A., Sterken, C. et al. (1994): A&AS, 106, 141

Štefl, S., Balona, L.A. (1996): A&A, 309, 787

van Leeuwen, F., van Genderen, A.M., Zegelaar, I. (1998): A&AS, 128, 117

Discussion

S. Shore: I have a question and a comment. Have you looked at any of the magnetic stars, i.e., He-strong stars, that overlap with the β Cep range? There could be the sort of trapped modes at the poles that are modulated on the rotational timescales.
As a comment: in the Be stars, the formation of a boundary layer would be expected and this should damp the pulsations. Is there a change in the amplitude of pulsation during outburst?
D. Baade: No, we have not done this. But it is an interesting suggestion.
Irrespective of the presence or absence of a disk, the driving of mass loss by pulsation would remove energy from the pulsation. We have not looked for a reduction in amplitude, but given the simultaneous presence of g modes and the short duration of the outbursts proper of only very few days, I suspect that our sampling of these events is insufficient to establish conclusive results.

G. Peters: Have you looked for closely spaced periods around the dominant 0.7-day one in λ Eri? Comment 1: I have examined the entire IUE database on λ Eri and looked for cyclical wind variability. By using Scargle's method to analyse the equivalent widths of C IV, I find the same 1.3-year period that Mennickent detected in his photometry. Interestingly, most if not all of the variability is in a DAC that appears at < -1000 km/s but slowly shifts to about $-700 - -800$ km/s as the outburst progresses. Comment 2: I would like to add that when μ Cen undergoes an outburst, it can be spectacular, with He I λ6678 variability on a time scale of < 1 m (see forthcoming paper in ApJL, 502, in press).
D. Baade: In the LTPV data of λ Eri the 0.7-day period is just barely detectable. The S/N is too low to search for multiple periods around this value. Reply to comment 2: We also find something similar (Rivinius et al. 1998, A&A 333, 125).

Theory of Pulsational Instabilities of Hot Stars

Alexei A. Pamyatnykh

Copernicus Astronomical Center, Bartycka 18, 00-716 Warsaw, Poland
Institute of Astronomy, Pyatnitskaya 48, 109017 Moscow, Russia

Abstract. Basic physical mechanism of the pulsational instability in OB stars is shortly reviewed. Similarities and differences in the driving effect acting in these objects and in the classical pulsating variables are emphasized. Updated theoretical instability domains in the H-R diagram, in the $\log g - \log T_{\text{eff}}$ and in the Period $- \log T_{\text{eff}}$ diagrams are presented and compared with the observational data on β Cephei, SPB and some other types of hot variables.

1 The opacity mechanism in Main Sequence stars

This short review is devoted to the new instability strip in the HR diagram which was determined theoretically approximately five years ago (see Gautschy & Saio 1996 and references therein). It is located along upper part of the Main Sequence and is populated by Slowly Pulsating B-type (SPB) stars, β Cephei stars and other types of OB variables. In its uppermost part, at luminosities of about 10^6 of the solar value, the OB-star instability strip extends well beyond the Main Sequence and probably merges with the instability strip of classical Cepheids. At these high luminosities both radial and nonradial oscillations correspond mostly to unstable strange modes which are discussed by W. Glatzel in these Proceedings.

In the upper Main Sequence, the pulsational instability is caused essentially by the same opacity (or κ–) mechanism which drives pulsations in the classical instability strip. An exception is that in OB stars this mechanism is connected with a different opacity maximum. Figure 1 illustrates the opacity behaviour at astrophysical conditions. The main local maxima (ridges or bumps) are due hydrogen ionization, second helium ionization and due to absorption by excited ions of metals. In various stellar models these three bumps occur at approximately the same temperatures ($1.0 - 1.2 \times 10^4$ K, $4.5 - 5.0 \times 10^4$ K and $1.5 - 2.0 \times 10^5$ K, respectively), but at different densities and geometrical depths. The third maximum was missing in the earlier data, it has been recognized recently by taking into account a huge number of absorption lines produced by intrashell transitions in iron-group elements (for newest versions of two independent series of the opacity tables, OPAL and OP, see Iglesias & Rogers 1996 and Seaton 1996, respectively). In regions around opacity maxima the radiative flux from the interiors may be partly blocked during compression and converted into kinetic energy of oscillations.

It can be seen from Fig. 1 that all stellar models cross the opacity bridges. However, not all of the models are pulsationally unstable and only certain kinds of pulsation are excited in different models.

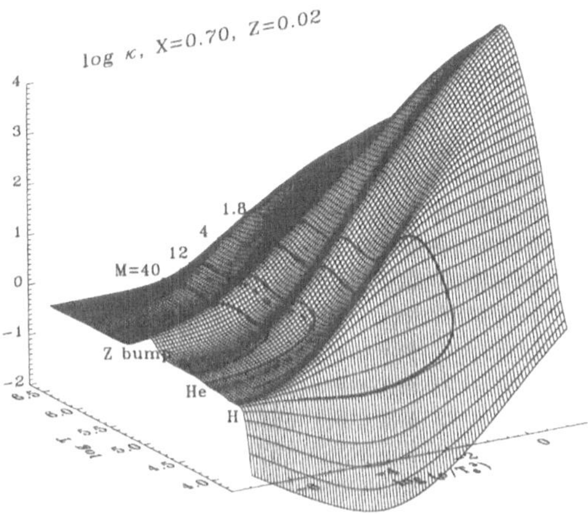

Fig. 1. Stellar opacity versus $\log T$ and $\log \rho/T_6^3$ where $T_6 \equiv T/10^6$. An OPAL table from Iglesias & Rogers (1996) for a standard chemical composition was used. Thick solid lines correspond to the ZAMS models of 1.8–40$M_\odot$, dashed line corresponds to the 12$M_\odot$ model on the TAMS (Terminal-Age Main Sequence). Hydrogen, helium and metal (Z) ridges are marked by corresponding symbols.

To explain the opacity mechanism, it is useful to consider contributions from different layers in stellar interiors to the thermodynamic work integral, W, which is the net energy gained by an oscillation mode during one cycle of pulsation (for detailed discussion, see Dziembowski (1995) and Gautschy & Saio 1995 and references therein). It is easy to show that a region in the stellar envelope in which the opacity derivative κ_T (more exactly, $\kappa_T + \kappa_\rho/(\Gamma_3 - 1)$) is increasing outwards, will contribute to the driving. Here $\kappa_T = (\partial \ln \kappa/\partial \ln T)_\rho$, $\kappa_\rho = (\partial \ln \kappa/\partial \ln \rho)_T$, and $\Gamma_3 - 1 = (d \ln T/d \ln \rho)_{\rm ad}$. If opposite is true, the region will contribute to the damping the oscillations. Note that spatial variation of $(\Gamma_3 - 1)$ due to ionization of abundant elements or due to radiative pressure effects near opacity maxima has only small influence on OB star instability.

Two additional conditions must be fulfilled to excite global stellar pulsation in a given mode: (i) the amplitude of oscillation must be relatively large and slowly varying in the potentially driving region, (ii) thermal timescale in the driving zone, $\tau_{\text{th}}(r) = \int_r^R Tc_p dM / L$, must by comparable or longer than the oscillation period. Otherwise, potentially driving region remains in thermal equilibrium during pulsation (neutral stability). The last condition means that to excite the oscillations, the opacity bump has to be located at an optimal geometrical depth in the stellar envelope.

The efficiency of the κ–mechanism in representative models of β Cephei, SPB and δ Scuti variables is demonstrated in Fig. 2. The stars are located in the middle of corresponding instability domains in the HR diagram. It is easily to see that main driving in all models takes a place in the layers with steep gradient of the opacity derivative near relevant opacity bump. Acoustic

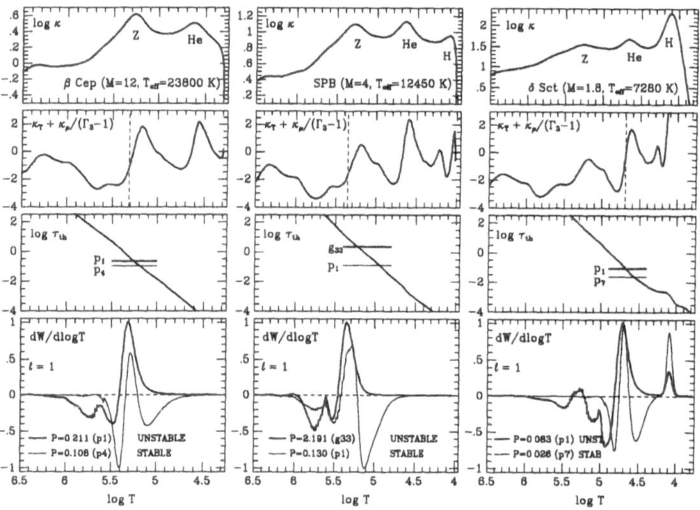

Fig. 2. Opacity, κ, opacity derivative, $\kappa_T + \kappa_\rho/(\Gamma_3 - 1)$, thermal timescale, τ_{th} (in days), and differential work integral, $dW/d\log T$ (arbitrary units, positive in driving zones), for selected pulsation modes (ℓ=1), plotted versus temperature for three representative models of β Cephei (left), SPB (middle) and δ Scuti (right) variables. All models have initial chemical composition $X = 0.70$ and $Z = 0.02$. Dashed vertical lines mark the position of the maximum driving for the unstable modes shown in the lower panels. Horizontal lines in the τ_{th} diagrams correspond to the periods of selected modes.

mode p_1 (as well as other low-order radial and nonradial modes) is excited in β Cep and in δ Sct model, but it is damped in SPB model. The helium bump region, where pulsations of δ Sct stars and other classical variables are driven, is neutrally stable in B stars due to very short thermal scale there. However, the timescale requirement can be satisfied at the position of the metal bump. In the β Cep model of $12M_\odot$, the thermal timescale there is comparable with periods of the low order p-modes. In the SPB model of $4M_\odot$, the thermal timescale at the metal bump is approximately 20 times longer because the bump is located deeper and due to smaller luminosity of the star. It is comparable now with periods of the high-order g-modes which become unstable. For low order g- or p-modes the damping between helium and metal bump is activated, therefore these modes are stabilized. More detailed discussion of the SPB versus β Cep–type instability in terms of the work integral is given by Dziembowski et al. (1993) and Dziembowski (1995).

The long-period, high-order g-mode instability extends to high luminosities, well beyond the observed SPB domain in the HR diagram (see Pamyatnykh 1998 and Fig. 4 below). In this region, two separate frequency ranges appear in the oscillation spectrum, which can be easily understood with help of Fig. 3. In upper panels, the MS evolution of the spectrum of quadrupole ($\ell = 2$) modes for a $30M_\odot$ star is shown. In a model of $\log T_{\rm eff} \approx 4.44$ both high-order $g_{14} - g_{19}$ modes with periods 2.4 – 3.1 days and low-order p_0, $g_3 - g_6$ modes with periods 0.6 – 1 days are excited by the metal opacity bump. (We classify acoustic and gravity modes according to distribution of their kinetic energy inside the star.) For these modes the oscillation amplitude is large only in the outer layers where driving occurs. In deep interiors, the amplitude behaviour for these two groups of modes is very different but the contribution to work integral is negligible in both cases. On contrary, intermediate modes which have relatively small amplitudes in the metal bump region, are stabilized in deep layers. A small driving at $r/R \approx 0.55$ caused by the local opacity maximum at $\log T \approx 6.25$ is not enough to outweigh the damping. Note that in hotter models, the high-order g-modes may also have large amplitudes in the potentially driving zone, but the thermal timescale is short there in comparison with the periods.

2 Instability domains

Figure 4 shows the instability domains along the upper Main Sequence. The models were computed using OPAL opacities and a standard stellar evolution code developed by B. Paczyński, M. Kozłowski and R. Sienkiewicz (private communication). The linear nonadiabatic analysis of low-degree oscillations ($\ell \leq 2$) was performed with a code developed by Dziembowski (1977). For general consideration, the effects of stellar rotation, mass loss and convective overshooting were neglected.

Slowly Pulsating B star models are unstable only to nonradial high-order gravity modes. Due to very dense frequency spectrum, a lot of modes are unstable simultaneously: for example, in the TAMS $4M_\odot$ model, the modes $g_{43} - g_{85}$ of $\ell = 1$ and $g_{36} - g_{81}$ of $\ell = 2$ are excited (the period ranges are 2.0–4.5 and 1.4–2.7 days, respectively). As it was discussed in the previous

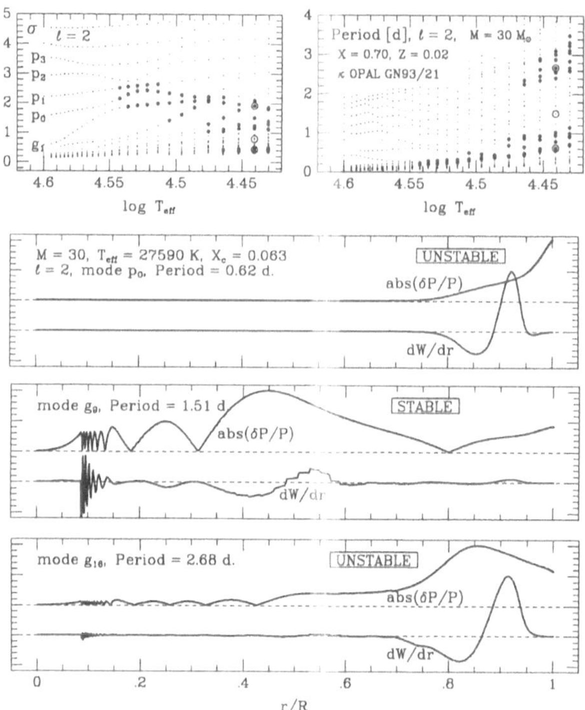

Fig. 3. *Upper panels:* Dimensionless frequencies, $\sigma = \omega/\sqrt{4\pi G\langle\rho\rangle}$, and periods of quadrupole modes for a $30M_\odot$ star in its MS evolution (from the left to the right). The small and large dots correspond to stable and unstable modes, respectively. Open circles at $\log T_{\mathrm{eff}} \approx 4.44$ mark modes shown in the lower panels.
Lower panels: The differential work integral, dW/dr, and the pressure eigenfunction, $\delta P/P$ (both in arbitrary units), for three selected modes. Zero-lines for the variables are shifted one relative to another. The metal opacity bump is located at $r/R \approx 0.916$.

section, this instability extends to high luminosities, the extension is continuous for the OP opacities (Pamyatnykh 1998) or for models with convective overshooting resulting in a widening of the MS band. For the SPB domain and its extension, the computed Red Edge coincides with the TAMS, which is explained by very strong damping of the high-order gravity modes in evolved post-MS star models.

In fact, the opacity driven domains in the HR diagram present a sequence of several instability strips shifted in effective temperature one after another, as it can be easily understood from the position of blue (hot-temperature) boundaries. From the right to the left, we have classical instability strip driven by helium bump, high g-mode instability strip, and β Cep instability

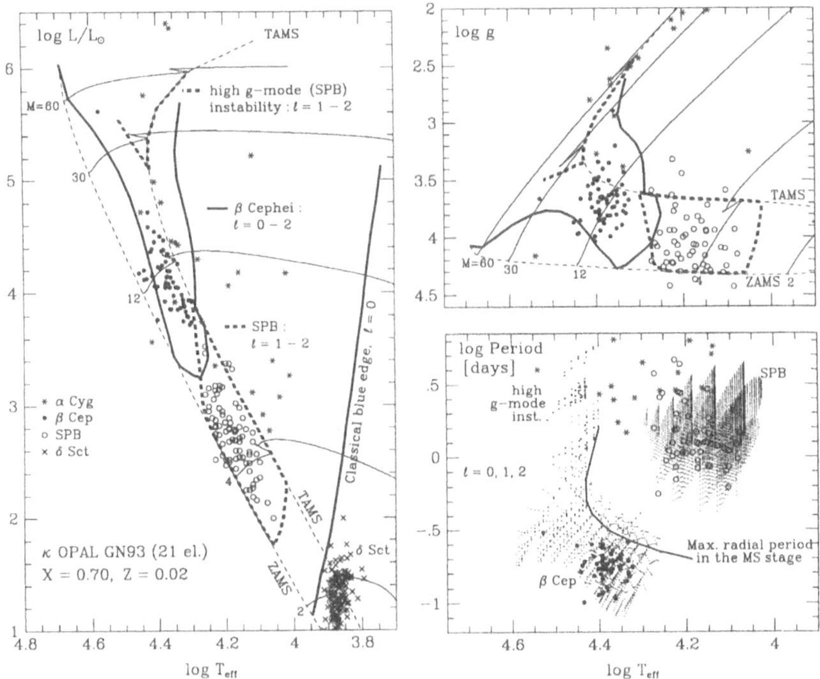

Fig. 4. *Left:* Instability domains in the upper Main Sequence. For the classical instability strip, only blue edge is shown. Thin solid lines show evolutionary tracks for a few stellar models with masses given in solar units. Thin dashed lines show ZAMS and TAMS. The positions of photometrically variable stars are due to L. A. Balona (private communication on β Cep in open clusters), Sterken & Jerzykiewicz (1993, β Cep), North & Paltani (1994, SPB), Waelkens et al. (1998, α Cyg and SPB), Garcia et al. (1995, δ Sct).
Right top: β Cep and high-order g-mode (SPB–type) instability domains in the the $\log g - \log T_{\rm eff}$ diagram
Right bottom: Periods of unstable low-degree modes in β Cep and SPB domains. For clarity, only MS models of $M \leq 40 M_\odot$ are considered in the β Cep domain (post–MS models result in a decrease of the gap between β Cep and SPB domains). Solid line marks the position of the TAMS line for radial pulsations in the β Cep domain. All points to the right of the line correspond to low-order nonradial modes in this domain. Only main observational periods are plotted for multiperiodic variables.

strip, both driven by Z bump. Different slope of the SPB and especially of the β Cep domain is caused partly by effects of large radiative pressure on structure of high luminosity stars.

Both β Cep and SPB instability domains are very sensitive to the assumed metallicity and shrink with decreasing value of Z (see Dziembowski 1998).

Observational data include results obtained with Hipparcos (Waelkens et al. 1998). Because of limited space, we do not go beyond a simple illustrative comparison. The theoretical instability domains fit almost all of the observed β Cephei and the SPB variables.

Acknowledgments. Most of the theoretical results presented here were obtained and discussed in a group headed by Wojciech Dziembowski (WD, Paweł Moskalik, Ryszard Sienkiewicz, AAP). I appreciate a support due to grants KBN 2-P03D-014-14, RFBR 98-02-16734 and a travel grant from IAU. Many thanks to the organizers of the meeting for their hospitality.

References

Dziembowski W. A. (1977): Oscillations of giants and supergiants. Acta Astron. **27**, 95–126

Dziembowski, W. A. (1995): Pulsation in hot stars. *Astrophysical applications of powerful new databases*, eds. Adelman S.J., Wiese W.L., ASP Conf. Ser., vol. 78, 275–289

Dziembowski, W. A. (1998): B star pulsation – theory and seismological prospects. *New eyes to see inside the Sun and stars*, Proc. IAU Symp. 185, eds. Deubner F.-L., Christensen-Dalsgaard J., Kurtz D.W., Kluwer, Dordrecht, 355–360

Dziembowski W. A., Moskalik P., Pamyatnykh A. A. (1993): The opacity mechanism in B-type stars – II. Excitation of high-order g-modes in main-sequence stars. MNRAS **265**, 588–600

Garcia J. R., Cebral J. R., Scoccimarro E. R., et al. (1995): A catalogue of variable stars in the lower instability strip. A&AS **109**, 201–262

Gautschy, A., Saio, H. (1995): Stellar pulsations across the HR diagram. I. ARA&A **33**, 75–113

Gautschy, A., Saio, H. (1996): Stellar pulsations across the HR diagram. II. ARA&A **34**, 551–606

Iglesias, C. A., Rogers, F. J. (1996): Updated OPAL opacities. ApJ **464**, 943–953

North P., Paltani S. (1994): HD 37151: a new "slowly pulsating B star". A&A **288**, 155–164

Pamyatnykh A. A. (1998): Pulsation instability domains in the upper main sequence. *A half-century of stellar pulsation interpretations*, eds. Bradley P.A., Guzik J.A., ASP Conf. Ser., vol. 135, 268–269

Seaton, M. J. (1996): Interpolations of Rosseland-mean opacities for variable X and Z. MNRAS **279**, 95–100

Sterken C., Jerzykiewicz M. (1993): β Cephei stars from a photometric point of view. Space Sci. Rev. **62**, 95–171

Waelkens, C., Aerts C., Kestens E., et al. (1998): Study of an unbiased sample of B stars observed with Hipparcos: the discovery of a large amount of a new slowly pulsating B stars. A&A **330**, 215–221

Discussion

J. Cassinelli: The star ϵ CMa lies just beyond the red edge of the β Cep instability zone and has a very large EUV flux (in the 500 – 700 Å region). Is it possible that pulsational energy that arises deep in the envelope is deposited as heat rather than as an organised pulsation?

A. Pamyatnykh: The red edge of the β Cep instability domain is determined in our computations in the same way as the blue edge: if the driving metal opacity bump is too close to the surface (a model to the left of the blue edge), a star will be stable because this potential driving zone will be in thermal equilibrium due to the very short thermal timescale. If the driving metal opacity bump is located too deep in the envelope (a model to the right of the red edge), then: (i) the damping above this region will be activated; (ii) the oscillation amplitude of acoustic-type modes in the driving zone is significantly smaller; and (iii) the oscillations in deep layers are nearly adiabatic. Due to all these factors, the star will be stable as well. So, formally, we do not need any additional energy losses to stabilise oscillations to the right of the red edge. However, the position of the red edge depends quite strongly on stellar parameters, especially on metal abundance and also on the model of the atmosphere, which is important to determine conditions of possible dissipation of pulsational energy in the outermost layers. Therefore, the EUV flux you mentioned may indeed be a manifestation of such an energy dissipation. More definitive answers require a detailed theoretical study of this star and its oscillations.

M. Marlborough: Is it possible in a rapidly rotating B star, e.g., for the polar regions to be stable while the equatorial regions are unstable due to opacity effects?

A. Pamyatnykh: In the evolutionary computations we take the rotational effect into account by adding a correction due to an averaged centrifugal force to the hydrostatic equilibrium equation, so the modified structure of a star remains spherically symmetric. Therefore, there is no difference in the driving effect between polar and equatorial regions in our models. Lee (1998, ApJ 497, 312) found that due to rotational deformation of the outer envelope the damping effect above the main driving metal-bump region can become significant in the equatorial zone. It may result in the stabilisation of some axisymmetric non-radial modes of a rapidly rotating B star which are unstable in the non-rotating case.

A. Maeder: You have shown the metal opacity peak at 200 000 K, but in your plots of $\log \kappa$ vs. $\log \mathrm{T}$ and $\log \mathrm{P}$ there was another peak at higher temperatures. Is it driving any pulsations?

A. Pamyatnykh: The deeper opacity bump near $\log \mathrm{T} \sim 6.3 - 6.4$ is produced by fine-structure transitions in iron-group elements as well as in carbon and oxygen. It may be important to helioseismology and to the lithium problem in low mass stars due to some modification of solar and stellar interior structure (see, e.g., the review by Rogers & Iglesias on astrophysical opacity, Science, 1994, 263, 50). As for the driving of pulsations: this bump may be potentially important for some very hot stars to the left of the upper main sequence, but I don't know any study about this matter.

Alosza Pamyatnykh (returns from being lost in the Odenwald)

Non-radially Pulsating Hot Stars: Non-radial Pulsations and Be Phenomenon

Yoji Osaki

Department of Astronomy, School of Science, University of Tokyo, Bunkyo-ku, Tokyo, 113, Japan

Abstract. We discuss a possible role of non-radial oscillations as a cause of mass-loss in hot stars. In particular, we propose a working model for the episodic mass-loss in Be stars. In this model, equatorial mass loss is thought to be driven by wave-breaking phenomenon of large-amplitude non-radial waves and a circumstellar disk could thus be formed around the equatorial plane of a rapidly rotating star. A kind of relaxation-oscillation cycle could be established between the Be phase and non-Be phase, in which an interplay between non-radial oscillations in stellar atmosphere and the circumstellar disk is essential. We also discuss a viscous decretion-disk model for the circumstellar envelope around Be stars.

1 Introduction

As more precise observations have been made for luminous hot stars both photometrically and spectroscopically, it becomes evident that pulsations and oscillation related phenomena are ubiquitous in these stars. Since the main objective of this colloquium is to discuss non-spherical and variable mass loss from these stars, it is my understanding that my task assigned is to discuss the connection between the non-radial pulsations (abbreviated as "NRP") and mass loss in hot stars. Thus I will discuss in this talk a possible role of NRP for equatorial mass loss in hot stars. In particular, I propose a working model for episodic mass loss in Be stars.

2 NRP and Mass Loss in Be Stars

The Be stars are those rapidly rotating B-type stars that exhibit from time to time emission lines in their spectra, most conspicuously in hydrogen Balmer lines (see, e.g., Slettebak, 1988). Emission lines in these stars are believed to be formed in the circumstellar disks around stars. More than a half century ago, Otto Struve (1931) had already proposed a simple model of Be stars where a rapidly rotating B-type star is surrounded by a rotating equatorial disk which is supported by centrifugal force against stellar gravity. This simple picture for Be stars still seems to be basically correct.

The most fundamental problem of Be stars is the origin of the Be phenomenon. That is, the basic question, why these stars have circumstellar envelopes in the first place, has still remained unanswered. It has been known that the rapid rotation in these stars is not sufficient to produce mass loss and equatorial disks. Some extra mechanism other than rapid rotation of a star seems to be needed. As discussed below, one of possible mechanisms seems to be "NRP" in these stars.

3 Basic Model for Mass Loss in Be Stars by NRP

A very interesting small workshop called "The connection between non-radial pulsations and stellar winds in massive stars" was held in Boulder, Colorado in 1985 and its summary of review talks appeared in PASP in 1986. In this workshop, Penrod (1986) suggested a model for episodic mass loss in Be stars. Since his results were very important and since my working model presented below basically follows his suggestion, here I summarize his conclusions in what follows. Penrod (1986) observed about 1500 spectra of a sample of 25 rapidly rotating Bn and Be stars. He pointed out in his talk that

1. "all but two of the program stars show obvious line-profile variations due to nonradial pulsation".
2. "The non-emission B stars are each pulsating in one or two short period high-degree ($\ell = 4$ to 10) modes while the Be stars are in all cases pulsating in a long period $\ell = 2$ mode, and usually in a short-period high-ℓ mode as well" (see, also Baade, 1987).
3. "There is no evidence of excessive winds or episodic ejection events in the large-amplitude radial pulsating β Cephei stars, or in the slowly rotating 53 Persei stars, which like the Be stars are energetic long-period $\ell = 2$ pulsators."

He suggested from these observational evidences that "both rapid rotation and a long-period nonradial pulsation mode are essential ingredients of a Be star."

Here we propose a working model for episodic mass loss in Be stars along the line suggested by Penrod (1986). It is a scenario of relaxation-oscillation cycle between Be phase and non-Be phase for a star. In fact, this kind of model is not new but it has existed for more than a decade as Penrod (1986), Osaki (1986), and Ando (1986) proposed models similar to the present model.

Our scenario of relaxation-oscillation cycle of Bn and Be phases goes in the following way.

1. Start of relaxation-oscillation cycle. We begin with non-emission phase in which no disk exists around the star. It is assumed in this model that the B stars in our interest are pulsationally unstable to nonradial g-modes. In fact, it is now well established that hot stars in a wide range of B spectral type are pulsationally unstable to NRP g-modes due to the opacity bump of Fe-peaked elements (Pamyatnykh, 1998).
   In particular, some of NRP g-modes with $\ell = 2$ are assumed to be unstable and grow in amplitude.

2. As (prograde-type) travelling waves of NRP g-modes with $\ell = 2$ and $m = -2$ grow in amplitude reaching some large amplitudes, angular momentum is transported within the star by these NRP modes. As discussed by Osaki (1986), this angular momentum transport occurs usually in a sense that angular momentum is transported from the deep interior to the stellar surface and thus the atmospheric layer is accelerated, if waves are prograde.
3. Ultimately, mass loss will occur from the equatorial region and the rotating disk will then be formed around the equatorial region above the star. As discussed below, mass loss may be driven by *breaking* NRP waves. This corresponds to the Be phase.
4. Once the circumstellar disk is formed around the stellar equator, the outer boundary condition for NRP waves has changed in a sense that NRP waves can leak from the stellar surface into the circumstellar envelope. Leakage of wave energy into the circumstellar envelope leads to damping of NRP waves. Thus the wave amplitudes decrease greatly.
5. Once the NRP waves are damped, the supply of mass and angular momentum to the circumstellar envelope stops, which results in the dissipation of the circumstellar disk. Most of mass and angular momentum of the disk is then accreted back onto the star once more while some small amount of mass will leave the system by carrying away extra angular momentum.
   This corresponds to the end of Be phase and the star returns to the Bn phase. This is the end of the relaxation-oscillation cycle and a new cycle starts by going back to the phase 1.

4 Critical Examination of the Model

Let us now scrutinize this scenario for the Be phenomenon. In particular, following three points must be cleared, in order for this scenario to work.

4.1 Angular Momentum Transport by NRP Waves

Problems of angular momentum transport by travelling-type NRP waves within stars have been discussed by Osaki (1986) and Ando (1986) and it need not be repeated here. In short, if progradely travelling NRP waves are excited in some finite amplitudes in stars, NRP waves can carry both energy and angular momentum from the exciting region to the dissipation region within the star. Since the stellar atmospheric regions usually act as dissipation zone in stellar NRP, angular momentum will be deposited in the stellar atmosphere by progradely travelling NRP waves, which may accelerate the stellar atmosphere toward the break-up velocity.

It may be noted here that Kambe et al. (1993) have found observationally that the stellar rotational velocity in a Be star, λ Eri, increased from $V_{\rm rot} =$

380 km/s in the non-emission Bn phase to $V_{\rm rot} = 480$ km/s during Be active phase.

4.2 Mass Loss from the Equatorial Region

Even if the stellar rotational velocity approaches close to the break-up velocity, we need some extra mechanisms for mass ejection to occur from the stellar surface and to form a circumstellar disk. Here we propose a wave "breaking" phenomenon as one of possible causes for mass ejection. In this picture it is assumed that progradely travelling NRP waves in the stellar equatorial region may attain large amplitudes so that non-linear phenomena become important. The wave breaking phenomenon is one of such non-linear effects.

It is well known in ocean waves that wave breaking occurs when amplitudes of waves become sufficiently large (see, e.g., Kinsman 1965). Figure 1 illustrates schematically the wave breaking phenomenon. It is known that when the wave breaking occurs, matter of the wave crest is detached from the main part of fluid. The condition for wave breaking to occur is known as Stokes criterion in that matter at the wave crest moves with the same speed as that of wave profile (i.e., the phase speed of waves). If the star rotates very rapidly, the additional velocity produced by the wave breaking phenomenon may be large enough for the matter to attain above the Keplerian velocity at the stellar surface and to leave the stellar surface by joining the circumstellar disk. In this picture mass is thrown out not vertically but instead horizontally in the direction of wave propagation (in our case in the direction of rotation) near the stellar equatorial region.

Fig. 1. Schematic wave profile when wave breaking occurs. Matter at the very crest moves at the same speed as the wave profile.

We may now understand Penrod's finding why the low-degree modes with $\ell = 2$ is favoured for the occurrence of the Be phenomenon. We note here that the phase velocity of waves is given by $R \times \omega/(-m)$, where R is the radius of the star and ω is the angular frequency of NRP waves and m is the azimuthal order of NRP modes. For a given angular frequency of NRP waves, the largest phase velocity is realized for the lowest order modes. The lowest order modes observed in early-type stars are usually NRP modes with $\ell = -m = 2$.

In this respect, the most interesting observations are those by Rivinius et al. (1998) of μ Cen, a well known Be star. These authors have discovered that emission outburst events occur in μ Cen, whenever the vectorial sum of amplitudes for three NRP modes having closely spaced periods around 0.505 d exceed some critical level by multi-mode beating.

Furthermore these authors (Rivinius et al. 1998) have noted that conspicuous sharp absorption spikes appear close to the line wings, particularly during precursor and early phases of a line emission outburst. This phenomenon could be well understood in term of wave breakings. As seen in figure 1, the NRP wave profiles will then be very steep and almost vertical near the wave crest, that is, they have saw-tooth forms. This means that the wave velocity almost jumps from negative minimum to positive maximum when the wave crest passes. When this wave velocity component is added to stellar rotational velocity, it gives rise to a sharp violet absorption spike near the line wing when the wave crest comes around the stellar limb. As waves propagate over the visible hemisphere, the same feature of absorption spikes will become visible near the red wing after about one-half cycle of oscillation period. Since the wave velocity vector of non-radial g modes is dominantly directed horizontally, these wave fronts are best visible near the line wings. These authors have also noted that "narrow absorption spikes can occur in both the blue and the red wing but are never seen in both wings simultaneously". This indicates that waves of our interest are produced by $\ell = -m = 2$ modes as only one wave crest is seen at any one instance on the visible hemisphere for these modes.

4.3 Damping of NRP Waves in Stars

In order for our scenario of the relaxation oscillation cycle to work, damping of NRP waves must occur before the dissipation of the circumstellar disk. In other words, we need a phase lag between the damping of NRP waves in stars and the dissipation of the circumstellar disk. Otherwise, a kind of steady state could be established in a sense that some low-amplitude NRP waves might be maintained with reduced leakage of wave energy into a remnant circumstellar disk.

The damping of NRP waves is supposed to occur in the dynamical timescale of the star which is on the order of days while if the supply of mass and angular momentum into the circumstellar disk ever stops, the dissipation of the circumstellar disk is supposed to occur in viscous timescale which is on the order of years. Thus we may expect reasonably a time lag between the occurrence of damping of NRP waves and that of dissipation of the circumstellar disk.

5 Circumstellar Disk in Be Stars

As discussed in the previous section, mass is ejected from the equatorial region by wave-breaking to form the circumstellar disk. The disk is basically supported against gravity by the centrifugal force of rotation and therefore its rotational velocity is Keplerian. As mass supply from the star to the disk continues, the matter drifts outwardly in the disk by viscous stresses. Such a disk is called either "decretion disk" or "excretion disk" .

The basic equations describing the decretion disk are basically the same as those of accretion disks which have been studied extensively for cataclysmic variables and X-ray binaries. A question may then naturally arise why a disk behaves sometime as "decretion disk" while sometime as "accretion disk" and what causes this difference. The answer to this question is the difference in boundary conditions !

In order for a decretion disk to be formed, strong torques must act at the inner edge of the disk (which is assumed to extend to the surface of the star). In our case, the NRP waves are supposed to work for this purpose. As travelling-type NRP waves propagate outward into the stellar atmosphere from much denser inner part, wave amplitudes increase and they ultimately begin to break. This wave breaking phenomenon deposit mass and angular momentum in the stellar surface (i.e., at the inner boundary of the decretion disk), which serves for the necessary boundary condition for the decretion disk.

The steady-state decretion disks (or the excretion disks) around Be stars have been studied by Lee, Saio, and Osaki (1991). It must, however, be cautioned here that the steady state decretion disk may never be realized in realistic situations because the viscous timescale is very long of the order of years to decades. The viscous timescale, $t_{\rm visc}$, and the radial drift velocity, v_r, in the decretion disk are estimated as

$$t_{\rm visc} \sim \frac{r^2}{\nu} \sim \frac{r}{\alpha c_s}\frac{r}{h}$$

$$v_r \sim \frac{r}{t_{\rm visc}} \sim \alpha c_s \,(h/r)\,,$$

where r and h are the radial coordinate and the half thickness of the disk, respectively, c_s is the sound speed, and ν and α are the kinematic viscosity and the Shakura-Sunyev viscosity parameter which has been used widely in the theory of accretion disks and it is thought to be $\alpha \lesssim 1$.

The viscous timescale of the decretion disk is estimated to be of the order of years to decades and the the drift velocity is of the order of 1 km/s or less. The matter in the decretion disk therefore drifts outward very slowly and thus the decretion disk model is in a good contrast with the wind-compressed disk model proposed by Bjorkmann and Cassinelli (1993) in which the typical expansion velocity is much higher.

The observational evidence for the existence of Keplerian disks in Be stars is provided by the long timescale variation in V/R ratio in emission lines. Some of Be stars exhibit the V/R variation in which the the intensity ratio of the "violet" and "red" components of double emission lines change quasi-periodically with a timescale of several years. Huang (1975) has shown that this curious phenomenon could be explained phenomenologically by the precessing elliptical ring model in which the disk takes an eccentric ring form with its apsidal line slowly precessing. Physical basis for the formation of the eccentric disk in Be stars has been proposed by Okazaki (1991) who has shown that decretion disks in Be stars may be unstable to non-axisymmetric perturbation of degree $m = 1$ (i.e., one-armed spiral mode) and the disk is then deformed into a lopsided shape.

6 Summary and Conclusion

In this talk, we have discussed a working model for episodic mass loss in Be stars in which a kind of relaxation oscillation cycle may be established between NRP waves in the stellar surface and the circumstellar envelope. In particular, I have proposed wave-breaking of non-radial g-modes as a possible mechanism for mass and angular-momentum injection from the stellar surface to the circumstellar envelope. It is concluded that NRPs could be responsible for episodic mass loss in Be stars.

Acknowledgement: I would like to thank Dr. Eiji Kambe for informative discussions on this subject.

References

Ando, H. (1986): A&A, 163, 97

Baade, D. (1987): in *IAU Colloq. 92, Physics of Be stars* ed., A. Slettebak and T. P. Snow (Cambridge University Press), p. 361

Bjorkmann, J.E., Cassinelli, J.P., (1993): ApJ, 409, 429

Huang, S.-S. (1975): Sky and Telescope, 49, 359

Kambe, E., Ando, H., Hirata, R., Walker, G.A.H., Kennelly, E.J., and Matthews, J. M. (1993): PASP, 105, 1222

Kinsman, B. (1965): *Wind Waves: their generation and propagation on the ocean surface* (Prentice Hall, New York)

Lee, U., Saio, H., and Osaki, Y. (1991): MNRAS, 250, 432

Okazaki, A. (1991): PASJ, 43, 75

Osaki, Y. (1986): PASP, 98, 30

Pamyatnykh, A. A. (1998) : in these proceedings

Penrod, D. (1986): PASP, 98, 35

Rivinius, Th., Baade, D., Stefl, S., Stahl, O., Wolf, B., and Kaufer, A. (1998): A&A, in press

Slettebak, A. (1988) : PASP, 100, 770

Struve, O. (1931): ApJ, 73, 94

Discussion

T. Rivinius: From the peak separation variability we have derived a radial drift velocity of the order of a few km/s.

J. Bjorkman: Is the dissipation zone where you deposit angular momentum by NRPs in the photosphere? If so, to obtain a viscous decretion disk, this zone must rotate at Keplerian velocities. Why then do you not observe evidence for such large rotation speeds in the photosphere line profiles?
Y. Osaki: Yes, it is. I suspect that the equator rotates at break-up speed. One possible explanation for why we do not observe such large rotation speeds might be that the rapid rotation is confined to the equator and matter at higher latitude rotates slower. The line profiles are produced by an average over the whole surface area.

K. Bjorkman: Can you comment on the time scale you expect for dissipation of the disk? And is your proposed cycle expected to be strictly periodic?
Y. Osaki: The viscous time scale depends on the size of the disk. I expect that the viscous time scale is on the order of several years for a disk of several stellar radii. The scenario I presented is an idealised model. In reality, many effects will enter. This gives rise to a quasi-periodic phenomenon.

S. Shore: Might you be able to drive (non-linear) breaking Rossby waves, depending on the rotation law, to drive higher latitude mass loss?
Y. Osaki: I am not able to answer your question. In my talk I discussed mass loss by centifugal acceleration. In such a case, we get a circumstellar ring or disk but not a stellar wind. A ring around the star at high latitude seems rather unlikely to me.

Pulsation Hydrodynamics of Luminous Blue Variables and Pulsation-Driven Winds

Joyce A. Guzik, Arthur N. Cox, Kate M. Despain, and Michael S. Soukup

Los Alamos National Laboratory, Los Alamos, NM 87545-2345 USA

Abstract. Many physical factors, including radial and nonradial pulsation, rotation, radiation pressure, convection, magnetic fields, or dynamical instabilities may play important roles in the hydrodynamics of Luminous Blue Variables. We review the current status of hydrodynamic modeling of LBV envelopes, and describe results of our models using the one-dimensional nonlinear hydrodynamics code of Ostlie and Cox. We find that the models pulsate in several simultaneous radial modes, driven by the helium and Fe ionization κ effect. The pulsations have quasi-periods between 5 and 80 days, with radial velocity amplitudes of 50-200 km/sec, and may be identified with the LBV microvariations. In some cases, depending on luminosity-to-mass ratio and helium abundance, deep layers in the model can periodically exceed the Eddington luminosity limit. The key to exceeding L_E is the inclusion of the time dependence of convection: Near the regions of opacity peaks produced by Fe and helium ionization, convection is turning on and off during each pulsation cycle. If convection cannot turn on rapidly enough to transport the required luminosity through the region, the Eddington limit is exceeded. If this region of the star is sufficiently adiabatic, an "outburst" may occur. In the hydrodynamic models, an outburst is indicated by the photospheric radial velocity suddenly becoming very large, and the photospheric radius increasing monotonically over several pulsation cycles. Such pulsation-triggered outbursts may be responsible for the driving of variable, nonspherical winds. If large and infrequent enough, these outbursts may be identified with the classic LBV eruptions accompanied by episodic mass loss.

1 Mechanisms Proposed to Destabilize LBV Envelopes

Many mechanisms have been proposed to destabilize the envelopes of Luminous Blue Variable stars, making them susceptible to prodigious mass outflow and sporadic S Doradus-type or Eta Carinae-like outbursts. It is likely that all of these mechanisms are relevant at some stage of a massive star's evolution, or at different levels in the stellar envelope.

The envelopes of LBV stars are very close to the Eddington luminosity limit, at which the outward acceleration due to radiation pressure equals the inward gravitational acceleration. As discussed by Lamers (1997) and Langer (1997), the classical Eddington luminosity $L_{edd} = 4\pi cGM/\kappa_e$ is modified for stellar envelopes by replacing the electron scattering opacity κ_e with the flux-mean opacity κ, and by reducing the gravitational acceleration by the centripetal acceleration for rotating stars. Thus the Eddington limit can vary within a stellar envelope, and as a function of latitude, and also varies with mass, metallicity, and evolutionary state. Lamers (1997) concludes that if a stellar envelope is within 5 or 10% of this "Atmospheric Eddington Limit", the envelope is loosely-bound enough that any small perturbation will cause the envelope to react violently. Langer (1998) shows that during main-sequence evolution of rotating massive stars, the surface layers reach the rotation-modified Eddington limit, called the Ω limit, and mass loss occurs first, and preferentially, along the equatorial plane. Angular momentum loss in the wind restricts the star's rotation rate to the Ω limit during subsequent evolution, and determines the maximum mass-loss rate, which is comparable to the observed LBV mass-loss rate.

Maeder (1989) finds that as massive stars evolve redward across the HR diagram, envelope opacities increase and the outer layers become supra-Eddington, resulting in a density inversion. Maeder suggests that such density inversions trigger instability, resulting in ejection of the outer layers above the density inversion. As the outer layers are ejected, the instability region moves downward into the star, with a rate characteristic of the local thermal timescale which is comparable to the dynamical timescale. Thus, the instability can encompass a significant fraction of the envelope during one episode, and result in mass loss of $0.1 - 1M_\odot$.

Stothers and Chin (1993) discuss the roles of the Fe-ionization opacity enhancement at 200,000 K and high radiation pressure in producing a region in the stellar envelope where $\Gamma_1 \leq 4/3$, the condition for dynamical instability. They suggest that the onset of dynamical instability is responsible for the mass expulsions seen in LBV outbursts; however, the models that Stothers and Chin find dynamically unstable are far redward of the LBV instability strip and the Humphreys-Davidson limit. Appenzeller (1989) emphasizes the role of radiation pressure in extending the envelopes of LBV stars, and in driving them into a state of dynamical instability, in which additional expansion results in further mass loss. The envelope of the star recovers on a thermal timescale, until conditions for expansion and onset of instability are again achieved (the "relaxation oscillation" scenario).

Pulsation also likely plays a role in destabilizing LBV envelopes and enhancing mass loss. According to linear nonadiabatic pulsation stability analysis, nearly all of the stars in the upper HR diagram are unstable to radial and nonradial pulsations, including p-modes, g-modes, radial modes, and so-called "strange modes" (see e.g. Glatzel and Kiriakidis 1993, Glatzel 1997; Cox et al. 1997). Some modes are predicted to have large growth rates in the linear analysis. Pulsation also has the potential to extend the stellar envelope. Glatzel and Kiriakidis (1993) suggest that resonant coupling between modes could trigger an LBV outburst. As discussed by Guzik et al. (1997) and Cox et al. (1998), the variations in physical conditions and radiation-matter interactions during a pulsation cycle can cause regions of the envelope to periodically exceed L_E.

2 Nonlinear Pulsation Hydrodynamic Modeling

Several groups have calculated nonlinear pulsation hydrodynamic models of Stars that have some physics in common with LBVs, such as strange mode instability, and high luminosity-to-mass (L/M) ratios. Kiriakidis et al. (1997) and Glatzel et al. (1998) discuss nonlinear evolution of strange modes in massive main sequence and Wolf-Rayet stars. They find that pulsation amplitudes can reach $\sim$ 100 km/sec, sufficient to exceed the escape velocity and to drive mass loss in massive main-sequence stars. Preliminary hydrodynamic models of LBV envelopes by Dorfi and Feuchtinger (1998, private communication) exhibit regular pulsations, and a rearrangement of the envelope structure as kinetic energy is deposited in outer layers, but no pulsation-driven mass loss. Bono and Marconi (1998) and Buchler et al. (1998, private communication) have examined nonlinear pulsations in Cepheid variables, with particular attention to the interaction between pulsation and convection. Aikawa and Sreenivasan (1996) examine the role of strange modes and a density inversion in nonlinear pulsations of 0.6 and $4M_{\odot}$ supergiant models, and comment on differences in envelope structure produced by the strange modes, compared to typical radial modes. Heger et al. (1997) considered 10-20 $M_{\odot}$ red supergiant models with high L/M ratios, and find large amplitude pulsations similar to those posited as precursors for the "superwind" of AGB stars.

Here we summarize our preliminary results for nonlinear hydrodynamic models of LBV envelopes (see also Guzik et al. 1997; Cox et al. 1997; Cox et al. 1998; Despain et al. 1998). The unique and essential ingredient in the Los Alamos models is the inclusion of a time-dependent convection treatment.

We base the composition, mass, luminosity, and effective temperature of our models on evolution models of initial mass 50 and 80 $M_{\odot}$, and $Z = 0.01$ or 0.02, on their first crossing to the red in the HR diagram. We vary the radius exponent in the mass-loss parametrization of Nieuwenhuijzen and de Jager (1990) to produce a range of surface helium abundance enhancements when the models reach the LBV instability region.

We next create hydrostatic envelope models with 60-120 zones, based on the evolution models. The models have extremely compact cores, and highly extended envelopes containing 95% of the stellar radius, but only 0.001% of the stellar mass. Linear nonadiabatic pulsation analysis shows that the models are unstable to pulsations driven by the κ mechanism, primarily in the Fe-bump opacity region for massive models, and in the He-ionization region for somewhat lower-mass models. The models are also unstable to multiple radial "strange modes", that have no counterpart in adiabatic analysis. These modes have large growth rates, typically more than 100% per period, and in one case exceeding 600% per period!

As discussed by Langer (1997), in a static model, the deeper layers of a stellar envelope avoid exceeding L_E because convection turns on to transport the excess luminosity. However, we note that convective energy transport in reality takes some time to turn on and build up efficiency, and this time can be a significant fraction of a pulsation cycle. During this time, the Eddington limit can be exceeded. To investigate this possibility, we use an updated version of the Lagrangian nonlinear hydrodynamics code described by Ostlie (1990), Cox (1990), and Cox and Ostlie (1993). The models use OPAL opacities and equation of state. This code includes a parametrized time-dependent convection treatment, based on the mixing-length theory of convection. The convective velocity of a zone at a given timestep is determined by a weighted interpolation between convective velocities of the past two timesteps, and the instantaneous velocity from mixing-length theory. Nonlocal effects are also included via a weighted average of local convective velocity with the convective velocities of neighboring zones.

The hydrodynamic models are initiated in the most unstable radial mode predicted by linear analysis, with radial velocity amplitude one km/sec inward. We find for all of our models that the pulsation amplitude grows rapidly. For models with sufficiently high L/M ratio and high envelope hydrogen abundance, the outward photospheric radial velocity suddenly becomes very large, and the photospheric radius monotonically increases during several pulsation periods. Eventually the outer zones of the model reach densities and temperatures beyond the lower limits of our opacity and equation of state tables, ending the hydrodynamic simulation. We define this behavior as an "outburst".

A closer examination of the models that outburst shows L_E is exceeded in zones near the Fe-opacity bump near 200,000 K each pulsation cycle. Figure 1 illustrates this behavior for a convecting zone (zone 20) near 200,000 K of a $47.28 M_\odot$ model, with envelope Y, Z = 0.02, 0.38. Convection turns on to carry nearly all of the luminosity in zone 20 at about 0.4×10^6 s, but just before this time, the Eddington luminosity of the zone has decreased (due to increased opacity), and the radiative luminosity has increased and exceeded L_E. This behavior occurs again at 0.8 and 1.6×10^6 seconds; however, this zone has also moved outward in radius and may no longer be the zone where L_E is exceeded by the largest amount.

For temperatures and densities typical of these LBV envelope models, the opacity decreases by about 5% for an increase in Y of 10%. If Y increases sufficiently due to additional mass loss, the models can avoid exceeding L_E. The hydrodynamic models then show pulsations in one or more simultaneous modes, with periods of 5-50 days, similar to the periods of observed LBV microvariations, and amplitudes of $\sim 50-150$ km/sec, decreasing systematically with increasing Y.

We need to examine the sensitivity of our results to the number and distribution of zones, amount of artificial viscosity, inclusion of turbulent pressure and energy, mixing length, time lagging and spatial averaging of convective velocity in the time-dependent convection treatment, pulsation mode of initiation, and initial radial velocity amplitude. We expect that any of

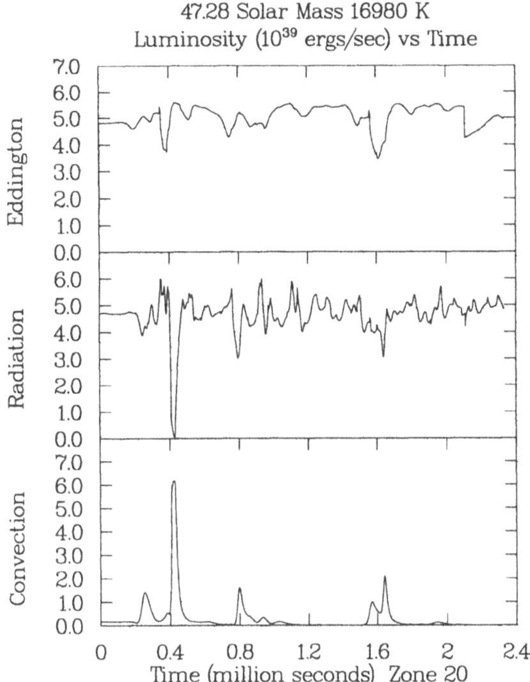

Fig. 1. Eddington, radiative, and convective luminosity of LBV model envelope zone near 200,000 K. As the opacity due to Fe ionization increases during the pulsation cycle, the Eddington luminosity increases, but convection does not turn on rapidly enough to transport the luminosity, and the radiative luminosity exceeds the Eddington limit. These conditions occur at 0.4, 0.8, and 1.6 million seconds.

these may change the quantitative details, but will not destroy the conclusion that regions of the stellar envelope can briefly exceed L_E each pulsation cycle while convection is becoming established to transport some luminosity. We also need to address with a separate code, or a code with enhanced capabilities, the fate of the mass outflow, development and propagation of shocks, replenishment of the envelope by material from the core, whether the envelope structure stabilizes or becomes periodically unstable, and how the envelope recovers from an outflow event.

We have shown only that pulsations can trigger instability, and cause the outer layers of the stellar model to move outward rather than oscillate with the pulsation cycle. However, the envelopes of these models contain only $10^{-4}M_\odot$, wherease in observed LBV outbursts much more mass can be lost. Furthermore, the predicted outflow is generated after a few pulsation cycles (less than one year), beginning from a near-static configuration, whereas the observed time between LBV eruptions is years or decades. Thus, we may have discovered a mechanism for a pulsation-modulated wind, whereby pulsation amplitude builds up and pushes off the outer $10^{-4}M_{\odot}$ of material every few pulsation cycles. On the other hand, if Davidson's (1989) "geyser" analogy/hypothesis applies, once an instability is generated, the "trigger" may be able to propagate deeper into the star on a dynamical timescale, and involve a larger fraction of the star. In this case, this mechanism may be able to initiate substantial episodic mass ejection as is seen in LBV outbursts.

Our models do not include rotation, and treat only one-dimensional radial pulsations. Rotation reduces the effective gravitational force and modifies L_E as a function of latitude, and may result in a latitude-dependent pulsation amplitude and envelope extension, or pulsation-modulated mass loss. In addition, high-degree nonradial modes are expected in the linear analysis to have larger growth rates than radial modes (Glatzel 1997). Do these modes grow to large enough amplitudes to impose a nonspherical perturbation on the outflows?

3 Addendum on η Carinae

Humphreys (these proceedings) presented the case that the most extreme LBV stars η Car, SN 1961V, and P Cyg experienced a giant eruption, followed by a second eruption of comparable strength (but obscured by ejecta from the first eruption), with an interval of 40-50 years. If this interval corresponds to a "recovery time" between giant outbursts, and can be associated with the thermal timescale of a portion of the LBV envelope, the amount of mass involved in the outburst is very large. For example, applying this argument for η Car, $\sim 12M_{\odot}$ participated in the eruption. This implies a deep-seated mechanism for the outbursts. It also implies that for a given amount of mass that escapes in the outburst (roughly $2M_{\odot}$ for η Car), several times this much mass readjusts back onto the star during the recovery, and sets up the conditions for another outburst. In the case of η Car, there is evidence that the first outburst generated mass loss in bipolar lobes at the poles, while the second generated mass outflow predominantly along the equator (N. Smith et al., these proceedings). How can we reconcile this observation with proposed mechanisms for giant outbursts?

Acknowledgments:

The authors gratefully acknowledge P. Bradley, W. Glatzel, H. Lamers, K. Oedegaard, R. Humphreys, N. Smith, J. Cassinelli, S. Owocki, S. Shore, G. Koenigsberger, N. Langer, T. Aikawa, S.R. Sreenivasan, A.M. van Genderen, G. Bono, J. Bjorkman, and C. Sterken for useful discussions.

References

Aikawa, T., Sreenivasan, S. R. (1996): PASJ **48**, 29

Appenzeller, I. (1989): in *Physics of Luminous Blue Variables*, ed. K. Davidson et al. (Kluwer Academic, Dordrecht), p. 195

Bono, G., Marconi, M. (1998): in *A Half Century of Stellar Pulsation Interpretations*, ed. P.A. Bradley and J.A. Guzik (ASP), p. 287

Cox, A. N. (1990): in *Nonlinear Astrophysical Fluid Dynamics*, ed. J.R. Buchler and S. T. Gottesman (New York Acad. Sciences), p. 54

Cox, A. N., Ostlie, D. A. (1993): Ap &SS **210**, 311

Cox, A. N., Guzik, J. A., Soukup, M. S., Despain, K. M. (1998): in *A Half Century of Stellar Pulsation Interpretations*, ed. P.A. Bradley and J.A. Guzik (ASP), p. 302

Cox, A. N., Guzik, J. A., Soukup, M. S. (1997): in *Luminous Blue Variables: Massive Stars in Transition*, ed. A. Nota and H.J.G.L.M. Lamers (ASP), p. 133

Davidson, K. (1989): in *Physics of Luminous Blue Variables*, ed. K. Davidson et al. (Kluwer Academic, Dordrecht), p. 101

Despain, K.M., Guzik, J.A., Cox, A.N. (1998): in *A Half Century of Stellar Pulsation Interpretations*, ed. P.A. Bradley and J.A. Guzik (ASP), p. 307

Glatzel, W., Kiriakidis, M. (1993): MNRAS **263**, 375

Glatzel, W. (1997): in *Luminous Blue Variables: Massive Stars in Transition*, ed. A. Nota and H.J.G.L.M. Lamers (ASP), p. 128

Glatzel, W., Kiriakidis, M., Chernigovskij, S., Fricke, K.J. (1998): MNRAS, in press

Guzik, J. A., Cox, A.N., Despain, K.M., Soukup, M. S. (1997): in *Luminous Blue Variables: Massive Stars in Transition*, ed. A. Nota and H.J.G.L.M. Lamers (ASP), p. 138

Heger, A., Jeannin, L., Langer, N., Baraffe, I. (1997): A&A **327**, 224

Humphreys, R. (1998): these proceedings

Jeannin, L., Fokin, A.B., Gillet, D. Baraffe, I. (1997): A&A **326**, 203

Kiriakidis, M., Chernigovskij, S., Fricke, K. J., Glatzel, W. (1997): in *Luminous Blue Variables: Massive Stars in Transition*, ed. A. Nota and H.J.G.L.M. Lamers (ASP), p. 150

Lamers, H. J. G. L. M. (1997): in *Luminous Blue Variables: Massive Stars in Transition*, ed. A. Nota and H.J.G.L.M. Lamers (ASP), p. 76

Langer, N. (1997): in *Luminous Blue Variables: Massive Stars in Transition*, ed. A. Nota and H.J.G.L.M. Lamers (ASP), p. 83

Langer, N. (1998): A&A **329**, 551

Maeder, A. (1989): in *Physics of Luminous Blue Variables*, ed. K. Davidson et al. (Kluwer Academic, Dordrecht), p. 15

Nieuwenhuijzen, H. and de Jager, C. (1990): A&A **231**, 134

Ostlie, D. A. (1990): in *The Numerical Modelling of Nonlinear Stellar Pulsations: Problems and Prospects*, ed. J.R. Buchler (Kluwer Academic, Dordrecht), p. 89

Smith, N., Guzik, J. A., Lamers, H. J. G. L. M., Cassinelli, J. P., Humphreys, R. (1998): Blitz Model for the Eruptions of Eta Carinae, these proceedings

Stothers, R.B., Chin, C.-W. (1993): ApJ **408**, L85

Discussion

J. Puls: How sensitively do your results depend on the outer boundary condition, e.g., the presence of a wind?
A. Guzik: The models do not include a wind, but do include zones beyond the photosphere. Since the place where this mechanism operates is deep in the interior, at 200 000 K, I do not think that including a steady wind would affect the result.

N. Langer: I was surprised to hear that you expect only quantitative changes from changing the coupling between the pulsations and convection. Is this coupling not one of the fundamental uncertainties of your model which – suitably chosen – could even make the "outbursts" go away?
A. Guzik: We have not explored all possibilities, but as long as convection does not adapt instantaneously to conditions in the zone, some convective zones in the model cannot avoid exceeding the Eddington limit for some fraction of each pulsation cycle.

S. Shore: These are neat results. You have put a characteristic time scale into the model because of the "switch on/off" convection scheme, so in a very non-linear model I am not surprised that you get this resonance-like behaviour. Have you explored the range of parameters ($\alpha = l/H_\text{p}$ and this time delay) to see if there are characteristic instability regimes?
A. Guzik: We did not tune the mixing length and time-dependent convection lag factor to find a resonance. It is a good suggestion to try varying these parameters relative to each other to see whether a resonance exists, or whether the conditions that lead to an "outburst" disappear for certain choices.

G. Koenigsberger: What determines the speed with which convection initiates in your code?
A. Guzik: The criterion for turning on convection is given by the standard mixing length theory criterion, $\nabla_\text{rad} > \nabla_\text{ad}$. The rate at which convection turns on in a particular zone is modified from the instantaneous values given by mixing-length theory according to the lag in time specified by the time-dependent convection parameterization. The lag time is the same fraction of the convection time scale = mixing length/convective velocity determined by some average of the condition in the previous time steps.

Linear Strange Modes in Massive Stars

Wolfgang Glatzel

Universitäts-Sternwarte, Geismarlandstraße 11, D-37083 Göttingen, Germany

Abstract. The occurrence and properties of strange modes and associated instabilities in massive stars are reviewed. If applicable, strange modes may be classified by the maximum of the opacity they are associated with. Whether they are still present in the limit of vanishing or infinite thermal timescales – which corresponds to the NAR and adiabatic approximations respectively – is another criterion for classification. A model for the instability mechanism of strange mode instabilities is discussed.

1 Methods and assumptions

Strange modes and associated instabilities can be identified on the basis of *standard* stellar physics (see, e.g., Baker & Kippenhahn 1962, Cox 1980 and Unno et al. 1989). Models to be tested for stability are either taken from ordinary stellar evolution calculations or may be stellar envelopes with given luminosity, effective temperature and mass. Stability of these models is tested with respect to infinitesimal (both radial and nonradial) perturbations, where - apart from the mechanical equations - also the equations of energy conservation and transport are taken into account, i.e., a linear nonadiabatic analysis (LNA) of the stability is performed.

Even if the input physics and the basic equations describing the problem are quite common, concerning the identification of strange modes (SMs) and associated instabilities (SMIs) for its solution not all standard methods are appropriate. In particular, relaxation methods relying on an adiabatic initial guess for the solution fail in general. A robust technique which allows for a reliable and accurate determination of any eigenfrequency (not only those of strange modes) is the Riccati method which was adapted to stellar stability problems by Gautschy & Glatzel 1990a (for the treatment of nonradial perturbations see Glatzel & Gautschy 1992).

2 Occurrence and definition of strange modes

Strange modes (and associated instabilities) are a common phenomenon in the envelopes of luminous stars (see, e.g., Saio & Jeffery 1988, Saio et al. 1984 and Gautschy 1993 for the case of less massive objects and Glatzel & Kiriakidis 1993a and 1993b for massive stars). The essential parameter which controls their occurrence and properties is the ratio L/M of luminosity to mass. More or less independent of the particular stellar model strange modes together with instabilities having growth rates in the dynamical range occur, if the value of L/M exceeds $\approx 10^4$ (in solar units). Thus the mass - luminosity relation becomes important, if domains of instability in the HRD are to be determined. In the upper HRD, for massive stars and given luminosity masses taken from stellar evolution calculations tend to be considerably higher than those indicated by observations. The results discussed are based on conservative high masses which favour stability. Up to now there is no precise definition of the term "strange mode". Some studies on the instability mechanism even suggest that an unambiguous definition might not be possible at all (see Glatzel, 1994). Therefore strange modes may in a loose way be regarded as additional modes not anticipated from experience with adiabatic spectra and neither fitting in the "ordinary spectrum" nor following its dependence on stellar parameters.

3 Classification

Phenomenologically, for massive stars SMs and SMIs come in three groups apparently related to the three opacity maxima (see, e.g., Iglesias et al. 1992 and Rogers & Iglesias 1992) caused by the contribution of heavy elements ("Fe") and the ionization of helium ("He") and hydrogen/helium ("H/He") respectively (see Kiriakidis et al. 1993). The run of the growth rates of the instabilities directly reflects the run of the opacity. Accordingly the opacity bump related to a strange mode may be used for classification. This seems to be justified also by the fact that some properties of SMs vary systematically within this classification scheme:

- *Sensitivity to the treatment of convection:*
  Fe-SMs: sensitive; He- and H/He-SMs: insensitive
- *Dependence on metallicity:*
  Fe-SMs: strongly dependent; He- and H/He-SMs: independent
- *Dependence on the harmonic degree l* (see Glatzel & Mehren 1996):
  Fe-SMs: instabilities up to $l \approx 10$, maximum growth rates are attained around $l \approx 0$;
  He- and H/He-SMs: instabilities up to $l \approx 500$, maximum growth rates are attained around $l \approx 100$
- *Instability mechanism:*
  Fe-SMs: partially κ- mechanism; He- and H/He-SMs: no κ- mechanism
- *Domain of existence in terms of the effective temperature:*
  Fe-SMs: $\log T_{\rm eff} > 4.3$; He- and H/He-SMs: $\log T_{\rm eff} < 4.3$
- *Adiabatic counterpart:*
  Fe-SMs: yes; He- and H/He-SMs: no
- *NAR counterpart (see Section 7):*
  An NAR counterpart exists in any case.

An alternative criterion for classification can be derived by considering various approximations. One of them implying an infinite thermal timescale is the adiabatic approximation. The opposite implies a vanishing thermal timescale and has been denoted by NAR approximation (see Section 7). Whether an adiabatic or NAR counterpart exists for a given mode may be used for its classification. Moreover, consideration of these limits can be helpful to identify its physical origin and the mechanism of an associated instability.

4 Strange modes in the adiabatic approximation

Stable adiabatic counterparts of strange modes are found only for the group associated with the opacity maximum due to the contribution of heavy elements (Fe-SMs). In this case the existence of strange modes can be attributed to the particular run of the sound speed which – together with the density – exhibits an inversion at the position of the opacity maximum. This inversion implies an acoustic barrier and splits the stellar envelope into two cavities, each of which develops an independent acoustic spectrum. The spectrum of the outer cavity is identified with strange modes, that of the inner cavity with ordinary modes. A shift of the barrier during stellar evolution causes the spectra to cross thus leading to multiple resonances which in an adiabatic environment unfold into avoided crossings (see Kiriakidis et al. 1993).

Except for extremely low irrelevant effective temperatures the stability analysis based on the adiabatic approximation does not reveal any instability (Glatzel & Kiriakidis 1998; see, however, Stothers & Chin 1993, 1994, 1995). Moreover, even if an instability in the adiabatic approximation would be present, this would require a confirmation by the nonadiabatic analysis, since the adiabatic approximation does not hold for the objects considered which are characterized by high L/M ratios and therefore by short thermal timescales.

5 Domains of instability in the HRD

On the basis of a standard LNA analysis for radial perturbations domains of instability in the HRD have been determined by Kiriakidis et al. (1993). Instabilities associated with He-SMs and H/He-SMs define a largely metallicity independent unstable region in the upper right corner of the HRD, whose boundaries approximately coincide with the observed Humphreys - Davidson limit (Humphreys & Davidson 1979). For solar metallicity (and above) it is continuously connected with the β Cepheid instability strip by an additional instability domain associated with Fe-SMs. It extends to the ZAMS for luminosities above $\log L/L_{\odot} \approx 5.9$ (corresponding to a ZAMS mass of $\approx 80 M_{\odot}$) and encompasses, e.g., the region of line profile variable O stars as well as the position where LBVs are found (see Fullerton et al. 1996).

6 Wolf - Rayet stars

With respect to growth rates and the number of simultaneously unstable modes the most extreme examples for SMIs are found in models for Wolf - Rayet stars (Glatzel et al. 1993, Kiriakidis et al. 1996). Largely independent of metallicity and opacity instability prevails for masses above $3...5 M_{\odot}$. In this case the NAR approximation may not only be used in a qualitative way to classify modes or to identify the instability mechanism: It provides even quantitatively correct results.

7 The NAR approximation

The NAR approximation (where NAR stands for Non - Adiabatic - Reversible, see Gautschy & Glatzel, 1990b) consists of neglecting the time derivative of the entropy in the perturbation equations. We emphasize that this approximation does not necessarily imply adiabatic changes of state which may be seen by considering the differential of the entropy s expressed in terms of the differentials of the density ρ and the pressure p:

$$ds = C\frac{\alpha}{\delta}\left(-\Gamma_1\frac{d\rho}{\rho} + \frac{dp}{p}\right) \tag{1}$$

where C, α, δ and Γ_1 denote the specific heat, the logarithmic derivative of the density with respect to the pressure at constant temperature, the negative logarithmic derivative of the density with respect to temperature at constant pressure, and the adiabatic index respectively. ds vanishes either, if the term in brackets is zero – which corresponds to the adiabatic limit – or, if the specific heat vanishes (NAR limit). Accordingly, the latter representing the opposite of the adiabatic limit is justified, if matter cannot store heat efficiently, e.g., in a tenuous, extended stellar envelope. The ratio of thermal and dynamical timescales varies from infinity in the adiabatic limit to zero in the NAR limit. The NAR approximation implies several consequences:

- As the matter cannot store heat, the luminosity perturbation has to vanish.
- Any instability mechanism relying on a Carnot - type process, which requires a finite specific heat – such as the classical κ- and ε- mechanisms – is excluded in the NAR limit.
- Due to vanishing thermal timescales thermal modes cannot exist in the NAR limit. Essentially only the mechanics of the system is considered, whereas its thermodynamics is disregarded. Although the NAR limit represents the opposite of the adiabatic limit, both limits are similar in this respect.
- The system is reversible with respect to time, which is common for purely mechanical configurations.

- Closely related to reversibility is the property that eigenfrequencies come in complex conjugate pairs.

Thus, due to its listed properties, the NAR approximation is a useful tool for the classification of modes and the identification of various instability mechanisms.

8 The mechanism of strange mode instabilities

Models for the mechanism of strange mode instabilities have meanwhile been proposed in three studies by Glatzel (1994), Papaloizou et al. (1997) and Saio et al. (1998). These investigations seem to agree upon the following issues:

- The strange mode phenomenon is found both for radial and nonradial perturbations. Geometry is therefore not essential for its occurrence and a plane - parallel model is sufficient for a physically correct description.
- Cowling's approximation does not have a significant influence on the strange mode phenomenon. For a model, the perturbation of the gravitational potential may therefore be neglected.
- As strange mode instabilities do also exist in the NAR limit, a model can be based on this approximation. In particular and as a consequence, the luminosity perturbation can be assumed to vanish. The existence of SMIs in the NAR limit proves them not to be related to a Carnot - type process. In particular, the classical κ- mechanism is thus excluded as their origin.
- A finite value or a particular run of the logarithmic derivative κ_T of the opacity κ with respect to temperature T is not crucial for the occurrence of SMIs. This again excludes the classical κ- mechanism as their origin.
- All objects suffering from SMIs are characterized by high ratios of luminosity to mass. Apart from small thermal timescales this has the consequence that radiation pressure contributes significantly to the total pressure in the envelopes of these stars. Therefore all models are based on considering the limit of small ratios β of gas pressure to total pressure: $\beta \to 0$.

Using these findings as basic assumptions of a model (for its description we closely follow Glatzel 1994), the mechanical equations (mass and momentum conservation) may be condensed into an acoustic wave equation:

$$\frac{\partial^2 \tilde{\rho}}{\partial t^2} - \frac{\partial^2 \tilde{p}}{\partial r^2} = 0 \tag{2}$$

where $\tilde{p}$ and $\tilde{\rho}$ denote pressure and density perturbation respectively. By adopting the NAR approximation the equation of energy conservation is satisfied. Moreover, within this approximation and in the limit $\beta \to 0$ of dominant radiation pressure the diffusion equation for energy transport reads:

$$\frac{\partial \tilde{p}}{\partial r} - \beta \frac{\bar{p}}{\bar{\rho}} \frac{\partial \tilde{\rho}}{\partial r} = \frac{\partial \bar{p}}{\partial r} (\kappa_\rho \frac{\tilde{\rho}}{\bar{\rho}} + 4\kappa_T \frac{\tilde{p}}{\bar{p}}) \tag{3}$$

where $\bar{p}$ and $\bar{\rho}$ denote pressure and density in the hydrostatic configuration respectively. κ_ρ and κ_T are the logarithmic derivatives of the opacity with respect to density ρ and temperature T. Equation (3) provides the – in general differential – relation between pressure and density perturbation, which is needed to close the wave equation (2).

In the following we perform a local analysis of equations (2) and (3), i.e., we consider solutions of the form $\exp(i\omega t + ikr)$ characterized by the frequency ω and the wavenumber k.

In the limit of high wavenumbers the r.h.s of equation (3) may be neglected and we are left with an algebraic relation between $\tilde{p}$ and $\tilde{\rho}$, where the factor of proportionality involves the isothermal sound speed c_T. Its consequence is the dispersion relation of isothermal sound waves:

$$\tilde{p} = \beta \frac{\bar{p}}{\bar{\rho}} \tilde{\rho} = c_T^2 \tilde{\rho} \quad ; \quad \omega^2 = (c_T k)^2 \tag{4}$$

In this limit pressure and density perturbation are in phase and the perturbation equations reduce to a self-adjoint problem of second order which does not allow for any instabilities.

In an intermediate range of wavenumbers k ($1 < k \cdot H_p < 1/\beta$; H_p: pressure scale height), which exists only for sufficiently small β, the second term on the l.h.s. and the second term on the r.h.s. of the diffusion equation (3) may be neglected. As a consequence, we obtain a differential relation between pressure and density perturbation and a dispersion relation of the form (g: gravity):

$$\frac{\partial \tilde{p}}{\partial r} = \frac{1}{\bar{\rho}} \frac{\partial \bar{p}}{\partial r} \kappa_\rho \tilde{\rho} = g \kappa_\rho \tilde{\rho} \quad ; \quad \omega^4 = -(g\kappa_\rho k)^2 \tag{5}$$

A phase lag of $\pi/2$ is now found between pressure and density perturbation, the perturbation operator is no longer self-adjoint and the dispersion relation (5) represents oscillatory damped and unstable modes with growth rates in the dynamical range. The latter resemble the strange mode instabilities. We emphasize that the perturbation problem is still of second order.

In general an unstable mode is characterized by a phase lag between pressure and density perturbation. In the model for SMIs presented here it is caused by the differential relation between pressure and density perturbation which is determined by the diffusion equation for energy transport. The model thus shows that in a highly nonadiabatic environment involving low heat capacity and short thermal timescales a phase lag leading to SMIs in an intermediate range of wavenumbers is inevitable, if radiation pressure is dominant ($\beta \to 0$) provided gravity and κ_ρ do not vanish, as indicated by equation (5).

The model presented was proposed by Glatzel (1994) and essentially confirmed by Saio et al. (1998). Although there are many correspondences also with the study of Papaloizou et al. (1997), some differences need to be noted: Essential in their model of the instability is the coupling of isothermal sound waves to a thermal mode which can be found only in the original third order NAR perturbation problem. On the other hand, the approach discussed here relies on second order perturbation problems only which do not allow for the existence of a third wave and the corresponding coupling. Moreover, in the isothermal limit instabilities do not exist here, whereas isothermal sound waves are involved in the instability mechanism proposed by Papaloizou et al. (1997).

References

Baker, N.H., Kippenhahn, R. (1962): Z. Astrophys. **54**, 114

Cox, J.P. (1980): *Theory of Stellar Pulsation* (Princeton Univ. Press, Princeton)

Fullerton, A.W., Gies, D.R., Bolton, C.T. (1996): ApJS **103**, 475

Gautschy, A. (1993): MNRAS **265**, 340

Gautschy, A., Glatzel, W. (1990a): MNRAS **245**, 154

Gautschy, A., Glatzel, W. (1990b): MNRAS **245**, 597

Glatzel, W. (1994): MNRAS **271**, 66

Glatzel, W., Gautschy, A. (1992): MNRAS **256**, 209

Glatzel, W., Kiriakidis, M. (1993a): MNRAS **262**, 85

Glatzel, W., Kiriakidis, M. (1993b): MNRAS **263**, 375

Glatzel, W., Kiriakidis, M. (1998): MNRAS **295**, 251

Glatzel, W., Mehren, S. (1996): MNRAS **282**, 1470

Glatzel, W., Kiriakidis, M., Fricke, K.J. (1993): MNRAS **262**, L7

Humphreys, R.M., Davidson, K. (1979): ApJ **232**, 409

Iglesias, C.A., Rogers, F.J., Wilson, B.G. (1992): ApJ **397**, 717

Kiriakidis, M., Fricke, K.J., Glatzel, W. (1993): MNRAS **264**, 50

Kiriakidis, M., Glatzel, W., Fricke, K.J. (1996): MNRAS **281**, 406

Papaloizou, J.C.B., Alberts, F., Pringle, J.E., Savonije, G.J. (1997): MNRAS **284**, 821

Rogers, F.J., Iglesias, C.A. (1992): ApJS **79**, 507

Saio, H., Jeffery, C.S. (1988): ApJ **328**, 714

Saio, H., Baker, N.H., Gautschy, A. (1998): MNRAS **294**, 622

Saio, H., Wheeler, J.C., Cox, J.P. (1984): ApJ **281**, 318

Stothers, R.B., Chin, C.W. (1993): ApJ **408**, L85

Stothers, R.B., Chin, C.W. (1994): ApJ **426**, L43

Stothers, R.B., Chin, C.W. (1995): ApJ **451**, L61

Unno, W., Osaki, Y., Ando, H., Saio, H., Shibahashi, H. (1989): *Nonradial Oscillations of Stars* (Univ. of Tokyo Press, Tokyo)

Discussion

H. Lamers: I understand that the periods you predict for B supergiants and LBVs are on the order of days or less. This is a factor of 10 to 100 shorter than observed. Which effect might be responsible for the long periods?
W. Glatzel: The results discussed are based on rather conservative values for the masses of the objects. If the masses are reduced, the time scales of variability may increase by a factor of 10. Moreover, in the non-linear regime of the evolution of the instabilities, we observe some kind of "period-doubling".

D. Massa: Do these models depend on the interior physics used to construct the model?
W. Glatzel: Strange-mode instabilities have their origin in the envelope of the objects; the interior does not have an influence. The phenomenon is perfectly described on the basis of envelope models.

A. Maeder: What effect limits the growth of the amplitudes of the strange modes in the non-linear regime?
W. Glatzel: The non-linear regime is characterised by the formation of shocks. The associated dissipation seems to limit the growth of perturbations.

Yoji Osaki and Atsuo Okazaki

Instabilities in LBVs and WR Stars

Knut Jørgen Røed Ødegaard

Institute of Theoretical Astrophysics, University of Oslo, P.O. Box 1029 Blindern, N-0315 Oslo, Norway

Abstract. Using a dynamical stellar code, detailed evolutionary models have been computed that follow the evolution and nucleosynthesis of massive stars from the pre-MS (accretion phase) up to the onset of O-burning. Various instabilities occur during the evolution. As an example, the evolution of a star with maximum mass 120 $M_{\odot}$ is discussed.

1 Ingredients of the models

The code is an extended version of the Göttingen dynamical stellar evolution code and includes:

- A large and flexible nuclear network consisting of up to 177 nuclear species from n and ^{1}H to ^{71}Ge linked by more than 1700 nuclear reactions. Initial metallicity Z = 0.02. The network and the diffusion equation are solved for each time step during the evolution.
- Modern rates for both strong and weak interaction processes as well as neutrino emission.
- Semiconvection, overshooting (0.2 H_P). The semiconvective diffusion parameter α (Langer et al. 1985) was set to 0.04.
- Mass loss using different prescriptions for O-stars (radiation driven), (Kudritzki et al. 1989), LBVs (parametric) and WR stars (Langer 1989).
- Improved grid distribution, a large number of grid points and artificial viscosity

1.1 Instabilities

In the present models instabilities occur in at least 3 different evolutionary stages. The overall evolution and the nucleosynthesis of the models were discussed in Ødegaard (1996).

The computations are started from a protostar with low temperature. The onset of nuclear burning can therefore be followed in detail. No nuclear species are assumed to be in equilibrium. Towards the end of the rapid growth of the core, pulses occur because light nuclear species are mixed into the core. The increased core luminosity causes spikes affecting both the size of the convective core (cf. Fig. 1) and the surface. In a new sequence computed according to the accretion scenario (Bernasconi and Maeder 1995), similar pulses occur when the stellar mass is exceeding 105 $M_\odot$. These pulses are even more violent and cause periodic appearances of one or more semiconvective layers above the convective core. The luminosity of the star varies 2 percent or less.

LBV pulsations are shown in the upper panel of Fig. 2. For a period of about 7800 years, $\dot{M}$ is very large, but variable, and reaches $\log \dot{M} = -2.15$. The time scales range from small variations in a few years or less to large variations in periods from 100 - 600 years.

The very heavy mass loss during the LBV stage and the succeeding WR-stages remove the outer layers completely, leaving a simple He-star. During subsequent burnings, the interior again becomes extremely complex with a large number of thin and active zones with convection as well as semiconvection. Local luminosities reach high values and the interior becomes rather dynamic. Waves seem to be excited and reach the surface, causing vibrations, as illustrated in the lower panel of Fig. 2. The total timespan of the plot is 1430 seconds.

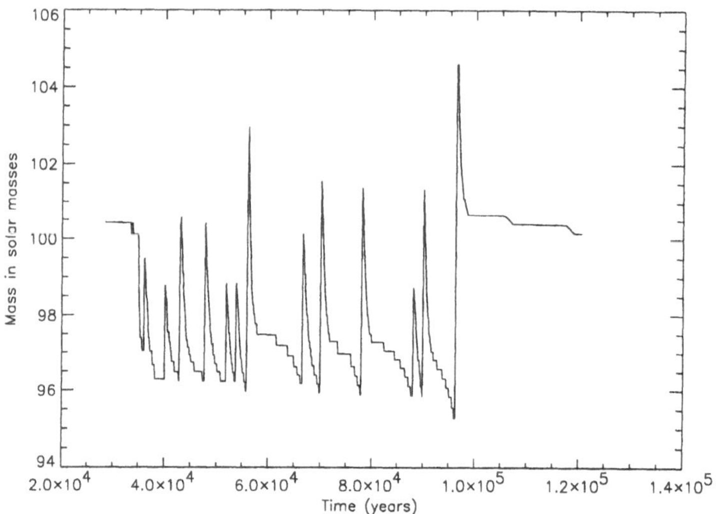

Fig. 1. Extention of convective core early in the evolution

References

Bernasconi, P.A., Maeder, A. (1995): About the absence of a proper zero age main sequence for massive stars. A&A, **307**, 829–839

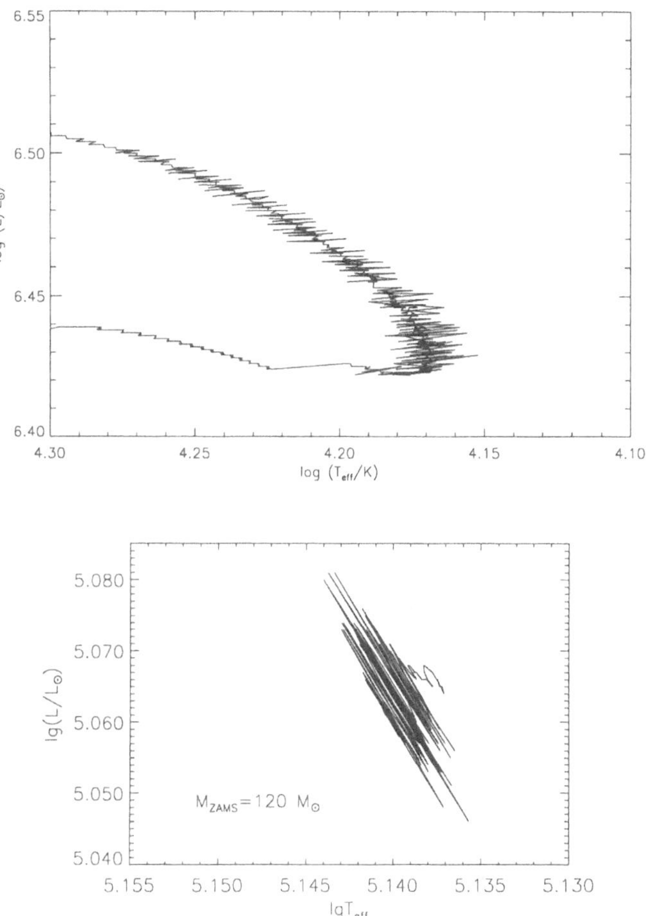

Fig. 2. Top panel: Pulsations during LBV stage. $M_{init} = 120 M_{\odot}$. **Lower panel:** Vibrations, probably caused by waves excited in the interior, affect the stellar surface during the transition from Ne-burning to O-burning.

Kudritzki, R.P., Pauldrach, A., Puls, Abbott, D.C. (1989): Radiation-driven winds of hot stars. VI. Analytical solutions for wind models including the finite cone angle effect. A&A, **219**, 205–218

Langer, N. (1989): Mass-dependent mass loss rates of Wolf-Rayet Stars. A&A, **220**, 135–143

Langer, N., El Eid, M.F., Fricke, K.J. (1985): Evolution of massive stars with semiconvective diffusion. A&A, **145**, 179–191

Ødegaard, K.J.R., (1996): Structure and evolution of single WR stars. Ap&SS, **238/1**, 107–111

Discussions

G. Koenigberger: What is the largest mass for a star in your model that can exist?
K. Ødegaard: The physics in my code (OPAL etc.) allow a mass of at least 200 $M_\odot$ at the MS for $Z = 0.02$. Pulsation mechanisms not taken into account in this case may strongly enhance mass loss for these very high masses.

Knuth Ødegaard and Joyce Guzik

Session VI

Evolutionary Aspects

chair: C. Sterken

The Evolution of Non-spherical and Non-stationary Winds of Massive Stars

Norbert Langer

Institut für Physik, Universität Potsdam, D–14415 Potsdam, Germany

Abstract. We describe present theoretical ideas about the time evolution of the winds of luminous stars with emphasis to effects of non-sphericity and non-stationarity. We discuss the evolution of the winds of rotating luminous stars during their main sequence evolution, in particular when they approach their Eddington-limit or any other surface instability. We then consider the winds of post-main sequence stars up to the immediate pre-supernova stage. We connect the giant outbursts of Luminous Blue Variables with luminous rotating post-main sequence stars in thermal disequilibrium. We further discuss the spin-up effect of Heger & Langer (1998) for post-red supergiants and describe its observational consequences. We compare theoretical models with observations of the winds of B[e] supergiants and Luminous Blue Variables in general, and with SN 1987A, VY CMa and η Car and the Pistol Star in particular.

1 Introduction

Massive stars ($M \gtrsim 10\,\mathrm{M}_\odot$) are very luminous ($L \gtrsim 10^4\,\mathrm{L}_\odot$) and have strong winds. During their evolution, their surface properties change which gives rise to changes in their wind properties. For example, the terminal wind speed is always of the order of the escape velocity at the stellar surface (Cassinelli & Lamers 1998). Therefore, as the massive stars move across the HR diagram during their post-main sequence evolution, the speed of their winds are changing dramatically. For contracting stars, i.e., for an increasing wind speed, this leads to swept up dense shells around stars which then can form bright visible nebulae — this concerns in particular LBVs and post-red supergiants; cf. García-Segura et al. (1996ab) — much like in the case of planetary nebulae around low-mass post-AGB stars.

As important as the time dependence of the stellar wind properties is their spatial structure. In fact, there is compelling evidence that the winds of massive stars are not generally isotropic (cf. below).

In this paper, we focus on non-sphericity and non-stationarity of massive star winds, but referring always to large spatial and time scales. I.e., the shortest variability time scales which we consider here is that of the thermal time scale of the stellar envelope — which can be as short as one year — while we generally do not consider variability due to stellar pulsations (although see Section 5) or due to rotation. And while we explicitly consider the effect of rotation on the spatial structure of the wind, we do not discuss small scale features as clumping or magnetic star spots.

2 Assumptions

As theoretical models for the envelopes and atmospheres of rapidly rotating luminous stars do not yet exist, we have to make several assumptions in the following. Let us consider a rotating star which, either due to increasing luminosity or — more realistically — due to increasing surface opacity, approaches its Eddington limit. Whether or not the stellar luminosity and the equatorial radiation flux are strongly affected by the rotation is unclear at the moment. Even though it is sometimes assumed that the von Zeipel (1924) theorem on rotational or "gravity darkening" can be applied (e.g., Owocki et al. 1998, Glatzel 1998), it may give spurious results for stars considered here, in which only the outer "skin" ($\sim 0.01\%$ of the stellar mass) is actually close to critical rotation, the energy transport in these layers is largely due to convection, and the internal circulation pattern is unknown. In fact, Kippenhahn (1977) has shown that even solutions for which the radiative flux increases with decreasing gravity ("gravity brightening") can be obtained. Thus, even though it can not be excluded that rapidly rotating stars blow a denser wind along their polar axis than in their equatorial plane as a result of "gravity darkening" (Owocki et al. 1996), we rather assume a constant surface brightness of the star, and consequently an outflow preferentially in the equatorial plane for the rapidly rotating stars considered here. Were the star non-rotating, its surface would become unstable when the Eddington factor $\Gamma = L/L_{\rm edd} = \kappa L/(4\pi c G M)$ approaches the critical value of one. However, when the star is rotating, the centrifugal force has to be considered in the force balance. Assuming an unchanged radiation flux as function of latitude, instability occurs at the stellar equator actually for $\Omega \rightarrow 1$ (with $\Omega = v_{\rm rot}/v_{\rm crit}$ and $v_{\rm crit}^2 = GM(1-\Gamma)/R$; cf. Langer 1997) at a value of $\Gamma < 1$ (cf. Fig. 1).

We want to emphasize that the principle of the so called Ω-limit (instead of Eddington-limit) is more general: Any surface instability in a non-rotating star — e.g. the turbulent pressure instability investigated by Nieuwenhuijzen and de Jager (1995) — will be affected by rotation such that the spherical symmetry is broken and mass outflow will be strongly latitude-dependent — i.e., critical rotation is reached (cf. Langer 1998).

A further assumption made in the following concerns the time dependence of the obtained mass loss. As mentioned in Sect. 1, we will ignore here variability on short time scales. Furthermore, we will assume that for a star which approaches the surface instability the mass loss increases drastically (Friend & Abbott 1986, Owocki et al. 1996), but remains limited to values of the order of $\dot{M} \sim M_{\rm env}/\tau_{\rm evol}$, where $M_{\rm env}$ is the mass of the H-rich stellar envelope and $\tau_{\rm evol}$ its evolutionary time scale; this relation has actually been verified for massive main sequence stars (Langer 1998; cf. Sect. 3).

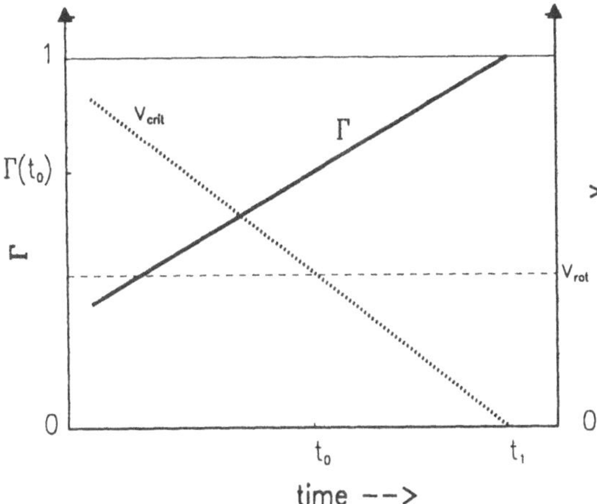

Fig. 1. Schematic time evolution of the Eddington factor $\Gamma = L/L_{\mathrm{Edd}}$ (thick line), with Eddington luminosity $L_{\mathrm{Edd}} = 4\pi cGM/\kappa$, where c is the speed of light, M and L are mass and luminosity of the star, and κ is the opacity at the stellar surface, and of the critical rotation velocity $v_{\mathrm{crit}} = [(1 - \Gamma)GM/R]^{1/2}$ (dotted line), where R is the radius of the star. For simplicity, we assumed a constant equatorial rotational velocity v_{rot} (dashed line); however, a time dependence of v_{rot} would not alter the conclusions. The curves shown are for a massive star expanding at constant luminosity L, which — in the non-rotating case — would reach the Eddington limit due to an opacity increase in its surface layers at $t = t_1$. However, since the rotational velocity v_{rot} must always be finite, the critical rotation, where the combination of radiation pressure and centrifugal force exceeds surface gravity at the equator, occurs at a time $t_0 < t_1$, i.e. *before* the Eddington factor reaches the critical value of $\Gamma = 1$. Thus, outflows occur before the star reaches its Eddington limit, acting to prevent further expansion.

3 Core hydrogen burning

Very luminous stars may approach their Ω-limit during core hydrogen burning. Langer (1998) has shown that this is *not* connected with any catastrophic mass loss, but that instead the coupling between mass and angular momentum loss leads to a stable, long-lasting evolutionary stage at the Ω-limit (cf. Fig. 2). During this stage, the mass loss is increased (by roughly a factor 10 for a $60\,\mathrm{M}_\odot$ example) and stable, but strongly unisotropic. Langer & Heger (1998) proposed that the luminous B[e] supergiants (Gummersbach et al. 1995) may correspond to this stage.

The question as to stars of which mass, metallicity and initial rotation rate do actually reach the Ω-limit during core hydrogen burning is far from settled (cf. Langer 1998). It is intriguing that the so called Pistol Star close to the Galactic Center (Figer et al. 1998), which appears to have an initial mass well above $150\,\mathrm{M}_\odot$, is best understood with the assumption that rotation did not affect its evolution. In fact, its location in the HR diagram, in particular

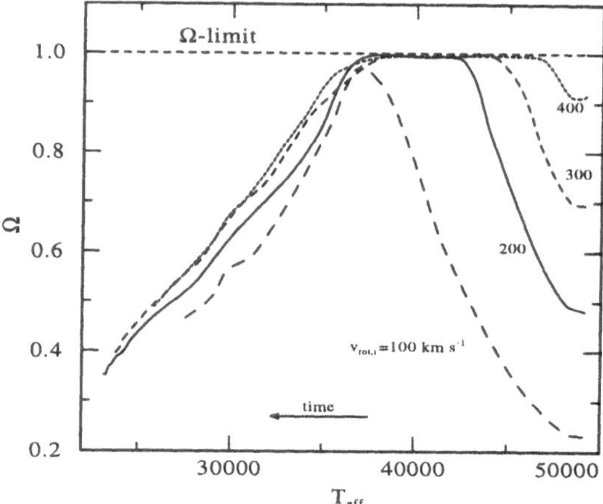

Fig. 2. Ω as function of the effective temperature during core hydrogen burning, for 60 $M_{\odot}$ sequences with different initial rotation rates (cf. Langer 1998).

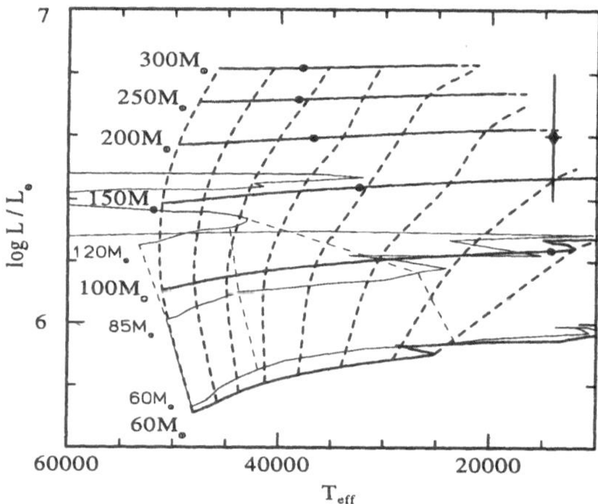

Fig. 3. Stellar evolutionary tracks in the HR diagram for core H-burning non-rotating stars in the initial mass range 60...300 $M_{\odot}$ and a metallicity Z of 2% (thick continuous lines). Thick dashed lines connect models with central helium mass fractions of 0.28, 0.4, 0.5, ..., 0.9, and 0.98. Black dots mark the first appearance of hydrogen burning products at the stellar surface. The tracks of the 200, 250, and 300 $M_{\odot}$ models end due to the occurrence of surface instabilities. Thin continuous lines show evolutionary tracks for 60, 85, and 120 $M_{\odot}$ and Z=0.02 obtained by Schaller et al. (1992) — without their effective temperature correction — who included convective core overshooting, with thin dashed lines connecting models with central helium concentrations of 0.30, 0.60, and 0.90. The position of the Pistol Star is marked by a diamond, together with the error bar (cf. Figer et al. 1998).

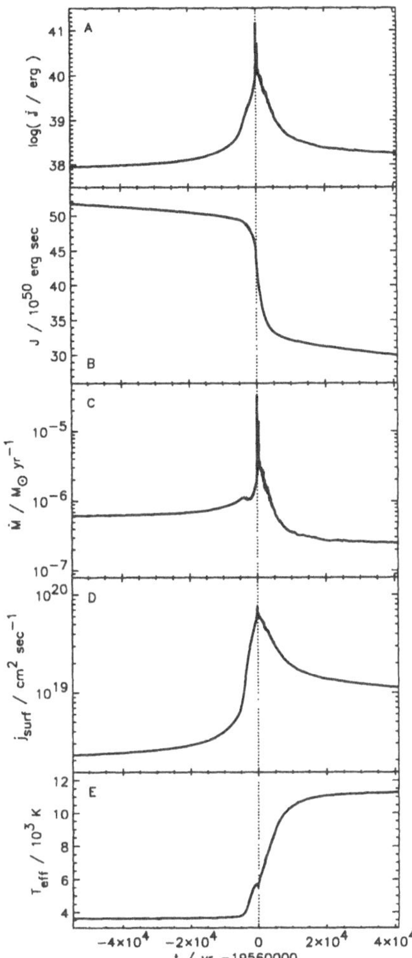

Fig. 4. Evolution of characteristic stellar properties as function of time, during the first part of the blue loop of a $12\,M_\odot$ model (cf. Heger & Langer 1998). The time zero-point is arbitrarily defined. Displayed are: the angular momentum loss rate $\dot{J}$ (**A**), the total angular momentum J (**B**), the mass loss rate $\dot{M}$ (**C**), the specific angular momentum loss rate $\dot{J}/\dot{J} = j_{\rm surf}$ (**D**), and the effective temperature $\log T_{\rm eff}$ (**E**).

its low effective temperature, seems to exclude any extra mixing as produced by rotation or even convective core overshooting (cf. Fig. 3).

4 Luminous Blue Variables

The scenario of the Ω-limit can lead to an understanding of several essential features of the Luminous Blue Variables (LBVs). First of all, it is at the least conceivable that very massive stars arrive at their Ω-limit directly after core hydrogen exhaustion, since the ensuing envelope expansion leads to stellar effective temperatures where the helium opacity peak appears close to the surface and thus the Eddington factor Γ becomes close to one. According to Sect. 2, we can then expect the star to produce very high mass loss rates due to the small evolutionary time scale of the envelope, and the mass loss

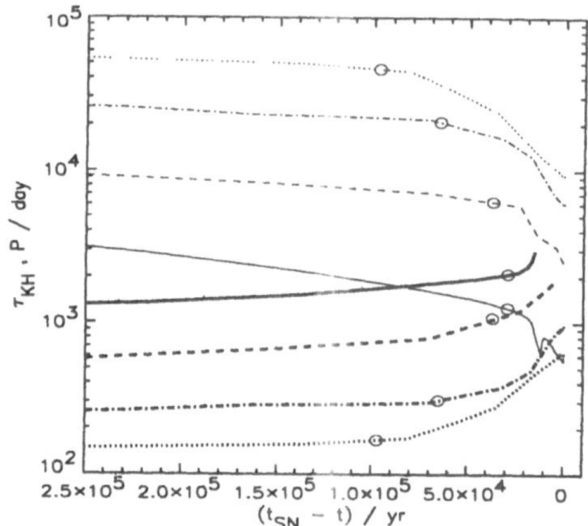

Fig. 5. Kelvin-Helmholtz time scale of the H-rich stellar envelope (thin lines) as function of the time left until the supernova explosion of the star, for the computed 10 (dot), 12 (dash-dot), 15 (dash), and 20 $M_\odot$ (solid line) sequences. The thick lines show the period of the fundamental mode as derived from a linear stability analysis. Small circles designate the time when the central helium mass fraction has reached 1%, i.e. roughly central helium exhaustion. It is evident that, at least for the 15 and 20 $M_\odot$ sequences, pulsation periods of the order of the Kelvin-Helmholtz time scale occur. Note that for the last models of these two sequences linear pulsation periods could not be derived due to strong departures from thermal and hydrostatic equilibrium.

may be better designated eruption than wind (see also Langer 1997). Due to rotation, i.e. at the Ω-limit, the circumstellar nebulae formed in this phase are expected to be strongly bipolar (Langer et al. 1998), which appears to be realized in practically all observed cases (Nota et al. 1995), η Carinae being a well known example. Finally, the dependence of the Ω-limit on the stellar rotation rate offers an explanation for the low luminosity of some LBVs: those would be the initially rapid rotators (Langer 1997, 1998).

5 Core helium burning and beyond

During core helium burning, massive stars would be red supergiants, blue supergiants or Wolf-Rayet stars. While most (though not all) Wolf-Rayet stars are expected to have spun down due to excessive mass and angular momentum loss (Fliegner & Langer 1994), highly anisotropic mass loss can be expected for blue and even for some red supergiants. Heger & Langer (1998) have found that the surface specific angular momentum (not only the surface rotation rate) can increase by more than one order of magnitude during the transition from the Hayashi line to hotter surface temperatures in the HR diagram (cf. Fig. 4). This can lead to circumstellar rings or bipolar structures around blue supergiants, perhaps such as found around the SN 1987A progenitor or B[e] supergiants (cf. Langer & Heger 1998), and may already lead to anisotropic mass loss on the RSG branch, for which VY CMa may be a candidate (Wittkowski et al. 1999).

Finally, note that even though the statistical probability to observe this is small, massive red supergiants may produce highly variable and even unstable mass outflows shortly before they explode as Type II supernova (cf. Fig. 5), as their pulsation periods become of the same order as the thermal time scale of their envelopes (Heger et al. 1997).

Acknowledgements The author is very grateful to Peter Conti, Guillermo García-Segura, Alexander Heger, André Maeder, and Stan Owocki for enlightening discussions. This work has been supported by the Deutsche Forschungsgemeinschaft through grant No. La 587/15-1 and La 587/16-1.

References

Cassinelli J.P., Lamers H.J.G.L.M., 1998, *Introduction to Stellar Winds*, Cambridge University Press

García-Segura G., Mac Low M.-M., Langer N. 1996a, A&A 305, 229

García-Segura G., Langer N., Mac Low M.-M. 1996b, A&A 316, 133

Glatzel W., 1998, preprint

Gummersbach C.A., Zickgraf F.-J., Wolf B. 1995, A&A, 302, 409

Figer D.F., Najarro F., Morris M., McLean I.S., Geballe T.R., Ghez A.M., Langer N.: 1998, ApJ, in press

Fliegner J., Langer N., 1994, in: *Wolf-Rayet Stars: Binaries, Colliding Winds, Evolution*, Proc. IAU-Symp. No. 163, K.A. van der Hucht et al., ed., p. 326

Friend D.B., Abbott D.C., 1986, ApJ 311, 701

Heger A., Jeannin, L., Langer, N., Baraffe, I. 1997, A&A, 327, 224

Heger A., Langer N. 1998, A&A 334, 210

Kippenhahn R. 1977, A&A 58, 267

Langer, N. 1997, in: *Luminous Blue Variables: Massive Stars in Transition*, ed. A. Nota, H.J.G.L.M. Lamers., ASP Conf. Series, 120, 83

Langer, N. 1998, A&A, 329, 551

Langer, N., Heger A., 1998, in *B[e] stars*, A.M. Hubert and C. Jaschek, eds., Kluwer, p. 235

Langer N., García-Segura G., Mac Low M.-M. 1998, ApJ, submitted

Nieuwenhuijzen H., de Jager C., 1995, A&A 302, 811

Nota A., Livio M., Clampin M., Schulte-Ladbeck R., 1995, ApJ 448, 788

Owocki S.P., Cranmer S.R., Gayley K.G. 1996, ApJ 472, L151

Owocki S.P., Gayley K.G. 1998, proc. 2nd Boulder-Munich Workshop on *Hot Stars*, ed. I. Howarth, A.S.P. Conf. Ser. Vol. 131, p. 237

Schaller G., Schaerer D., Meynet G., Maeder A., 1992, A&AS 96, 269

Wittkowski M., Langer N., Weigelt G., A&A, submitted

von Zeipel, H. 1924, MNRAS, 84, 665

Discussion

R. Kudritzki: Your scenario of evolution at critical rotation in LBV-phases or the blue-supergiant phase could be checked by looking at observed rotational velocities obtained from echelle spectra. Have you already looked at this?
N. Langer: I have tried, but with ambiguous results. The problem is complicated by the fact that a critical rotator does not have to be a rapid rotator since, e.g., close to the Eddington limit the critical rotational velocity goes to zero. However, more observational determinations of $v_{\rm rot}$ could shed light on this question in a statistical sense.

F.-J. Zickgraf: You showed that the first time a star reaches the ω-limit, it might stay there for up to 10^5 years. Interestingly, this agrees with an estimate for the duration of the B[e] phase based on the ratio of main sequence stars to B[e] stars. Is the time scale at the second time of reaching the ω-limit significantly shorter?
N. Langer: Yes it is, since the post-main sequence evolutionary time scale is much shorter than that on the main sequence.

G. Koenigsberger: In the case of a star in a binary system, one gets even larger possible mass loss rates. Could these rates be so large that the star is not able to go to cooler effective temperatures and remains at an early spectral type even during the LBV phase?
N. Langer: Yes. For a given luminosity, the minimum effective temperature can be estimated through the "radius" of the Roche lobe via the Stefan-Boltzmann law. Even though the mass-loss mechanism is different (i.e., Roche lobe overflow) from the mechanism for single stars, one may still expect an LBV-type outburst behaviour if the primary star (the mass loser) is in the immediate post-core H-burning phase.

S. Owocki: A general comment: I suggest we think of gravity darkening not just in terms of a simple parameterisation, but in terms of the fundamental physical effect: namely, diffusing photons will always tend to escape in the direction of least resistance, which for an oblate star is from the pole. Even if one were to avoid gravity darkening, it would be hard to use radiation to support an equatorial "ring" of higher density, simply because the radiative flux will avoid such high density.

R. Humphreys: Regarding the origin of the bipolar rings around SN 1987A: an evolved, mass-losing M supergiant – VY CMa – is bipolar and probably has an equatorial disk of dust.
N. Langer: Yes. In a recent paper with M. Wittkowski and G. Weigelt (see references), we argue that VY CMa may in fact be at the stage of leaving the Hayashi line, i.e., an immediate progenitor of IRC +10 420 – which has

bipolar outflows – and already shows the spin-up effect recently found by Heger & Langer (1998).

R. Schulte-Ladbeck: Could you comment on the implications of your results for the B/R supergiant ratio of galaxies? I have an approved HST programm to revisit this problem.
N. Langer: In a recent paper, A. Maeder and I (1995, A&A 195, 685) have shown that present stellar evolution models cannot predict the B/R ratio as a function of metallicity correctly. Whether this changes when the effects of rotation are included is not yet known. However, it would be nice to have improved observational data to investigate this question.

Peredur Williams and Norbert Langer

Rotation and Anisotropic Losses of Mass and Angular Momentum

André Maeder

Geneva Observatory, CH - 1290 Sauverny, Switzerland

Abstract. The expressions of the radiative flux at the surface of a non–uniformly rotating star are revised and this leads to a small extra-term in the von Zeipel theorem. The Eddington factor needs also to be carefully defined in a rotating star, as well as the critical break-up velocity. This leads us to reconsider the so-called Ω–limit. The most massive stars reach the Ω and Γ limits almost simultaneously.

We also examine the latitudinal dependence of the mass loss rates $\dot{M}(\vartheta)$ in rotating stars and find two main effects: 1) the "$g_{\rm eff}$" effect which enhances the polar ejection; 2) the "opacity effect" (or "κ–effect"), which favours equatorial ejection. In O–stars, the $g_{\rm eff}$ effect is expected to largely dominate. In B– and later type stars the opacity effect should favour equatorial ejection and the formation of equatorial rings. Possible relations with η Carinae and the inner and outer rings of SN 1987 A are mentioned. Opacity peaks produce some extrema in $\dot{M}(\vartheta)$ and this may also lead to the formation non-equatorial symmetrical rings.

Anisotropic stellar winds remove selectively the angular momentum. For example, winds passing through polar caps in O–stars remove very little angular momentum, an excess of angular momentum is retained and rapidly redistributed by horizontal turbulence. These excesses may lead some Wolf–Rayet stars, those resulting directly from O–stars, to be fast spinning objects, while we predict that the WR–stars which have passed through the red supergiant phase will have lower rotation velocities on the average. We also show how anisotropic ejection can be treated in numerical models by properly modifying the outer boundary conditions for the transport of angular momentum.

1 Introduction

Rotation has effects in the deep stellar interiors and also at the stellar surfaces. The main effects in the interior results from the transport of chemical elements and of angular momentum by shear turbulence and meridional circulation (cf. Zahn, 1992; Maeder, 1997; Maeder and Zahn, 1998). Here, in accordance with the thema of the meeting, we shall examine some of the interesting problems concerning the coupling of mass loss and rotation, which are important for stellar evolution. We may mention in particular the distribution of the radiative flux at the stellar surface, the correct expressions of the Eddington factor Γ and of the critical velocity in rotating stars, the latitudinal dependence of the mass loss rates $\dot{M}(\vartheta)$, the loss and gain of angular momentum in the remaining star as a result of the anistropies in the mass loss rates, etc.

2 The flux and the Eddington factor Γ

On the surface of a rotating star, the gravity, $T_{\rm eff}$ and the flux are not constant. The flux is generally given by the von Zeipel theorem (von Zeipel, 1924), which states that a given colatitude ϑ

$$\mathbf{F} = -\frac{L(P)}{4\pi G M_\star(P)}\,\mathbf{g}_{\rm eff} \tag{1}$$

with

$$M_\star = M\left(1 - \frac{\Omega^2}{2\pi G \rho_m}\right) \tag{2}$$

where ρ_m is the average density inside the considered surface level. This allows us to know the local $T_{\rm eff}$

$$T_{\rm eff}(\vartheta) \sim g_{\rm eff}^{1/4}(\vartheta) \tag{3}$$

The von Zeipel theorem usually applies to constant or cylindrical rotation. However, stars are likely to have more complicated rotation law, such as the "shellular rotation"proposed by Zahn (1992). Such a law of the form $\Omega = \Omega(r)$ results from the strong horizontal turbulence which rapidly damps the horizontal differences of the rotation. A more general expression of the von Zeipel theorem has been obtained (cf. Maeder, 1999), it is

$$\mathbf{F} = \frac{-L(P)}{4\pi G M_\star(P)}\mathbf{g}_{\rm eff}(1 + \zeta(\vartheta)) \tag{4}$$

with

$$\zeta(\vartheta) = \frac{H_T}{\delta}\frac{d\Theta}{dr}P_2(cos\vartheta) \tag{5}$$

$$\Theta = \frac{1}{3}\frac{r^2}{\bar{g}}\frac{d\Omega^2}{dr} \tag{6}$$

The term ζ is of the order of a few 10^{-2}, with $P_2(cos(\vartheta)) = 1$ at the pole and -0.5 at the equator. Thus, the term $\zeta(\vartheta)$ contributes to enhance the radiative flux at the pole and to decrease it at the equator.

The Eddington factor Γ is usually taken as

$$\Gamma = \frac{L\kappa}{4\pi c G M}. \tag{7}$$

It needs to be redefined more precisely in the case of a rotating star, since both the flux and the gravity are varying over the stellar surface. We consider a local $\Gamma(\vartheta)$ taken as the ratio of the stellar flux to the local limiting flux, i.e.

$$\Gamma(\vartheta) = \frac{F(\vartheta)}{F_{\text{lim}}(\vartheta)} \tag{8}$$

The limiting flux is obtained from the condition

$$\mathbf{g_{tot}} = \mathbf{g_{eff}} + \mathbf{g_{rad}} = \mathbf{g_{grav}} + \mathbf{g_{rot}} + \mathbf{g_{rad}} \tag{9}$$

which shows that the limiting $\mathbf{g_{eff}}$ and $\mathbf{g_{rad}}$ have the same direction, which introduces a major simplfication. Thus, the limiting flux is

$$\mathbf{F_{lim}} = \frac{-c}{\kappa}\mathbf{g_{eff}} \tag{10}$$

where $\mathbf{g_{eff}}$ is obtained from a Roche model, valid at the stellar surface. Thus, we finally have for the Eddington factor in a rotating star

$$\Gamma(\vartheta) = \frac{\kappa(\vartheta)L(P)}{4\pi cGM_\star(P)}\left[1 + \zeta(\vartheta)\right] \tag{11}$$

We note three differences between expressions (7) and (11). 1) The ζ–term favours a higher Γ at the pole. 2) The opacity $\kappa(\vartheta)$ applies only locally, which means that if κ grows for lower T, Γ is larger at the equator. 3) Γ depends also on rotation.

3 The Ω and Γ limits

These results are very useful in relation with the concept of an "Ω–limit" recently introduced by Langer (1997) and Heger and Langer (1998). The central idea is that if one has, for the critical break–up velocity

$$v^2_{\text{crit}} = \frac{GM}{R}(1 - \Gamma) \tag{12}$$

the value of v^2_{crit} would then tend towards zero for a star approaching the Eddington limit. Thus, Langer's conclusion was that for any non–zero rotation a star would reach its critical rotation velocity (the "Ω–limit") before the formal Eddington limit.

Indeed, one has, for the total gravity taking into account von Zeipel's theorem,

$$\mathbf{g_{tot}} = \mathbf{g_{eff}}\left[1 - \frac{\kappa(\vartheta)L(1+\zeta(\vartheta))}{4\pi cGM(1 - \frac{\Omega^2}{2\pi G\rho_m})}\right] \tag{13}$$

The break–up velocity occurs when rotation is such that $\mathbf{g_{tot}} = 0$ at the equator. We see that expression (13) has formally two roots, one given by $\mathbf{g_{eff}} = 0$ and the other obtained by nulling the bracket term in (13). The condition $\mathbf{g_{eff}} = 0$ at the equator gives the usual expression

$$v_{\mathrm{crit}}^2 = \Omega^2 r_{eb}^2 = \frac{GM}{r_{eb}} \tag{14}$$

where r_{eb} is the equatorial radius at break–up velocity. Thus we see that expression (12) is not correct since the current expression v_{crit}^2 does not contain any Γ. The reason for this is simply that close to break–up the effective gravity goes down to zero, and so does the radiative flux according to von Zeipel's theorem. Solving the surface equation, we find for the second root of (13)

$$v_{\mathrm{crit}}^2 = 3.3256\,\frac{GM}{r_{eb}}\left[1 - \frac{\kappa(\vartheta)L(P)}{4\pi cGM}\left(1 + \zeta(\frac{\pi}{2})\right)\right] \tag{15}$$

which is generally bigger than the first root given by (13), except when $\frac{\kappa(\vartheta)L(P)}{4\pi cGM}\,(1+\zeta(\vartheta))$ is bigger than 0.6993. When the rotation velocity grows, the root which is met first determines the physically significant zero of (13) and thus the critical velocity. We can conclude that (14) is the general expression for the critical velocity, except if $\frac{\kappa(\vartheta)L}{4\pi cGM}(1+\zeta(\vartheta))$ happens to become larger than 0.6993, a situation which, if it does occur, will only concern stars very close to the Eddington limit. Thus, the Ω and Γ limits are met almost simultaneously.

4 Mass loss in rotating stars

4.1 Latitudinal variations

According to the radiative wind theory (cf. Pauldrach et al. 1986; Kudritzki et al. 1989; Puls et al. 1996) the mass loss rates are essentially scaling like

$$\dot{M} \sim (k\alpha)^{1/\alpha}\left(\frac{1-\alpha}{\alpha}\right)^{\frac{1-\alpha}{\alpha}} F(\vartheta)^{1/\alpha} g_{tot}^{1-\frac{1}{\alpha}}(\vartheta) \tag{16}$$

where k and α are the force multiplier parameters. $F(\vartheta)$ is the local flux as given by the revised von Zeipel theorem. The values of k and α are changing with T_{eff} (cf. Pauldrach et al. 1986); for $T_{\mathrm{eff}} = 50\,000$, $40\,000$, $30\,000$ and $20\,000$ K one has respectively $k = 0.124$, 0.124, 0.17, 0.32, and $\alpha = 0.64$, 0.64, 0.59, 0.565. Then, taking into account the expression for the flux with the revised von Zeipel theorem, as well as g_{tot}, we obtain for the mass flux,

$$\dot{M}(\vartheta) \sim (k\alpha)^{\frac{1}{\alpha}}\left(\frac{1-\alpha}{\alpha}\right)^{\frac{1-\alpha}{\alpha}}\left[\frac{L(P)}{4\pi GM_\star(P)}\right]^{\frac{1}{\alpha}}\frac{g_{\mathrm{eff}}\,(1+\zeta(\vartheta))^{\frac{1}{\alpha}}}{(1-\Gamma(\vartheta))^{\frac{1}{\alpha}-1}} \tag{17}$$

This relation expresses the dependence of the mass loss rates on colatitude. If α and k are constant in latitude (as normally expected in O–stars), we see that $\dot{M}(\vartheta)$ mainly depends on g_{eff} (cf. also Owocki et al. 1996, 1998). This means that in a rotating hot star the mass loss rates by unit surface are much larger over the polar caps than at the equator. The terms $\zeta(\vartheta)$ and $\Gamma(\vartheta)$ in (17) slightly reinforce the polar mass loss.

As mentioned above k and α vary with T_{eff}. This means that over the surface of a rotating star k and α also vary. The term $(k\alpha)^{\frac{1}{\alpha}}\left(\frac{1-\alpha}{\alpha}\right)^{\frac{1-\alpha}{\alpha}}$ increases by a factor of three from $T_{eff} = 50\,000$ K to $20\,000$ K. The term in brackets in (17) is also larger for lower α–values. The term containing Γ increases with the value of Γ, this growth being much faster in case of lower α–values. On the whole the following picture emerges: 1. In hot, rotating stars the mass flux is higher at the poles and lower at the equator (the respective surface areas must of course be accounted for in numerical models). Let us call this the " g_{eff}–effect" in rotating stars. 2. This behaviour is also present near the $\Omega\,\Gamma$–limit where the mass flux is strongly increased as shown by (17). 3. In B and later type stars the enhanced polar ejection is progressively compensated by effects of larger line opacities (higher k and lower α), which favour progressively larger mass flux in the cooler equatorial regions. We shall call this the "κ–effect" in rotating stars. 4. For B and later type stars near the $\Omega\,\Gamma$–limit the mass flux is strongly enhanced, particularly in the equatorial regions. The wind density also has important latitudinal variations (cf. Maeder, 1999).

A so–called bistability of stellar winds has been found by Lamers (1997) in non–rotating stars: near $T_{eff} = 20\,000$ K and also close to $10\,000$ K, large and rather abrupt changes of the force multiplier parameters k and α modify the relations between v_∞ and v_{esc} and the mass loss rates. These important changes of k and α should also occur in B– and A–type rotating stars, as a result of the decrease of T_{eff} between the pole and the equator.

4.2 The case of η Carinae and of the rings in SN 1987 A

The HST pictures of η Carinae show two broad polar ejections and an equatorial skirt (cf. Ebbets et al. 1997; Davidson 1997). η Carinae is clearly a hot star close to the $\Omega\Gamma$–limit. Among the various explanations possible for the observed geometry of the ejections from η Carinae we might point out the possibility that polar ejections result from the " g_{eff}–effect" in (17) while the equatorial skirt is more likely to stem from the "κ–effect".

The complex structure around SN 1987 A consists of a bright elliptical inner ring and of two outer rings moved away from the central ring (cf. Crotts et al. 1989; Burrows et al. 1995). Currently the two outer rings are interpreted as real rings and not as rings due to the limb brightening of an hour glass shell (cf. Burrows et al. 1995). Their location and CNO composition suggest (cf. Panagia et al. 1996) that they were ejected at an earlier stage of evolution than the inner ring, perhaps when the SN progenitor was a blue supergiant. The bright inner ring is generally associated to the red supergiant stage in view of its composition, location and the timescales involved. We notice that an equatorial ring could be consistent with the "κ–effect" in cool stars while symmetrical outer rings would better correspond with peaks in the function $\dot{M}(\vartheta)$ due for example to some opacity peaks.

We must really wonder about the possibility of sharp extrema of the functions $\dot{M}(\vartheta)$ given by (17). This appears as a likely possibility. Indeed, if for some ϑ, the T_{eff} is such that there is a peak or a discontinuity in the opacity, $\dot{M}(\vartheta)$ will also show corresponding features at this colatitude. The net result will be the formation of rings. As a matter of fact, the very strong variations of α (cf. Lamers et al. 1995), which changes abruptly at $T_{eff} = 20\,000$ and 10 000K, may produce steep enhancements in $\dot{M}(\vartheta)$ and in the wind density and this may lead to the formation of symmetrical rings in nebulae. Future numerical models will tell us what are the features in the observed anisotropic nebulae which can be accounted for by the above rotational effects.

5 Change of angular momentum as a result of anisotropic mass loss

The anisotropic mass loss demonstrated above may have major consequences for stellar evolution. For example, polar ejection, as in O-type stars, removes very little angular momentum, and in particular much less than if the mass loss rates would be the same over the stellar surface. This implies that the angular momentum not embarked by polar winds remains as an excess $\mathcal{L}_{excess}$ in the outermost layers. This excess is rapidly redistributed within these layers by strong horizontal turbulence which operates on short timescales. Conversely equatorial mass loss will remove more angular momentum.

The above considerations imply that we must carefully rediscuss the surface boundary conditions applicable to the equation expressing the conservation of angular momentum in a rotating star. In particular, we find that the change of the angular momentum of the last shell between r and R is (Maeder, 1999)

$$\frac{\partial}{\partial t}\left[\bar{\Omega}\int_r^R \rho r^4 dr\right] = -\frac{1}{5}\rho r^4 \bar{\Omega}\,[U(r) - 5\dot{r}] + \mathcal{L}_{excess} \tag{18}$$

The second member expresses the transport by the meridional circulation (term with $U(r)$), the effect of contraction or expansion (term with $\dot{r}$ and the effect of the anisotropic mass loss.

Let us now express the excess of angular momentum $\mathcal{L}_{excess}$ applied to the last remaining shell as a result of the upper inhomogeneous mass removal. This excess is the difference between the angular momentum $\mathcal{L}(\Omega)$ of the last shell and the angular momentum $\mathcal{L}_{anis}(\Omega)$ anisotropically removed by mass loss. One has

$$\mathcal{L}_{excess}(\Omega) = \mathcal{L}(\Omega) - \mathcal{L}_{anis}(\Omega) = J(\Omega)\Omega\left[1 - \frac{J_{anis}(\Omega)}{J(\Omega)}\right] \tag{19}$$

$J(\Omega)$ is the moment of inertia of a shell at the surface of a rotating star of angular velocity Ω while $J_{anis}(\Omega)$ is the moment of inertia of the mass which is ejected by stellar winds. $\mathcal{L}_{excess}$ will be positive for a polar ejection an negative for an equatorial one. The application of this new boundary condition will modify the angular momentum during the evolution' of mass losing stars.Thus, it is likely that some WR stars, which after the O–phase have always stayed on the blue side of the HR diagram are fast rotators, since polar ejection is likely to have been dominant. At the opposite, WR stars resulting from an evolution through the red-supergiant phase, where equatorial mass loss dominates, may show lower rotational velocities on the average.

References

Burrows C.J., Krist J., Hester J.J. et al. 1995, ApJ 452, 680
Crotts A.P.S., Kunkel W.E., McCarthy P.J. 1989, ApJ 347, L61
Davidson K. 1997, Bull. American Astron. Soc. 188, 50.04.
Ebbets D.C., Walborn N.R., Parker J. 1997, ApJ 489, L161
Heger A., Langer N. 1996, A&A 315, 421
Kudritzki R.P., Pauldrach A., Puls J., Abbott D.C. 1989, A&A 219, 205
Lamers H.J.G.L.M. 1997, in Luminous Blue Variables: Massive Stars in Transition, ed. A. Nota, H. Lamers, ASP Conf. Ser. 120, 76
Lamers H.J.G.L.M., Snow T.P., Lindholm D.M. 1995, ApJ 455, 269
Langer N. 1997, in Luminous Blue Variables: Massive Stars in Transition, ed. A. Nota, H. Lamers, ASP Conf. Ser. 120, p. 83
Maeder A. 1997, A&A 321, 134 (paper II)
Maeder A. 1999, A&A in press (paper IV)
Maeder A., Zahn J.P. 1998, A&A 334, 1000 (paper III)
Owocki S.P., Cranmer S.R., Gayley K.G. 1996, ApJ 472, L115
Owocki S.P., Gayley K.G., Cranmer S.R. 1998, in Boulder–Munich II: Properties of Hot, Luminous Stars, ASP Conf. Ser. 131, p. 237
Panagia N., Scuderi S., Gilmozzi R., Challis P.M., Garnavich P.M., Kirshner R.P. 1996, ApJ 459, L17
Pauldrach A., Puls J., Kudritzki R.P. 1986, A&A 164, 86
Puls J., Kudritzki R.P., Herrero A. et al. 1996, A&A 305, 171
von Zeipel H. 1924, MNRAS 84, 665
Zahn J.P. 1992, A&A 265, 115

Discussion

J. Puls: In the model of ζ Pup ($\Gamma = 0.6$) Peter Petrenz presented in his poster paper, we included both the effects of gravity darkening and a latitude-dependent Γ, as well as a latitude-dependent α. As a result, the mass loss was still larger over the poles, since α remained roughly constant. The major change was in the velocity field, which was faster at the equator due to the lower Γ (higher $v_{\rm esc}$).

A. Maeder: I agree that for O-type stars not at break-up velocity, gravity darkening is insufficient to reduce the equatorial $T_{\rm eff}$ enough to make a lower α. Thus polar ejection dominates. For B stars, however, the reduction of α (i.e., the increase of opacity) at the equator may overcome the gravity darkening effect, so as to produce equatorial ejection. Concerning the velocity field, I would like to remind you of my remark that in general $v_{\rm esc}$ does not depend on Γ.

H. Lamers: Concerning the question of mass loss increasing toward the pole or the equator, you should not forget that observations tell us that there are disks around hot stars. These observations are, to name just a few: polarisation; the relation between polarisation and $v_{\rm rot}$ for Be stars; the double-peaked line profiles at low velocity of low ionisation lines; and the X-ray variability of Be-X-ray binaries. Although it may be possible to explain some of these effects by different models (e.g., polarisation by polar outflows) it is in my opinion impossible to explain all of these together by polar flows. For instance, polar flows might give double-peaked profiles, but that will be in high ionisation species (hot polar caps) and high wind velocity (high $v_{\rm esc} \rightarrow$ high v_∞).

F.-J. Zickgraf: Could your models also explain N enhancement on the main sequence for a $20\,M_\odot$ star? This would be of importance for the B[e] star R 4 for which we have found an N overabundance. We had interpreted this with a post-red supergiant evolutionary stage. In connection with the results of Norbert Langer, the N overabundance at an early stage in the evolution could indicate that the first arrival at the Ω-limit is important for the origin of B[e] stars.

A. Maeder: The first models show that $20\,M_\odot$ stars on the main sequence may undergo N-enhancement due to rotational mixing. I do not know which is the minimum mass for which this is possible. This may be an interesting point of comparison between models and observations.

S. Owocki: A general comment: I think we must be careful not to think that evidence for a "disk" must mean an equatorially enhanced stellar wind. A disk might not be outflowing at all, but in a stationary Keplerian orbit. Such Keplerian disks would likely have an origin quite independent of the wind.

R. Schulte-Ladbeck: I have to contradict what Henny said, at least from the point of view of spectropolarimetry of massive stars – it is not possible to tell from the data whether the material is in the equator or over the poles of the star. I do think the evidence for disks is strong only in the case of Be stars, owing to the existence of a large sample of stars observed with polarimetry, for which we can analyse the statistical probability of the distribution of projection angles on the plane of the sky. A possibility to distinguish an equatorial disk from a distribution of matter over the poles exists in principle (with polarimetry) in the case of binaries. Analysing a polarimetric time series can tell you whether material is located mainly in the binary (and hence equatorial) plane or over the pole (1st harmonic vs. 2nd harmonic). Finally, the behaviour of the polarisation across emission lines (see my paper on EZ CMa) might also become a useful tool to detect disks: unlike polar plumes, they rotate.

H. Henrichs: How do your evolutionary tracks depend upon the inclination of the star? If one wants to derive evolutionary masses from the observed luminosity and effective temperature, the inclination angle is apparently an important factor.
A. Maeder: I am pleased to say that we know the changes in M_V and (B−V) produced by rotation for various orientation angles rather well (cf., Maeder & Peytremann 1970, A&A 7, 120). As an example, at extreme rotation for pole-on stars, the excess brightness obtained by integration of the local fluxes over the visible part of the star is about 0.5 mag in M_V, the change in (B−V) is close to zero; for equator-on stars, the decrease of brightness is about 0.15 mag for a $5\,M_\odot$ and 0.4 mag for a $2\,M_\odot$ star. There is, however, an important reddening of the colour for equator-on stars. Other values can be obtained from the tables.

André Maeder and Rens Waters (Valeri Hambaryan in the background)

Rotation and Wolf-Rayet Star Formation

Georges Meynet

Geneva Observatory, CH-1290 Sauverny, Switzerland

Abstract. New results for 60 $M_\odot$ stellar models with different rotation rates are presented. Some effects of rotation on Wolf-Rayet star formation are discussed.

In these calculations we supposed that the angular momentum is locally conserved in radiative zones. The local conservation of angular momentum results from a balance between the outward flux of angular momentum essentially driven by shear instabilities and the inward flux resulting from meridional circulation. We checked by resolving the complete set of equations describing the transport of angular momentum (Maeder & Zahn 1998) that this stationary state is reached in a relatively short timescale compared to the evolutionary timescale. This result may not hold for the whole stellar mass range (see Denissenkov et al. A&A, in press).

The chemical species are diffused under the effects of both shear turbulence and meridional circulation. The diffusion coefficients used are given in Maeder (1997). We used the Ledoux criterion for setting the border of the convective core and, during the main sequence, we took into account semiconvective mixing. The effect of rotation on the mass loss rates is taken into account by using the Friend & Abbott's (1986) formula.

Figure 1 presents the evolution of the structure of the star during the H and He-burning phases for two 60 $M_\odot$ models: one is non-rotating and the other is rotating with an initial angular velocity Ω equal to 60% the critical angular velocity Ω_c (Ω_c is the angular velocity at which the star begins to loose matter at the equator). Many differences can be seen :

1) The evolutionary scenarios are not the same. Before becoming a WR star, the non-rotating 60 $M_\odot$ model goes through a short LBV phase (see also Fig. 2) which intervenes after the H exhaustion in its core. When the hydrogen surface mass fraction becomes inferior to about 0.4, the star enters into the WR phase as a WNL star. In the case of the fast rotating model[1], the star enters into the WR phase while still burning hydrogen in its core. The star thus skips the LBV phase and enters the WR phase at an earlier stage. Such an evolutionary scenario was also found by Fliegner & Langer (1994).

2) The mechanism of WR star formation is different. Surface abundances characteristic of the WNL star appear in the rotating model, not as a result of the mass loss which uncovers core layers, but as a result of diffusive mixing

[1] Similar evolutions occur for 60 $M_\odot$ stellar models with $\Omega/\Omega_c = 0.30$ and 0.40 and for 40 $M_\odot$ with $\Omega/\Omega_c = 0.30$, 0.40 and 0.60.

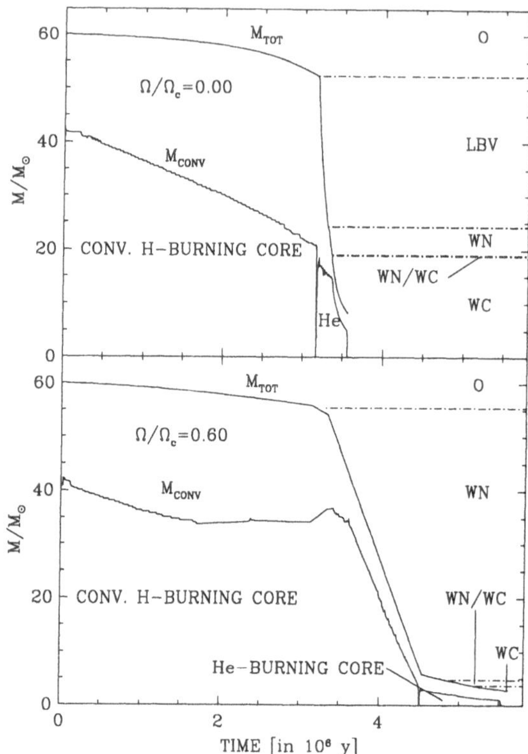

Fig. 1. Evolution of the total mass $M_{\rm TOT}$ and of the mass of the convective core $M_{\rm CONV}$ as a function of time. Various evolutionary stages are indicated on the right at the corresponding values of $M_{\rm TOT}$.

in the radiative zones. One observes that the same occurs for the entrance into the WC phase.

3) <u>During the WN phase, the surface abundances are different</u>. Indeed as a consequence of point 2) above, the N/C, N/O ratios obtained at the surface of the rotating WN model have not yet reached nuclear equilibrium in contrast with the no rotation case where nuclear equilibrium is reached as soon as the star enters the WN phase.

4) <u>The lifetimes are different</u>. When Ω/Ω_c increases the WR lifetime increases. The WN phase, as well as the transition WN/WC phase, become much longer. The WN/WC phase is characterized by the presence at the surface of both nitrogen and carbon. Non rotating models have difficulty in reproducing the observed number of such stars. Rotation might be the mechanism responsible for the formation of these stars (see also Langer 1991).

The evolutionary tracks in the HR diagram are shown on Fig. 2. We can see that higher is the angular velocity, bluer are the tracks during the main sequence. This is a consequence of the greater efficiency of mixing at higher rotation. We can see also, that when the rotation rate increases, less luminous WC stars are produced. This is due to the fact that when the star rotates, it enters the WR phase at an earlier stage and thus begins to loose high amounts of mass early in its evolution. Therefore, the star enters the WC phase with a small mass and a low luminosity.

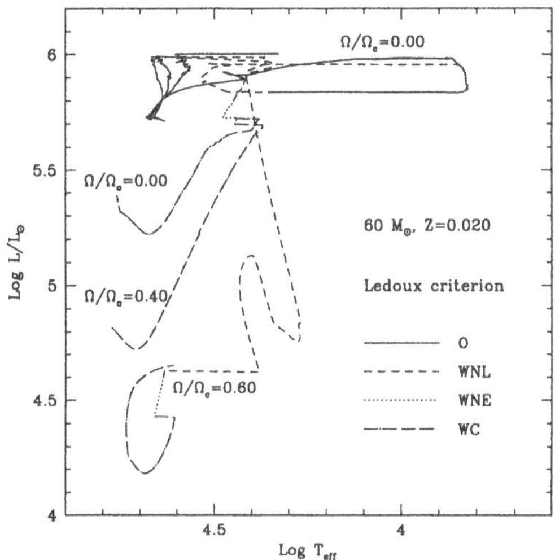

Fig. 2. Evolutionary tracks for 60 $M_\odot$ stellar models at solar metallicity for different rotational velocities. Ω/Ω_c is the angular velocity on the ZAMS expressed as a fraction of the critical angular velocity Ω_c . On the main sequence, the higher the rotational velocity, the bluer is the track. Different WR subphases are indicated with different types of lines. Only part of the $\Omega/\Omega_c = 0.30$ is shown.

References

Fliegner, J., Langer, N. (1994): IAU Symp. **163**, p. 326
Friend, D.B., Abbott, D.C. (1986): ApJ **311**, 701
Langer, N. (1991): A&A **248**, 531
Maeder, A. (1997): A&A **321**, 134
Maeder, A., Zahn, J.P., (1998): A&A **334**, 1000

Discussion

R. Kudritzki: You are predicting a new lower mass limit for WR progenitors. I thought that there are clear observational limits on this mass. Are the new calculations in agreement with the observations?
G. Meynet: In fact, I do not propose a new lower mass limit, merely that some WR stars could originate from low-mass progenitors rotating very fast. But it remains to be estimated what proportion of low-mass stars rotate sufficiently rapidly for this scenario to occur. They certainly belong to the upper end of the rotational velocity distribution.
On the other hand, it must be added here that if we accept an observational lower initial mass limit for WR star formation of around, let's say, 25 $M_\odot$ at solar metallicity, it is difficult to account for this low value without enhanced mass-loss rates. It may be that rotation can help to explain this fact without enhanced mass loss, even if (due to clumpiness) the mass-loss rates are to be reduced.

S. Shore: In laboratory shear flows, the factor 1/4 found from the classical Richardson instability criterion can be different; this could change the diffusion coefficient you have shown by a factor of 2 or more. So, what would such a change of "D" produce in your models? Also, if I understand correctly, Zahn's theory for sloping convection was developed in the perturbation limit; what happens at critical rotation? I would expect that the motion could be of high enough energy that the mechanical contribution, which is missing from the von Zeipel physical conditions, could be an important contributor to the energy balance at or near the equator.
A. Maeder: The factor 1/4 is the usual classical value (e.g., Chandrasekhar 1961). If we were to assume that the excess energy is not $1/4\,(\mathrm{d}U/\mathrm{d}z)^2$, but the same form with another numerical coefficient, the results would certainly be modified, like a shift in rotational velocities would do. Concerning von Zeipel's theorem, it applies to radiative envelopes. For stars with convective envelopes there is also some gravity darkening, but the results may be expected to show some quantitative differences. This is a problem which needs further investigation.

I. Appenzeller: Your models predict a significant fraction of H-rich WR stars. How does this compare with observations?
G. Meynet: H-rich WN stars are observed (see the paper by Hamann et al.). The interesting point to be checked is to estimate the extent to which the H-rich WN stars result from rotating progenitors. More rotating stellar models must be computed before careful comparisons with observations can be performed.
W.-R. Hamann: According to our spectral analysis (e.g., Hamann et al. 1995, A&A 299, 151) late-subtype WN atmospheres have typical hydrogen mass fractions of 10 – 40 %. Thus it seems that these new evolutionary calculations are now in good agreement with the empirical abundances.

Dusty LBV Nebulae: Tracing the Mass Loss History of the Most Massive Stars

Laurens B.F.M. Waters[1,2], Robert H.M. Voors[3], Patrick W. Morris[3,1], Norman R. Trams[4], Alex de Koter[1] and Henny J.G.L.M. Lamers[3]

[1] Astronomical Institute 'Anton Pannekoek', University of Amsterdam, Kruislaan 403, NL-1098 SJ Amsterdam, The Netherlands
[2] SRON Laboratory for Space Research Groningen, P.O. Box 800, NL-9700 AV Grogingen, the Netherlands
[3] SRON Laboratory for Space Research Utrecht, Sorbonnelaan 2, NL-3584 CA Utrecht, the Netherlands
[4] Astrophysics Division, ESA Space Science Department, ESTEC, Keplerlaan 1, Noordwijk, The Netherlands

Abstract. We present spectra obtained with the Infared Space Observatory (ISO) of the dust shells surrounding several Luminous Blue Variables (LBVs), both in our galaxy and in the LMC. The 20-45 μm spectra of R71, AG Car and Wra 751 show prominent emission features from crystalline silicates. The composition of the crystalline silicates in LBV dust shells is compared to that found in other types of objects, such as (post)-AGB stars and red supergiants (RSG). Both Wra 751 and AG Car have a high ratio of pyroxenes to olivines. This suggests that the grains in both stars experienced very similar processing, but that this processing has lead to a higher abundance of pyroxenes compared to RSG. The dust composition of the three LBVs discussed here suggests dust formation in a cool outflow not unlike those of RSG.

1 Introduction

Luminous Blue Variables (LBVs) are massive ($M_{zams} > 50\ M_{\odot}$), hot supergiants with dense, radiation-driven ionized winds. They show remarkable variations in temperature and radius, at roughly constant luminosity, with timescales of several years. LBVs are believed to be a short-lived post-main-sequence evolutionary phase of the most massive stars in galaxies; they will probably evolve to the He-rich Wolf-Rayet (WR) phase. There are on the order of 10 LBVs and LBV candidates known in our galaxy (Humphreys & Davidson 1994); well studied cases are P Cygni, η Car, and AG Car. For recent reviews on LBVs see *Luminous Blue Variables: massive stars in transition*, eds. A. Nota & H.J.G.L.M. Lamers (1997).

Almost all LBVs are associated with dusty ring nebulae (Nota et al. 1995). These nebulae are the result of previous phases of very high mass loss and/or of LBV outbursts, such as that of η Car around 1840. Thus, the shape, kinematics and composition of LBV ring nebulae can be used to trace their integrated mass loss history. Recently, the origin of LBV nebulae has received increased attention because the nebular gas-phase abundances of N and O in some nebulae are not in agreement with evolutionary predictions (Smith 1997; Smith et al. 1998). The mild N enhancements found in e.g. AG Car are better explained by a short Red Supergiant (RSG) phase during which the star was fully convective. Independently, Waters et al. (1997; 1998a) have shown that the dust in several LBV ring nebulae (among which AG Car) contains small amounts of crystalline silicates; these grains are also seen in the winds of RSG and AGB stars, suggesting similar physical and chemical conditions in the dust forming layers.

This paper discusses infrared spectroscopy taken with the Short Wavelength Spectometer (SWS, de Graauw et al. 1996) and with ISOCAM (Cesarsky et al. 1996) on board of the Infrared Space Observatory (ISO, Kessler et al. 1996) of two galactic LBVs (AG Car and Wra 751) and one LMC LBV (R71). These spectra yield new information about the composition of the dust through their solid state emission features. We compare the LBV dust spectra to those of NML Cyg (a RSG) and IRC+10 420 (a post-RSG).

2 Infrared spectroscopy of LBVs

2.1 AG Car and Wra 751

AG Car and Wra 751 are among the best studied galactic LBVs. Both stars have a prominent bipolar ring nebula expanding at 65-70 km/s (AG Car; Nota et al. 1992) and about 25 km/s (Wra 751; Hutsémekers & van Drom 1991; Garcia-Lario et al. 1998), and have strong infrared excess due to thermal emission from dust grains in their shell (e.g. McGregor et al. 1988; Hu et al. 1990). The LBV nature of Wra 751 is still somewhat uncertain, since it does not show the characteristic variations in temperature (Garcia-Lario et al. 1998). ISO-SWS spectra of both objects were discussed previously by Lamers et al. (1996) and Waters et al. (1997, 1998); these observations show the presence of weak, narrow emission bumps in the 30-45 μm region due to crystalline silicates. In the case of Wra 751, also the 9.7 μm amorphous silicate feature is present. In figure 1, we show new ISO-SWS observations of AG Car, as well as a ground-based 12.5 μm narrow-band N3 image taken with TIMMI (Käufl et al. 1992) at the 3.6m telescope, ESO, La Silla. The SWS spectrum is better centered on the nebula than the spectrum presented previously.

The better quality data for AG Car allows us to re-determine the shape and strength of the emission bands in the 30-38 μm window region. In figure 2, we show continuum subtracted 30-38 μm spectrum of AG Car. The presence of relatively narrow emission bands near 33, and 35 μm is evident. Note that the 33.8 μm band reported by Waters et al. (1997) is not seen in the speed 4 spectrum, nor in the (re-reduced) speed 1 spectrum (not shown). The differences between the present paper and previous results is due to improved data reduction and continuum subtraction procedures. We have

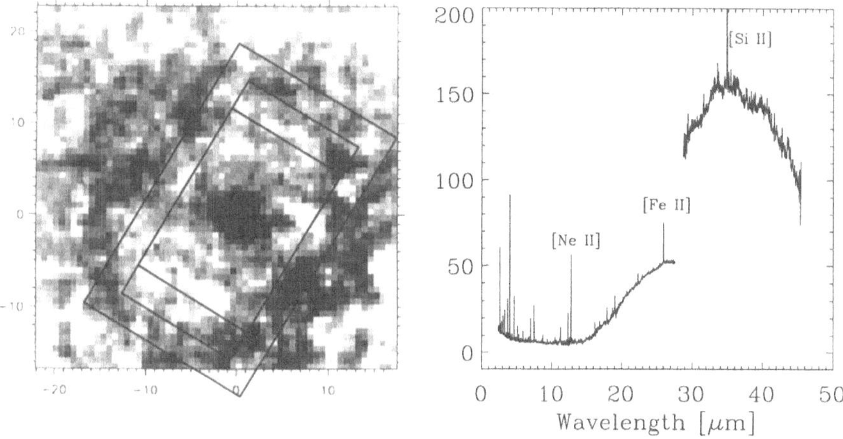

Fig. 1. left panel: TIMMI 12.5 μm image of AG Car. The scale is in arcsec. Overplotted is the position and size of the ISO-SWS apertures. Note the bi-polar structure of the nebula. right panel: ISO-SWS AOT01 speed 4 spectrum of AG Car. The jump at 30 μm is due to a change to a larger aperture of the SWS. Flux scale is in Jy.

compared the emission bands to laboratory spectra of crystalline silicates, and find a good match with the pyroxene ortho-enstatite ($MgSiO_3$) spectrum published by Koike and Shibai (1998). The laboratory results of Jäger et al. (1998) for a slightly Fe-enriched ortho-enstatite also give a reasonable match to the wavelengths of the bands, but the band strengths of these laboratory data do not agree with the AG Car data. It is remarkable that the 33.8 μm crystalline olivine forsterite (Mg_2SiO_4) peak, which is prominent in almost all oxygen-rich dust shells with crystalline silicates (see e.g. Waters et al. 1996), is weak or absent in AG Car. This suggests that the abundance ratio olivines to pyroxenes is much lower in AG Car than in other sources. We will return to this point below.

The ISO-SWS spectrum of Wra 751, and ground-based 10 μm imaging using TIMMI, have already been presented by Waters et al. (1998). The continuum subtracted spectrum is shown in figure 2. The pattern of emission features resembles that of AG Car, and we can draw similar conclusions concerning the lack of forsterite (see also Waters et al. 1998). Also the width of the emission bands is comparable in both objects. The similarity in solid state emission features in both stars implies that the formation and thermal

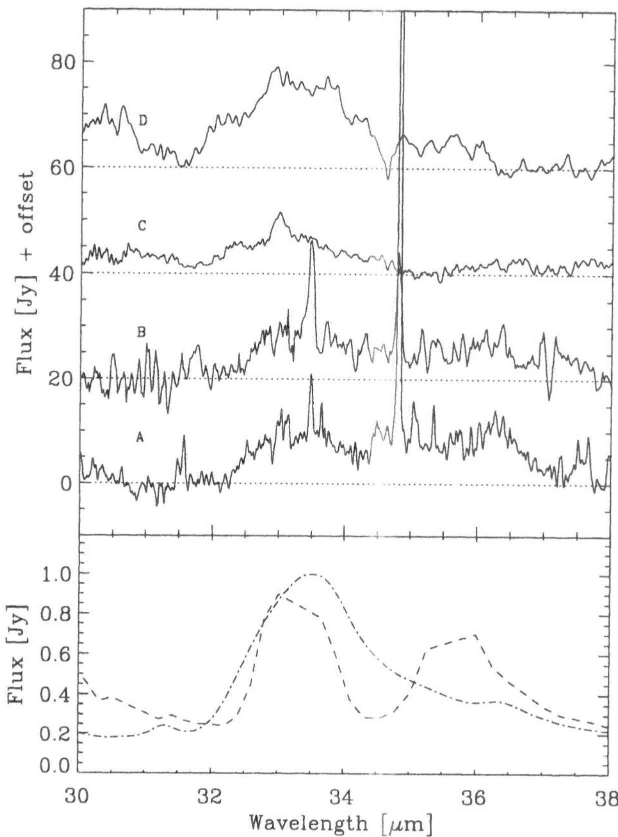

Fig. 2. Top: continuum subtracted 30-38 μm spectra of AG Car (slow scan) (A), Wra 751 (B), IRC+10 420 (C), and NML Cyg (D). Also shown are laboratory measurements of forsterite (dash-dotted line; Jäger et al. 1998) and enstatite (dashed line; Koike & Shibai 1998). The LBVs have little or no forsterite

processing history of the crystalline silicate grains is similar for both stars. AG Car and Wra 751 share other properties: the shape and kinematics of the nebulae, and the evolutionary phase in which both stars presently are. Although we are dealing with only two objects, these similarities are encouraging since it strengthens the diagnostic value of the solid state emission bands.

2.2 R 71

The LMC LBV R71 is one of the brightest stellar sources in the LMC at 25 μm (Wolf & Zickgraf 1986), due to the presence of a warm, detached dust shell with a dust mass of about 5 10^{-3} $M_{\odot}$ (Hutsemékers 1997), corresponding to a total mass of about 1 $M_{\odot}$ (depending on the uncertain gas/dust ratio). Roche et al. (1993) detected 10 μm silicate emission and conclude that the dust is oxygen-rich. We have observed R71 with ISO-SWS and with ISOCAM and will report on these observations elsewhere (Waters et al. in preparation). Here we mention the detection of a prominent 23 μm feature due to crystalline silicates. To our knowledge this is the first extragalactic detection of this band. We have fitted the dust spectrum using a 1D radiative transfer code, and assuming T_{eff} = 17,000 K and L = 7 10^5 $L_{\odot}$ (Lennon et al. 1994). We find a best-fit model with only large grains (0.1 to 1 μm) to fit the amorphous silicate feature, and derive a dust mass of 2 10^{-2} $M_{\odot}$, resulting in a total mass of 2 $M_{\odot}$ (for a gas/dust ratio of 100, which is very uncertain). The time-averaged total mass loss rate is 7 10^{-4} $M_{\odot}$/yr for a period of about 3 10^3 years and stopped 6000 years ago. We have neglected the effects of the interaction between the present-day fast wind and the slowly expanding dust shell in these calculations. Also, the outer radius of the dust shell is poorly constrained, which affects the estimate of the duration of the mass loss and the mass loss rate. However, the characteristics of the dust shell agree well with those expected for a RSG wind. Therefore R71 could have been a RSG when it produced the dust shell.

The ISOCAM data reveal the presence of a warm dust component which cannot be fitted with our radiative equilibrium dust model. This points to the presence of a dust component close to the star, or to a separate population of small grains out of thermal equilibrium.

3 PAH emission in LBVs

To our surprise, the new high quality SWS spectrum of AG Car shows clear evidence for the presence of the well-known family of emission bands near 3.3, 6.2, 7.7, 8.6 and 11.3 μm. The 5-12 μ CAM CVF spectrum of R71 also shows these bands (see figure 3). These bands are usually attributed to Polycyclic Aromatic Hydrocarbons (Leger & Puget 1984). PAH emission was previously found in the nebula of the candidate LBV HD168625 (Skinner 1997). The warm dust continuum present in both stars near 10 μm is probably related to a population of small grains as well.

The band strengths of the PAHs can be used to derive constraints on the nature of the grains and on their excitation (e.g. Allamandola et al. 1989). AG Car has a rather prominent 11.3 μm C-H out-of-plane bending mode peak compared to the strength of e.g. the 7.7 μm C-C stretch band strength; such a strong band can be due to e.g. a population of very large PAHs, or to predominantly neutral PAHs. The ratio of the 3.3 μm C-H stretch to 11.3

μm band strength is between 3 and 4, which rules out the presence of large PAHs, and suggests that the PAHs are neutral.

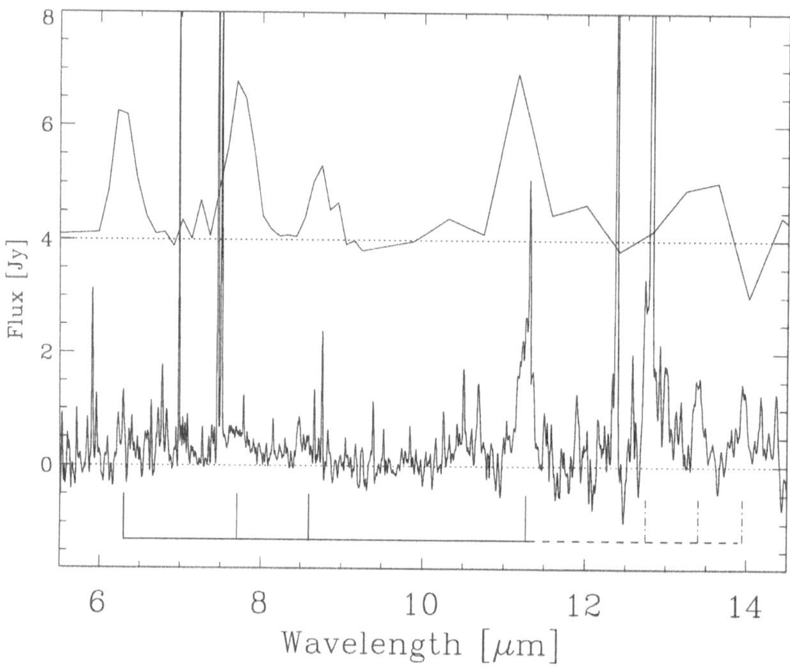

Fig. 3. PAH emission bands in R 71 (top) and AG Car (bottom). For R 71 we multiplied the spectrum by a factor 50 and we removed the prominent amorphous silicate band for clarity. Tickmarks indicate the expected position of the bands

In R71, the 11.3 μm band is strong but our data do not cover the 3.3 μm region; In addition, the 11.3 μm PAH band may blend with the 11.3 μm forsterite band. Given the strength of the 23 μm forsterite band and the relatively high temperature of the dust, the flux in this feature could be significant. Therefore, we cannot conclude much about the size of the PAHs in R71.

The presence of PAH emission in circumstellar envelopes is strongly correlated with C-rich chemistry: PAHs are abundant in C-rich planetary nebulae (PNe) and post-Asymptotic Giant Branch (AGB) stars. It is therefore not obvious that such C-bearing grains are present in the O-rich environment of the LBV nebulae. It is interesting to note that ISO spectroscopy of evolved C-rich objects with prominent PAH emission has shown that this 'mixed chemistry' is more widespread than previously thought: several [WC] central stars of PNe as well as the Red Rectangle nebula show a combination of oxygen-rich crystalline silicates and strong PAH emission (Waters et al. 1998b, 1998c; Cohen et al. 1999). In low mass stars, this can be understood in terms of the chemical evolution of the AGB star, which towards the end of the AGB changes from O-rich to C-rich chemistry. The LBVs discussed here are not yet C-rich. However, evolutionary calculations (Meynet et al. 1994) indicate that for an M_{ZAMS} 85 $M_{\odot}$ LBV the photospheric C/O ratio is about 0.8, i.e. above solar. Perhaps the grains are produced in the shock that occurs when the present-day C-enhanced fast wind collides with the slower dusty ejecta. This shock could drive a non-equilibrium C-rich chemistry which would result in the production of PAH molecules. Unfortunately, we have no information about the spatial distribution of the PAHs in the nebulae to verify this hypothesis.

In several galactic RSG weak PAH emission at 11.3 μm has been reported (Sylvester et al. 1994), proving that these molecules can be produced in an O-rich environment.

4 Comparison with other evolved dusty envelopes

In figure 2 we compare the 30-38 μm solid state emission bands of Wra 751 and AG Car to those of the M supergiant NML Cyg and the peculiar A-F type supergiant IRC+10 420. The latter object is believed to be a transition object, rapidly evolving to the blue part of the HR diagram (e.g. Jones et al. 1993). It is very well possible that IRC+10 420 is a WR progenitor. Both NML Cyg and IRC+10 420 have or have had mass loss rates in excess of 10^{-4} $M_{\odot}$/yr (Justtanont et al. 1996; Oudmaijer et al. 1996). It is evident that all four sources have crystalline silicates, but the ratio of the 33.0 (pyroxene) to 33.8 μm (olivine) peaks is different: in NML Cyg the olivine and pyroxene peaks are similar in peak strength, while in IRC+10 420 the olivine peak is quite weak. The olivine peaks seen in AGB and post-AGB stars (Waters et al. 1996) are significantly more prominent. Indeed, RSG and AGB stars have more olivines than Wra 751 and AG Car (Molster et al., in preparation).

In order to understand the differences in composition of the crystalline silicates between Wra 751 and AG Car on the one hand, and RSG and AGB stars on the other hand, it is useful to consider the physical and chemical conditions in the dust forming layers of dusty outflows. Crystalline silicates form predominantly in winds with very high mass loss rates, i.e. high densities in the dust forming layers (the LBVs are no exception to this trend). In order for the crystal lattice to become ordered after condensation from the gas phase, the temperature of formation must be higher than the glass temperature, which is about 1050 K for silicates. Therefore the crystalline grains must be among the first to condense from the gas phase. An important point to note is that crystalline silicates are Mg-rich and Fe-poor, while amorphous grains are Fe-rich and Mg-rich. A possible scenario for the dust condensation could be that first crystalline grains condense, and that most, but not all, of these grains *absorb* Fe at temperatures of about 1000 K or slightly lower, which destroys the ordered lattice structure (Tielens et al. 1998). The reaction with Fe is not possible for higher temperatures.

The condensation temperature for the Mg-rich olivine forsterite is about 1500 K. Slow cooling of this material results in the formation of the Mg-rich pyroxene enstatite. Therefore, the thermal processing timescale of crystalline grains after formation may determine the abundance ratio between olivines and pyroxenes. For the lower mass AGB stars and for RSG, the higher abundance of olivines may therefore indicate a more rapid cooling of the grains after formation compared to the LBVs Wra 751 and AG Car. This may be related to differences in stellar temperature, to the presence or absence of a chromosphere, and to the optical depth of the dust shell.

Perhaps the most surprising conclusion to be drawn from the ISO spectra is the similarity in physical and chemical conditions that must have occurred during the formation of the dust in RSG and LBVs. A picture emerges of a slow outflow at high density around a cool star. Simple models for the dust shell surrounding R71, AG Car and Wra 751 (Waters et al. 1997, 1998a), suggest that these shells could have been produced during a brief (2 10^3 yrs for AG Car) phase of very high mass loss. A more difficult question to answer is at what T_{eff} the star produced the dust shell. The high abundance of pyroxenes may point to an on average higher T_{eff} for the LBVs compared to RSG, but this is by no means certain. A possible scenario for the production of LBV ring nebulae could therefore be a brief period as yellow or red supergiant with a very high mass loss rate. Observationally, such objects would be very rare. For several LBVs it has now been established that the grain size is large (1 μm). Such grains produce *grey* extinction at optical wavelengths, i.e. the star would not appear to be very reddened. Perhaps the signature of LBVs during the ejection of the ring nebula is a yellow supergiant with modest circumstellar reddening but significant extinction, surrounded by a warm dust shell. This would not be in conflict with the observed lack of very luminous red supergiants in galaxies (Humphreys & Davidson 1979).

Acknowledgements. The authors thank Xander Tielens, Frank Molster, Teije de Jong, Jeroen Bouwman and Douwe Beintema for their help. TdJ kindly made the NML Cyg data available to us. LBFMW and AdK acknowledge financial support from an NWO *Pionier* grant.

References

Allamandola, L.J., Tielens, A.G.G.M., Barker, J.R.: 1989, ApJS 71, 733
Cesarsky, C.J. et al.: 1996, A&A 315, L32
Cohen, M., Barlow, M., Sylvester, R.J., Liu, X.-W., Cox, P., Schmitt, B., Speck, A.K.: 1999, ApJ (submitted)
De Graauw, Th. et al.: 1996, A&A 315, L49
Garcia-Lario, P., Riera, A., Manchado, A.: 1998, A&A 334, 1007

Hu, J.Y., De Winter, D., Thé, P.S., Pérez, M.R.: 1990, A&A 227, L17
Humphreys, R.M., Davidson, K.: 1979, ApJ 232, 409
Hutsemékers, D., van Drom, E.: 1991, A&A 251, 620
Hutsemékers, D.: 1997, in *Luminous Blue Variables: Massive Stars in Transition*, eds. A. Nota & H.J.G.L.M. Lamers, ASP Conference series vol. 120, p . 316
Jäger, C., Molster, C., Dorschner, J., Henning, Th., Mutschke, H., Waters, L.B.F.M.: 1998, A&A, in press
Jones, T.J., et al.: 1993, ApJ 411, 323
Justtanont, K., et al.: 1996, A&A 315, L217
Käufl, H-.U., Jouan, R., Lagage, P.O., et al., 1992, ESO messenger 70, 67
Kessler, M.F., et al., 1996 A&A 315, L27
Koike, C., Shibai, H.,: 1998, in press
Lamers, H.J.G.L.M. et al.: 1996a, A&A 315, L225
Leger, A., Puget, J.L.: 1984, A&A 137, L5
Meynet, G., Maeder, A., Schaller, G., Schaerer, D., Charbonnel, C.: 1994, A&AS 103, 97
McGregor, P.J., Finlayson, K., Hyland, A.R., Joy, M., Harvey, P.M., Lester, D.F.: 1988, ApJ 329, 874
Nota, A., Leitherer, C., Clampin, M., Greenfield, P., Golimowski, D.A.: 1992, ApJ 398, 621
Lennon, D.J., Wobig, D., Kudritzki, R.-P., Stahl, O., 1994, Space Sc. Rev. 66, 207
Nota, A., Livio, M., Clampin, M., Schulte-Ladbeck, R.: 1995, ApJ 448, 788
Oudmaijer, R.D., Groenewegen, M.A.T., Matthews, H.E., Blommaert, J.A.D.L., Sahu, K.C.: 1996, MNRAS 280, 1062
Roche, P.,F., Aitken, D.K., Smith, C.H.: 1993, MNRAS 262, 301
Skinner, C.J.: 1997, in *Luminous Blue Variables: Massive Stars in Transition*, eds. A. Nota & H.J.G.L.M. Lamers, ASP Conference series vol. 120, p . 322
Smith, L.J.: 1997, in *Luminous Blue Variables: Massive Stars in Transition*, eds. A. Nota & H.J.G.L.M. Lamers, ASP Conference series vol. 120, p . 311
Smith, L.J., Stroud, M.P., Esteban, C., Vilchez, J.M.: 1998, MNRAS 290, 265
Sylvester, R.J., Barlow, M.J., Skinner, C.J.: 1994, MNRAS 266, 640
Tielens, A.G.G.M., Waters, L.B.F.M., Molster, F.J., Justtanont, K.: 1998, Astrophys. Sp. Sc. 255, 415
Waters, L.B.F.M., et al.: 1996, A&A 315, L361
Waters, L.B.F.M., Morris, P.W., Voors, R.H.M., Lamers, H.J.G.L.M.: 1997, in *Luminous Blue Variables: Massive Stars in Transition*, eds. A. Nota & H.J.G.L.M. Lamers, ASP Conference series vol. 120, p . 326
Waters, L.B.F.M., Morris, P.W., Voors, R.H.M., Lamers, H.J.G.L.M., Trams, N.R.: 1998a, Astrophys. Sp. Sc. 255, 179
Waters, L.B.F.M., et al.: 1998b, Nature 391, 868
Waters, L.B.F.M., et al.: 1998c, A&A 331, L61
Wolf, B., Zickgraf, F.-J.: 1986, A&A 164, 435

Discussion

R. Schulte-Ladbeck: In η Car, as I showed in my talk, there is a different grain population in the UV/optical versus the IR. For the optical polarisation, we need small grains. The IR polarisation requires large, elongated grains with ferromagnetic inclusions.

S. Shore: That is very interesting since one does not see these, at least not in AG Car. But they are present at some stage in dust forming novae.

Georges Meynet and Andreas Korn

Joachim Dachs and Geraldine Peters

Wolf-Rayet and LBV Nebulae as the Result of Variable and Non-spherical Stellar Winds

Mordecai-Mark Mac Low

Max-Planck-Institut für Astronomie, Königstuhl 17, D-69117 Heidelberg, Germany

Abstract. The physical basis for interpreting observations of nebular morphology around massive stars in terms of the evolution of the central stars is reviewed, and examples are discussed, including NGC 6888, OMC-1, and η Carinae.

1 Introduction

The nebulae observed around massive, post-main sequence stars appear to consist of material ejected by the central stars during earlier phases of their evolution, rather than ambient interstellar matter. Models of these nebulae can be used to constrain the mass-loss history of the stars, giving an important input for stellar evolution models. Understanding the structure of these nebulae also clarifies the initial conditions for the resulting supernova remnants, which will interact with the circumstellar material for most of their observable lifetimes before encountering the surrounding interstellar medium.

A strong stellar wind sweeps the surrounding interstellar gas into a stellar wind bubble as shown in Figure 1. The stellar wind expands freely until it reaches a termination shock. If this shock is adiabatic, the hot gas sweeps up the surrounding ISM into a dense shell, forming a pressure-driven or energy-conserving bubble that sweeps up the surrounding ISM into a dense shell growing as $R \propto t^{3/5}$ in a uniform medium (Castor, McCray, & Weaver 1975). Should the termination shock be strongly radiative due to high densities or low velocities in the wind, the bubble only conserves momentum and will grow as $R \propto t^{1/2}$ (Steigman, Strittmatter, & Williams 1975). For more general discussions of blast waves in non-uniform media, see Ostriker & McKee (1988), and Bisnovatyi-Kogan & Silich (1995), as well as Koo & McKee (1990).

When these stars leave the main sequence, they pass through phases of greatly increased mass loss. These slow, dense winds expand into the rarefied interior of the main sequence bubble until their ram pressure $\rho_w v_w^2$ drops below the pressure of the bubble. (Should the main sequence bubble have cooled relatively quickly, this may never occur.) As the mass loss rate and velocity of the central wind vary during the post-main sequence evolution of the central star, these denser winds can in turn be swept up, producing the observed ring nebulae around evolved massive stars.

During their evolution, these nebulae are subject to a number of hydrodynamical instabilities, as well as thermal instabilities (*e. g.* Strickland &

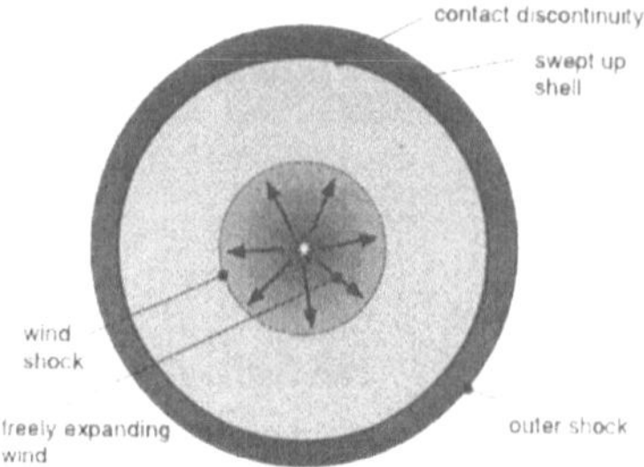

Fig. 1. Stellar wind bubble structure.

Blondin 1995). I explain how understanding the physical basis of the hydrodynamical instabilities gives insight into the dynamics of observed nebulae. High-resolution observations of nebular morphology can thus be used to constrain the mass-loss history of the central star.

I then describe how typical stellar evolutionary scenarioes can generate observed nebular morphologies, and show semi-analytic and numerical models derived from these scenarioes. For example, a star with a stellar wind varying from fast to slow and back again will have a clumpy circumstellar nebula due to hydrodynamical instabilities in the shell (García-Segura, Langer & Mac Low 1996). Nonspherical winds and stellar motion can add to the morphological richness of the resulting nebulae, as in the nebula around η Car (Langer, García-Segura, & Mac Low 1998). A recent review of this topic is Frank (1998).

2 Shell Instabilities

Gas swept up by a stellar wind will usually be subject to different instabilities. An adiabatic, decelerating shell with a density contrast across the shock of less than 10 in a uniform medium is stable. However, relaxing any of these constraints will lead to instabilities as I now describe.

2.1 Rayleigh-Taylor Instability

If the swept-up shell is denser than the stellar wind, as will be true in virtually all cases where a shell exists at all, the shell will be subject to RT instabilities if the contact discontinuity between the shocked stellar wind and the shell accelerates. This can be due to an external density gradient steeper than r^{-2} or to a sufficiently fast increase in the power of the central stellar wind, though these two mechanisms will lead to different shell morphologies, as I discuss below.

The RT instability occurs when the effective gravity due to acceleration points from a denser to a more rarefied gas. We can understand its driving

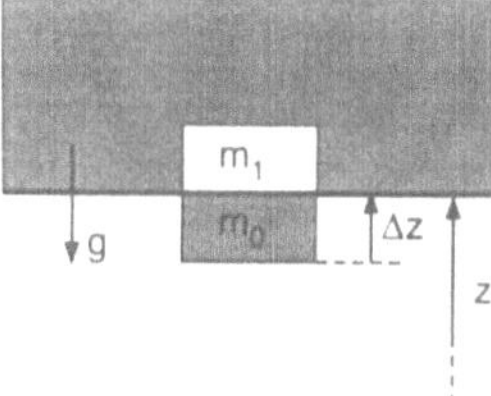

Fig. 2. Rayleigh-Taylor instability

mechanism by considering how the potential energy will change if we interchange a parcel of dense gas having mass m_1 with a parcel of more rarefied gas having $m_0 < m_1$, as shown in Figure 2. The potential energy before the interchange is given by $E_i = m_1 gz + m_0 g(z - \Delta z)$ which is greater than the potential energy after the interchange $E_f = m_0 gz + m_1 g(z - \Delta z)$ due to the difference in the masses. The decrease in potential energy drives an exponentially growing interchange of the two fluids.

When a RT instability occurs due to an external density gradient, dense fragments of shell are left behind as the less dense interior expands out beyond them, creating the characteristic bubble and spike morphology seen, for example, in models of superbubble blowout (e. g. Mac Low, McCray, & Norman 1989). The Wolf-Rayet ring nebula NGC 6888 shown in Figure 3 provides another example. Here a fast, rarefied Wolf-Rayet wind has swept up the slow, dense red supergiant wind that preceded it. While it was still sweeping up the slow wind, it was marginally stable to RT instabilities. However, at the outer edge of the slow wind, the sharp density gradient triggers RT instabilities, as modelled by García-Segura & Mac Low (1995) with the astrophysical gas dynamics and magnetohydrodynamics code ZEUS[1] (Stone & Norman 1992).

On the other hand, when a RT instability occurs due to an increase in power of the driving wind, some of the dense fragments of shell actually get shot out ahead of the bulk of the fragmenting shell, producing a markedly different morphology (Stone, Xu, & Mundy 1995). Although these fragments represent only a small fraction of the total mass of the shell, they can produce a very striking set of bow shocks in their wake. An example of this occurring around one or more pre-main sequence stars is given by the "bullets" observed around OMC-1 (Lane 1989, Allen & Burton 1993) in the Orion star forming region, as confirmed by McCaughrean & Mac Low (1997).

2.2 Vishniac Instability

If a pressure-driven shell is decelerating, but thin, with a density contrast across the shock of at least 25 for a stellar-wind bubble expanding into a

[1] Available by registration with the Laboratory for Computational Astrophysics at lca@ncsa.uiuc.edu

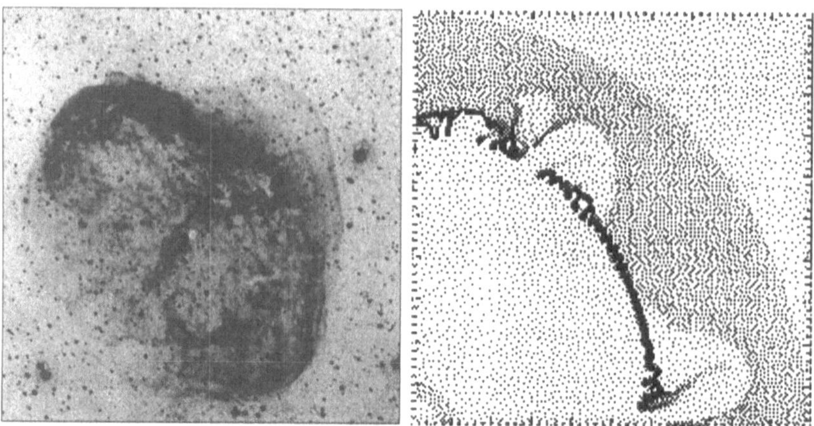

Fig. 3. Comparison between morphology of NGC 6888 in O [III] (in an image taken by K. B. Kwitter with the Burrell-Schmidt telescope of the Warner and Swasey Observatory, Case Western Reserve University), and a numerical simulation of RT instability due to a fast wind sweeping over the end of a slow, dense wind (García-Segura & Mac Low 1995). The model image shows a cross-section of the density structure in grayscale with black indicating high density and white low density.

uniform medium (Ryu & Vishniac 1988), or 10 for a point explosion (Ryu & Vishniac 1987), it will be subject to the Vishniac overstability (Vishniac 1983). This has been confirmed experimentally using blast waves generated by high-powered lasers propagating into gases with low and high adiabatic indexes (Grun et al. 1991).

The mechanism of the Vishniac overstability can be understood by considering a thin, decelerating shell driven from within by a high-pressure region, as shown in Figure 4. From within it is confined by thermal pressure acting normal to the shell surface, as adjacent regions can communicate with each other by sound waves, while from outside it is confined by ram pressure acting parallel to the velocity of propagation, as the shell moves supersonically into the surrounding gas. In equilibrium, these two forces remain in balance. Should the shell be perturbed, however, the thermal pressure will continue to act normally, but the ram pressure will now act obliquely, giving a transverse resultant force that drives material from "peaks" into "valleys" of the shell. The denser valleys will be decelerated less than the rarefied peaks, however, so that the positions of peaks and valleys are interchanged after some time. Vishniac (1983) showed that this overstable oscillation can grow as fast as $t^{1/2}$. It saturates when the transverse flows in the shell become supersonic and form transverse shocks, so that the end result of Vishniac instability is a shell with transonic turbulence and moderate perturbations (Mac Low & Norman 1993).

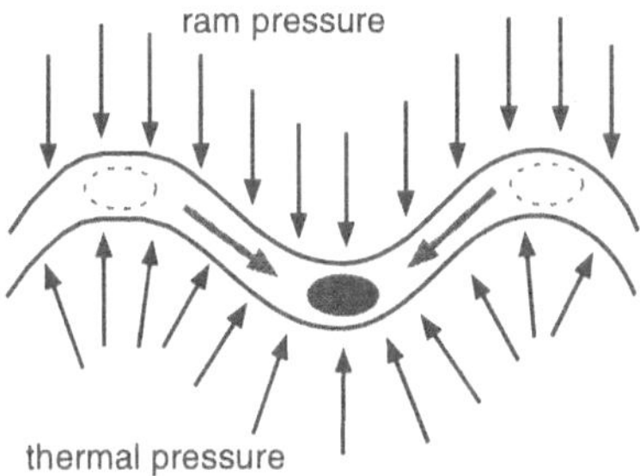

Fig. 4. Vishniac instability mechanism

2.3 Nonlinear Thin Shell Instability

Should the driving wind cool immediately behind its termination shock, for example because of exceptionally high mass-loss rates, it can form a decelerating shell that is momentum-driven rather than pressure-driven, so that it is effectively confined on both sides by ram pressure from shocks. Such a shell is not subject to the Vishniac instability, and is, in fact, linearly stable. However Vishniac (1994) has shown that if the shell is strongly perturbed, it will still be subject to a nonlinear thin shell instability (NTSI). When the shocks are oblique enough to the direction of flow, they will bend the streamlines passing through them, so that mass is driven towards the extrema of the perturbation. Numerical simulations by Blondin & Marks (1996), using a piecewise parabolic hydrocode called VH-1, have shown that the end result is a catastrophic breakup of the shell into a turbulent layer that grows in time.

3 A Final Example: Eta Carinae

As an example of how knowledge of these different instabilities can be used to constrain the evolution of a star, consider the example of the Homunculus Nebula around η Car. Langer, García-Segura, & Mac Low (1998) computed several two-dimensional models of it using ZEUS, following a basic scenario in which a luminous blue star with a fast stellar wind undergoes an outburst during which it has a much slower and denser wind strongly shaped by rotation, as described by Bjorkman & Cassinelli (1993), but then reverts to its previous state with a fast, rarefied wind. They chose two different values for the post-outburst wind, one consistent with current observed values of $\dot{M} = 1.3 \times 10^{-3} M_\odot$ yr^{-1} and $v_w = 1300$ km s^{-1}, and one with a faster, lower mass loss wind having $\dot{M} = 1.7 \times 10^{-4} M_\odot$ yr^{-1} and $v_w = 1800$ km s^{-1}. As shown in Figure 5, the slower, denser wind cools upon shocking, forming a momentum-driven shell that fragments due to the NTSI, producing a sharp, spiky shell morphology. On the other hand, the faster wind does not cool completely, and forms a bubble subject to Vishniac instabilities, giving it

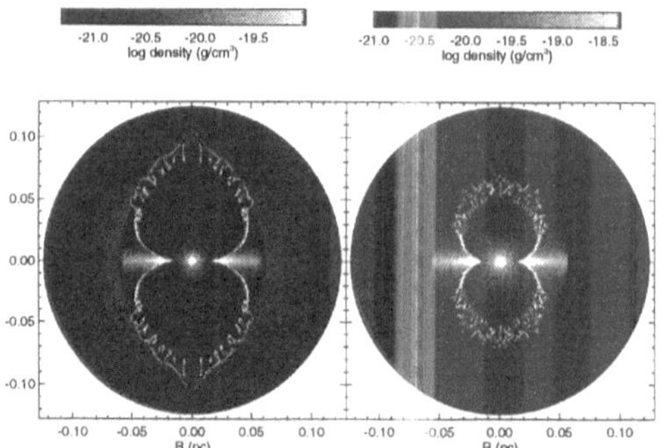

Fig. 5. Two-dimensional density distributions from models of η Car by Langer, García-Segura, & Mac Low (1998) with faster and slower post-outburst wind showing Vishniac instabilities and the NTSI respectively. Note how the faster wind model resembles a cross-section through the cauliflower-like observed lobes.

a much more curved, cauliflower-like appearance. Comparison to the high-resolution observations (Humphreys & Davidson 1994; Morse, Davidson, & Ebbets 1997) reveals that the actual morphology strongly resembles a three-dimensional version of the model with the faster wind. Langer et al. (1998) suggest that this reflects the typical behavior of the wind over the century since the outburst, and that the current wind properties are actually exceptional, and perhaps even indicative of another outburst on its way. This suggestion is supported by the gradual brightening of η Car over the last decades (Humphreys & Davidson 1994).

This work has made use of the NASA Astrophysical Data System Abstract Service. I thank the organizers for their invitation and their support of my attendence at this meeting.

References

Allen, D. A., & Burton, M. G. (1993): Explosive ejection of matter associated with star formation in the Orion nebula. *Nature*, **363**, 54–56

Bjorkman, J., & Cassinelli, J. (1993): Equatorial disk formation around rotating stars due to ram pressure confinement by the stellar wind. *Astrophys. J.*, **409**, 429–449

Bisnovatyi-Kogan, G. S., & Silich, S. A. (1995): Shock-wave propagation in the nonuniform interstellar medium. *Rev. Mod. Phys.*, **67**, 661–712

Blondin, J. M., & Marks, B. S. (1996): Evolution of cold shock-bounded slabs. *New Astron.*, **1**, 235–244

Castor, J., McCray, R., & Weaver, R. (1975): Interstellar Bubbles. *Astrophys. J. (Letters)*, **200**, L107–L110

Frank, A. (1998): Bipolar Outflows and the Evolution of Stars. *New Astron. Rev.*, in press (astro-ph/9805275)

García-Segura, G., Langer, N., & Mac Low, M.-M. (1996): The hydrodynamic evolution of circumstellar gas around massive stars. II. The impact of the time sequence O star → RSG → WR star. *Astron. Astrophys.*, **316**, 133–146

García-Segura, G., & Mac Low, M.-M. (1995): Wolf-Rayet Bubbles. II. Gasdynamical Simulations. *Astrophys. J.*, **455**, 160–174

Grun, J., Stamper, J., Manka, C., Resnick, J., Burris, R., Crawford, J., & Ripin, B. H. (1991): Instability of Taylor-Sedov blast waves propagating through a uniform gas. *Phys. Rev. Lett.*, **66**, 2738–2741

Humphreys, R. M., Davidson, K. (1994): The luminous blue variables: Astrophysical geysers. *Publ. Astron. Soc. Pacific*, **106**, 1025–1051

Koo, B.-C., & McKee, C. F. (1990): Dynamics of adiabatic blast waves in media of finite mass. *Astrophys. J.*, **354**, 513–528

Lane, A. P. (1989): Near Infrared Imaging of H2 Emission from Herbig-Haro Objects and Bipolar Flows. *Proceedings of the ESO Workshop on Low Mass Star Formation and Pre-main Sequence Objects* (ESO, Garching bei München), 331

Langer, N., García-Segura, G., & Mac Low, M.-M. (1998): "Giant Outbursts of Luminous Blue Variables and the Formation of the Homunculus Nebula Around η Carinae. *Astrophys. J. (Letters)*, submitted

Mac Low, M.-M., & McCaughrean, M. J. (1997): The OMC-1 Molecular Hydrogen Outflow as a Fragmented Stellar Wind Bubble. *Astron. J.*, **113**, 391–400

Mac Low, M.-M., McCray, R., & Norman, M. L. (1989): Superbubble blowout dynamics. *Astrophys. J.*, **337**, 141–154

Mac Low, M.-M., & Norman, M. L. (1993): Nonlinear growth of dynamical overstabilities in blast waves. *Astrophys. J.*, **407**, 207–218

Morse, J., Davidson, K., Ebbets, D. (1997): Multi-band WFPC2 Imaging of Eta Carinae. *Bull. Amer. Astron. Soc.*, **29**

Ostriker, J. P., & McKee, C. F. (1988): Astrophysical blastwaves. *Rev. Mod. Phys.*, **60**, 1–68

Ryu, D., & Vishniac, E. T. (1987): The growth of linear perturbations of adiabatic shock waves. *Astrophys. J.*, **313**, 820–841

Ryu, D., & Vishniac, E. T. (1988): A linear stability analysis for wind-driven bubbles. *Astrophys. J.*, **331**, 350–358

Steigman, G., Strittmatter, P. A., Williams, R. E. (1975): The Copernicus observations—Interstellar or circumstellar material. *Astrophys. J.*, **198**, 575–582

Stone, J. M., & Norman, M. L. (1992): ZEUS-2D: A radiation magnetohydrodynamics code for astrophysical flows in two space dimensions. I. The hydrodynamic algorithms and tests. *Astrophys. J. Suppl.*, **80**, 753–790

Stone, J. M., Xu, J., & Mundy, L. G. (1995): Formation of Bullets by Hydrodynamical Instabilities in Stellar Outflow. *Nature*, **377**, 315–316.

Strickland, R., & Blondin, J. M. (1995): Numerical Analysis of the Dynamic Stability of Radiative Shocks. *Astrophys. J.*, **449**, 727–738

Vishniac, E. T. (1983): The dynamic and gravitational instabilities of spherical shocks. *Astrophys. J.*, **274**, 152–167

Vishniac, E. T. (1994): Nonlinear instabilities in shock-bounded slabs. *Astrophys. J.*, **428**, 186–208

Discussion

A. Feldmeier: Do you see interactions between the Vishniac instability and the Rayleigh-Taylor instability in your thin-shell simulations?
M. MacLow: Yes. For example, in García-Segura & MacLow (1995) we modelled the behaviour of a Vishniac-unstable shell as it ran off the edge of a red supergiant wind and became Rayleigh-Taylor unstable. The Vishniac instability tends to act as a seed for R-T, by determining at least the initial wavelengths on which it acts.

L. Kaper: Could you comment on the time scales for growth of the different instabilities? Would it be possible that when the nonlinear, thin-shell instability operates the materialinvolved becomes so fragmented that it becomes difficult to observe?
M. MacLow: The Rayleigh-Taylor instability has an exponential growth rate, while both the Vishniac instability and the nonlinear, thin-shell instability have power-law growth rates. Time constants depend on many things, but will usually be somewhat shorter than the dynamical time. As far as observability goes, fragmentation per se will not reduce observability, except insofar as it creates a thicker, less dense shell. On the other hand, if the fragments are denser, their emission measure might actually increase.

H. Lamers: There is evidence based on a comparison between the nebular abundances and the abundances of the central object of η Car (obtained from high-resolution HST/UV spectra) that the star we see now (the LBV) is not the star that exploded. So one cannot derive the post-ejection wind properties from observations of the central object (Lamers et al., ApJ Letters, August 1998).
M. MacLow: This will naturally change the interpretation of our results, but the constraint of a rarified, fast wind having been dynamically important for the last century remains.

S. Shore: What happens in a full 3D model of the Vishniac instability? Would you expect enhanced filamentation when vorticity is not a constraint?
M. MacLow: This is an interesting question that is waiting for someone to put the time in to answer. We have the technology available to do these models now. I expect some enhanced filamentation as seen in the Grun et al. experiment. However, they also observed saturation suggesting that saturation by transverse shocks as described by MacLow & Norman (1992) will still occur.

G. Mellema: Can you explain why the polar lobes in your η Car models appear to be more unstable than the equatorial regions?
M. MacLow: Numerical dissipation is probably suppressing some Kelvin-Helmholtz instabilities. Secondly, other instabilities are not appearing strongly

because the shell is decelerating much more strongly at the equator due to the blowout of the polar lobes.

A. Moffat: All your calculations of wind interactions assume initially smooth winds. Then follows the obvious question as to the impact of clumped winds for which growing evidence suggests high degrees of multi-scale clumping. Even if you have not done any such calculations, can you guess what might happen?
M. MacLow: Yes, all of our models assume smooth winds. I am by no means satisfied with this assumption, so much so that I have devoted most of my time for the past year or so to modelling hydrodynamic and MHD turbulence with the ultimate intention of blowing winds into it to understand the behaviour of blast winds in real turbulent media. Simple models of dense clumps in a uniform density background will not be enough, that is already clear. The effects should include enhanced instabilities and thicker shells, perhaps also faster expansion.

Kerstin Weis, Mordecai-Mark MacLow and Lars Koesterke

Ring Nebulae Abundances: Probes of the Evolutionary History of Luminous Blue Variable Stars*

Linda J. Smith[1], Antonella Nota[2], Anna Pasquali[3], Claus Leitherer[2], Mark Clampin[2], and Paul A. Crowther[1]

[1] Dept. of Physics and Astronomy, UCL, Gower St., London WC1E 6BT, UK
[2] STScI, 3700 San Martin Drive, Baltimore, MD 21218, USA
[3] ST-ECF/ESO, Karl Schwarzschild Str. 2, D-85748 Garching, Munich, Germany

1 Introduction and Observations

The ring nebulae that surround most Luminous Blue Variable (LBV) stars are believed to be the relics of one or more giant eruptions (cf. Nota, these proc.). The nebulae thus represent the stellar surface layers at the time of the eruption(s) and by analysing their chemical composition and dynamics, it is possible to infer the past evolutionary state of the star.

Observations with the *Hubble Space Telescope* (HST) and the Faint Object Spectrograph (FOS) were obtained for the nebulae around the two LMC LBVs R127 and R143, and the Ofpe/WN9 star S119 for the purpose of obtaining abundances. The spectra cover the wavelength range 3235–6818 Å and a slit of dimensions $1''.7 \times 0''.2$ was placed on the brightest portion of each nebula. Full details of these observations are given in Smith et al. (1998).

2 Analysis and Results

Interstellar reddenings, electron densities N_e and temperatures T_e were derived for the three nebulae using the Balmer series, [S II] $\lambda6717/\lambda6731$ and [N II] $\lambda6584/\lambda5755$ ratios. For the R127 nebula, we derive $E(B-V) = 0.16 \pm 0.03$, $N_e = 720 \pm 90\,\mathrm{cm}^{-3}$, and $T_e = 6420 \pm 300\,\mathrm{K}$. For the S119 nebula, we find $E(B-V) = 0.05$, $N_e = 680 \pm 170\,\mathrm{cm}^{-3}$, and an upper limit to T_e of 6800 K. In contrast to the similar physical parameters we derive for the R127 and S119 nebulae, the R143 nebula has $N_e < 100\,\mathrm{cm}^{-3}$ and $T_e = 12\,000\,\mathrm{K}$. These parameters and the abundances (see below) indicate that the region observed with FOS is part of the 30 Dor H II complex. The real ejecta nebula is located just $2''$ from the star (see Smith et al., 1998; Nota, these proc.)

* Based on observations with the NASA/ESA Hubble Space Telescope, obtained at the Space Telescope Science Institute, which is operated by AURA for NASA under contract NAS5-26555.

Nitrogen and oxygen abundances were determined for the ejecta nebulae directly from the strengths of the [N II] λ6584 and [O II] λ3727 lines. Correction for unseen ionization stages is unnecessary because the nebulae have low values of T_e. For the S119 nebula, [N II] λ5755 was not detected; approximate upper and lower limits to the abundances were derived using the measured upper limit to T_e, and the value of T_e which reproduces the observed S^+/H in the R127 nebula. For the R143/30 Dor nebula, [O III] λ5007 is observed, and the abundances have been determined using the ionization correction factors of Kingsburgh & Barlow (1994).

In Table 1, we list the derived N/O ratios, the N enrichment factor Δ N, and the O depletion factor $1/\Delta$ O for the three nebulae. The mean of the LMC H II region abundances of Dufour (1984) and Russell & Dopita (1990) were used to calculate these factors. For comparison, we list the abundances for other objects containing processed stellar material in the LMC; non-type I PNe (nebulae which show CN-processed material), and the inner ring of SN 1987A which is believed to consist of red supergiant (RSG) wind material.

Table 1. Nebular abundances

Object	SpT	N/O	Δ N	$1/\Delta$ O	Ref.
R127	LBV	0.89 ± 0.40	10.7 ± 2.2	2.0 ± 1.0	1
S119	Ofpe/WN9	1.41–2.45	11.5–24.5	1.4–5.1	1
R143/30 Dor	LBV	0.04 ± 0.04	0.5 ± 0.6	2.2 ± 1.1	1
Non-Type I PNe		0.19 ± 0.09	4.6 ± 2.9	1.0 ± 0.5	2
SN 1987A	RSG?	1.55 ± 0.40	11.5 ± 2.3	3.2 ± 0.7	3
⟨H II⟩		0.04	1.0	1.0	4

1. this paper; 2. Barlow (1991), Walton et al. (1991); 3. Panagia et al. (1998, & in prep.); 4. Dufour (1984), Russell & Dopita (1990).

3 Discussion and Conclusions

For the R127 nebula, we derive an N/O ratio of 0.89 with N enriched by a factor of 11, and O showing little, if any, depletion. For the S119 nebula, we derive a similar N/O ratio of 1.41–2.45, where the range reflects the uncertainty in T_e. The R143/30 Dor abundances agree well with the mean H II region abundances listed in Table 1. Comparing first the R127 nebular abundance pattern with non-type I PNe, we find that while N is a factor of ≈ 2 more enriched, O is the same within the errors. Comparison with the SN 1987A abundances shows that the N enrichment is identical and that the O abundances agree reasonably well. We conclude that the high N enrichment and minimal O depletion in the R127 nebula are consistent with material that has been CN-processed only. The remarkable agreement with the SN 1987A inner ring abundances and the low expansion velocity of the nebula (Smith et al. 1998) suggests that the R127 nebula was once the CN-processed convective envelope of a RSG.

Comparison of the nebular abundances with those determined for LBVs from atmospheric analyses (e.g. Lennon et al. 1994; Venn 1997) shows that the atmospheres should consist of CNO-processed material. It thus appears that the R127 nebular abundances do not reflect the current surface composition assuming it has CNO-equilibrium abundances, and suggests that the nebula was ejected before, or at the very start of, the LBV phase. Interestingly, the η Car nebula shows extreme CNO-processing (Dufour et al. 1997).

We have examined whether the R127 nebular N/O abundance ratio can be reproduced by the evolutionary tracks of Meynet et al. (1994). We find that the amount of mass loss is the critical parameter. If it is too high, the observed N/O ratio occurs while the star is still on the main-sequence and the evolutionary timescale is then too long to produce a small nebula in the LBV phase. We find best agreement for a $60\,M_{\odot}$ model where the pre-LBV mass loss is low enough to allow the star to evolve redward, and $N/O \approx 1$ occurs when the star is a cool supergiant inside the Humphreys-Davidson limit.

Our finding that the R127 nebula was once a RSG convective envelope is also supported by recent *ISO* observations. Waters et al. (these proc.) find that the crystalline dust structure in LBV nebulae is suggestive of formation in RSG envelopes. How can we reconcile these findings with the observed absence of RSGs above the Humphreys-Davidson limit? First, the RSG phase must be brief. The model of Stothers & Chin (1996 and refs. therein) has the LBV eruption occurring in a brief RSG phase. Second, it is possible that the RSG phase does not correspond to an evolutionary phase in the usual sense, but rather to a pseudo-RSG phase, occurring as a result of encountering the Eddington limit. What happens at this point is very complicated. According to the model of Owocki & Gayley (1997), it is possible that the star will respond to the super-Eddington condition by developing a convective envelope and becoming very bloated such that it will have the appearance of a RSG. The outer envelope may then become detached due to a density inversion (Owocki & Gayley 1997).

On the basis of our observations we suggest the following picture for the formation of LBV nebulae. The pre-LBV does not lose enough mass while on the main sequence, and evolves redward. At some point it encounters the Eddington limit, and develops a deep convective envelope which is gently ejected to reveal the LBV underneath. LBVs should therefore be surrounded by massive RSG envelopes which in most cases, will be neutral because of the low ionizing fluxes.

References

Barlow M.J. 1991, in The Magellanic Clouds, IAU Symp. 148, ed. R. Haynes & D. Milne (Kluwer, Dordrecht), p. 291

Dufour R.J., 1984, Structure and Evolution of the Magellanic Clouds, IAU Symp. 108, ed. S. van den Bergh & K.S. de Boer (Kluwer, Dordrecht), p. 353

Dufour, R.J., Glover, T.W., Hester, J.J., Currie, D.G., van Orsow, D. & Walter, D.K., 1997, in Luminous Blue Variables: Massive Stars in Transition, ed. A. Nota & H.J.G.L.M. Lamers (ASP Conf. Ser.), 120, p. 255

Kingsburgh, R.L. & Barlow, M.J., 1994, MNRAS, 271, 257

Lennon, D.J., Wobig, D., Kudritzki, R.-P. & Stahl, O., 1994, Space Sci. Rev., 66, 207

Meynet, G., Maeder, A., Schaller, G., Schaerer, D. & Charbonnel, C. 1994, A&AS, 103, 97

Panagia, N., Scuderi, S., Gilmozzi, R. & the SINS Collaboration, 1998, in ESO/CTIO/LCO Workshop *SN 1987A: Ten Years After*, eds. M. Phillips and N. Suntzeff, A.S.P. Conf. Ser., in press.

Owocki, S.P. & Gayley, K.G., 1997, in Luminous Blue Variables: Massive Stars in Transition, ed. A. Nota & H.J.G.L.M. Lamers (ASP Conf. Ser.), 120, p.121

Russell, S.C. & Dopita, M.A., 1990, ApJS, 74, 93

Smith, L.J., Nota, A., Pasquali, A., Leitherer, C., Clampin, M. & Crowther, P.A., 1998, ApJ, 503, 278

Stothers, R.B. & Chin, C.-w., 1996, ApJ, 468, 842

Venn, K.A., 1997, in Luminous Blue Variables: Massive Stars in Transition, ed. A. Nota & H.J.G.L.M. Lamers (ASP Conf. Ser.), 120, p. 95

Walton, N.A., Barlow, M.J., Monk, D.J. & Clegg, R.E.S., 1991, in The Magellanic Clouds, IAU Symp. 148, ed. R. Haynes & D. Milne (Kluwer, Dordrecht), p. 334

Discussion

J. Cassinelli: About 10 years ago Kris Davidson found that stars could appear to the right of the Humphreys-Davidson limit at $\sim$ 8000 K. So, are your "pseudo red supergiants" really at the far right side (T_{eff} ~3000 K) of the HR diagram or is there a large uncertainty concerning T_{eff} at the cool phase? This is important for deciding whether dust can from or not.
L. Smith: There is a large uncertainty regarding T_{eff} since the amount of mass loss is the critical parameter in determining how far to the red a star will evolve.

M. Magalhães: Were the evolutionary tracks you used for non-rotating stars?
L. Smith: Yes, the tracks are those of Meynet et al. (1994) and do not include rotation.

R. Humphreys: How short do you expect the time scale to be for the red supergiant phase? We observe several stars now in different galaxies

(IRC +10420, Var A) undergoing evolutionary changes that take only 2000–3000 years. So for your proposed red supergiants to exist, they have to be < 1000 years and not be observed. Alternatively, the stars could be optically obscured. The LMC has been surveyed in the infrared. The two most luminous M supergiants are IRAS sources; they are OH/IR supergiants. They have $M_{bol} \sim -9.5 - -9.7$. There are no more luminous M supergiants known in the LMC.
So your proposed red supergiant phase will have to be extremely brief (R 127 is very unlikely given the size of red supergiants). HR Car is below the Humphreys-Davidson limit.
For a pseudo-RSG phase, take a look at Var A in M 33 (Humphreys et al. 1987) and ρ Cas – two F supergiants that temporarily became M supergiants.

Rolf Kudritzki and Henny Lamers

The Wind Momentum – Luminosity Relationship of Blue Supergiants

Rolf-Peter Kudritzki

Institut für Astronomie und Astrophysik der Universität München, Scheinerstr.1, D-81679 München, Germany
Max-Planck-Institut für Astrophysik, D-85740 Garching bei München, Germany

Abstract. The prediction of the **Wind Momentum – Luminosity Relationship** (WLR) based on the theory of radiation driven winds is verified by quantitative spectroscopy of winds of A-, B- and O-supergiants. The relationship depends on spectral type. New stellar wind calculations are presented reproducing the observed spectral type dependence. The impact of spectral variability on the WLR is investigated by an analysis of some hundred spectra of the luminous A0Ia-supergiant HD92207 obtained within the Heidelberg Spectral Variability Survey and found to be small.

Finally, the WLR is discussed as a tool for the determination of extragalactic distances. Recent results obtained for the Galaxy, M31 and M33 are presented. The potential of the method is discussed with the conclusion that it **may allow independent distance moduli to be obtained with an accuracy of ten percent out to the Virgo and Fornax clusters of galaxies**.

1 The Wind Momentum – Luminosity Relationship

The **Wind Momentum – Luminosity Relationship** (WLR) provides a new independent tool to derive extragalactic distances with an accuracy comparable to the Cepheid method. The basic concept is simple. Since the winds of the blue supergiants are a result of radiation pressure, we expect the mechanical momentum flow of a stellar wind $\dot{M}v_\infty$ to be a function of the photon momentum rate L/c provided by the stellar photosphere and interior. Indeed, a straightforward analytical solution of the hydrodynamic equations of radiation driven winds yields (see Kuddritzki et al. 1989, Kudritzki et al. 1996a, Puls et al. 1996, Kudritzki 1998)

$$\dot{\mathbf{M}}\mathbf{v}_\infty \propto \frac{1}{R_*^{0.5}}\mathbf{L}^{1/\alpha}, \tag{1}$$

where $\alpha(\approx 2/3...1/2)$ is well determined by atomic physics and represents the power law exponent of the distribution function of line strengths of the many thousands of lines driving the wind.

Thus, by measuring the rate of mass-loss and the terminal velocity directly from the spectrum we are – in principle – able to determine the luminosity

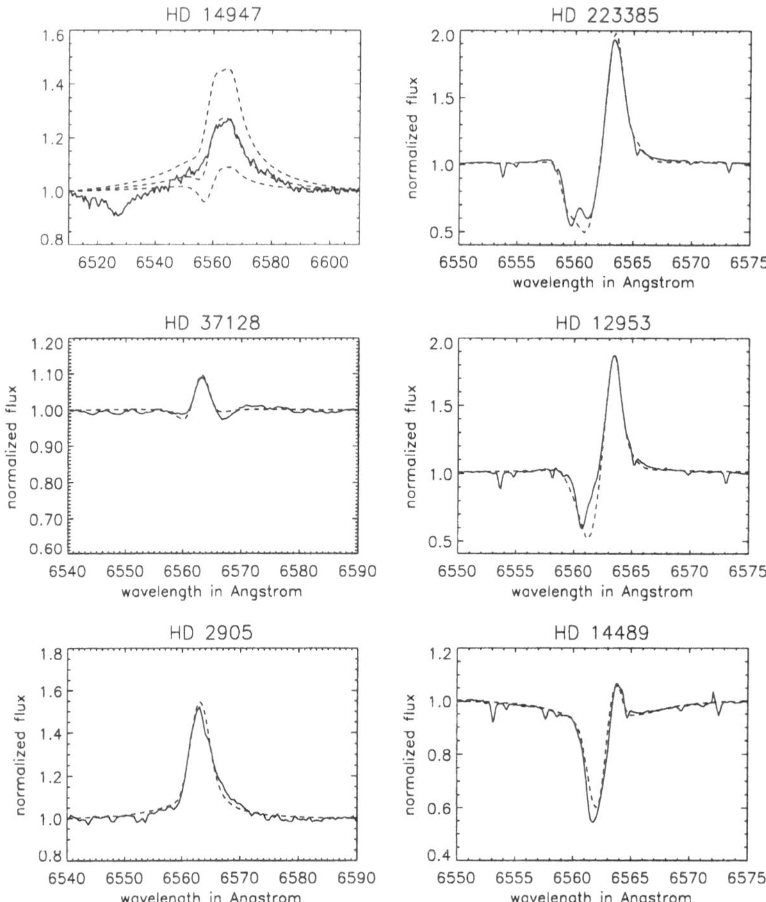

Fig. 1. H_α line profile fits of the galactic supergiants HD 14947 (O5 Ia), HD 37128 (B0 Ia), HD 2905 (B0.7 Ia), HD 223385 (A3 Ia), HD 12953 (A1 Ia), HD 14489 (A2 Ia). The fits of HD 14947 are taken from Puls *et al.*, 1996 and the three different curves indicate adopted changes in the model calculations of 25 percent in $\dot{M}$. The fits of all other objects are from Kudritzki et al. *et al.*, 1998, who used an improved version of the unified model atmosphere code developed by Santolaya-Ray, Puls and Herrero (1997) for the line profile fits.

of a blue supergiant. This is an exciting perspective, because it would give us a purely spectroscopic tool to determine stellar distances. Quantitative spectroscopy would yield T_{eff}, gravity, abundances, intrinsic colours, reddening, extinction, $\dot{M}$ and v_∞. With the luminosity from the above relation one could then compare with the dereddened apparent magnitude to derive a distance.

It is, of course, challenging to compare the prediction of radiation driven wind theory with the observations to see whether an observed WLR does

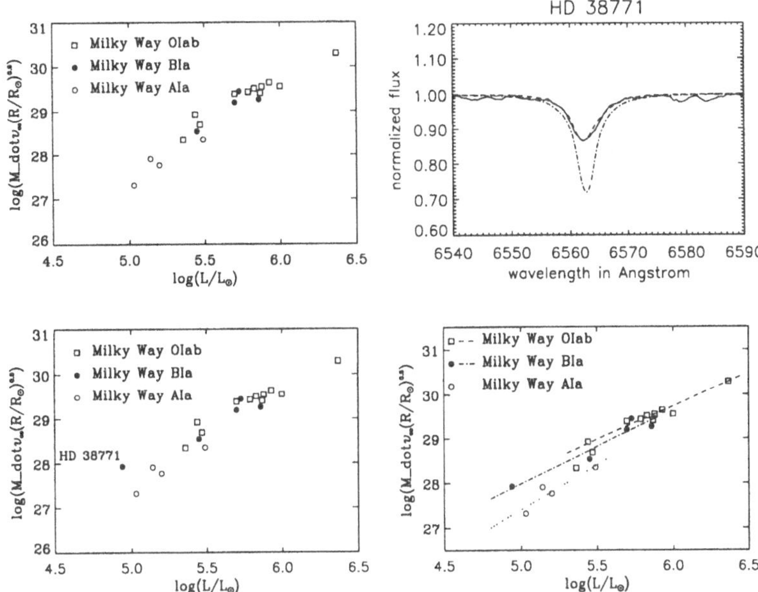

Fig. 2. Upper left: The observed WLR of the brightest galactic supergiants of spectral type O, early B and early A. Data from Puls *et al.*, 1996 and Kudritzki *et al.* 1998. Wind momenta are given in cgs-units. Upper right: H_α profile fit of the low luminosity supergiant HD 38771 (κ Ori, B0.5 Ia) with $\dot{M} = 2.8$ and $1.0\ 10^{-7}$ $M_\odot$/yr (dotted and dashed-dotted, respectively). Lower left: The observed WLR including HD 38771. Lower right: Three independent linear regressions fitted to the three spectral types assuming a temperature dependence of the WLR.

exist. This requires the determination of v_∞ and $\dot{M}$. While the measurement of the former is easy and straightforward by means of the blue edges of the P–Cygni profiles of UV resonance lines (for O-, B-, and A-supergiants) or H_α (for A-supergiants with winds of significant strength; for examples and references, see for instance Kudritzki 1998), the determination of the latter with sufficient precision used to be problematic. However, the development of NLTE Unified Model Atmospheres made it possible to use the strength of H_α as a wind line to determine $\dot{M}$ very precisely (see again Kudritzki 1998 for an overview). In this way, Puls et al. 1996 obtained accurate mass-loss rates for a large sample of galactic O-stars and provided a detailed discussion of the H_α line formation process and the possible error sources. Kudritzki et al. 1998b extended this work on the brightest galactic B- and A-supergiants with well determined distances. Fig. 1 indicates how accurately the observed H_α stellar wind line profiles can be reproduced by the theoretical models yielding mass-loss rates with a precision better than 0.1 dex. Fig. 2 (upper left) combines the results from Puls et al. 1996 and Kudritzki et al. 1998b with regard to the observed wind momentum and reveals the clear existence of a WLR.

There is the indication of a slight curvature of the observed WLR or of the existence of three independent relations of slightly different height and slope for the three spectral types. The latter interpretation is supported by the investigation of the wind momentum of the B0.5 supergiant HD 38771 (κ Ori). This object is usually considered as a twin of HD 37128 (ϵ Ori, H_α profile displayed in Fig. 1), however, the recent astrometric results obtained with the Hipparcos satellite show that it is in the foreground of the Orion association with a de-reddened absolute magnitude of only M_V = -4.8^m. This explains, why – contrary to HD 37128 - H_α is in absorption and is fitted with a significantly smaller rate of mass-loss (see Fig. 2, upper right). The wind momentum of HD 38771 – a low luminosity B-supergiant – is still larger than the wind momentum of the A-supergiants of comparable luminosity (see Fig. 2, lower left). We take this as an indication of a temperature dependence of the WLR and adopt three independent linear regression for the three spectral types (Fig. 2, lower right). We also note that the WLR of Central Stars of Planetary Nebulae, which are as hot or even hotter than O-star, follows the WLR of O-stars (see Kudritzki *et al.* 1997). This confirms our conclusion that the WLR is temperature dependent.

2 New wind models for supergiants of spectral type O, B and A

To be able to investigate the systematic behaviour of line driven winds across the whole HRD of hot stars (where "hot" means that hydrogen is ionized and can be as cool as T_{eff} = 8000 K for A-supergiants) in all different stages of evolution (including massive stars as well as post-AGB) we have developed a new approach to calculate the wind dynamics. This approach is based on the improvements achieved during the last decade with regard to atomic physics and line lists (see Pauldrach et al. 1998). We use the line list of 2.5 10^6 lines of 150 ionic species and apply analytical formulae (see Springmann 1997, Springmann and Puls 1998) for a fast approximation of NLTE occupation numbers to calculate the radiative line acceleration, which is then represented by a new parameterization using depth dependent force multiplier parameters (see Kudritzki et al. 1998a; note that now as an improvement real model atmosphere fluxes are used instead of Planck functions for the photospheric irradiation of the wind). Because of the depth dependent force multipliers a new formulation of the critical point equations is introduced (Kudritzki et al. 1998c). In this way, wind models can be calculated within a few seconds on a workstation for every hot star with specified effective temperature, mass, radius and abundances.

We have used this new approach to calculate wind models for typical parameters of A-, B- and O-supergiants. The results are compared with the observations in Fig. 3. Although much more work has yet to be invested in the future to compare theory and observation on a star by star basis, a first conclusion is that the theory is roughly able to reproduce the observed trends with regard to wind momentum and terminal velocity.

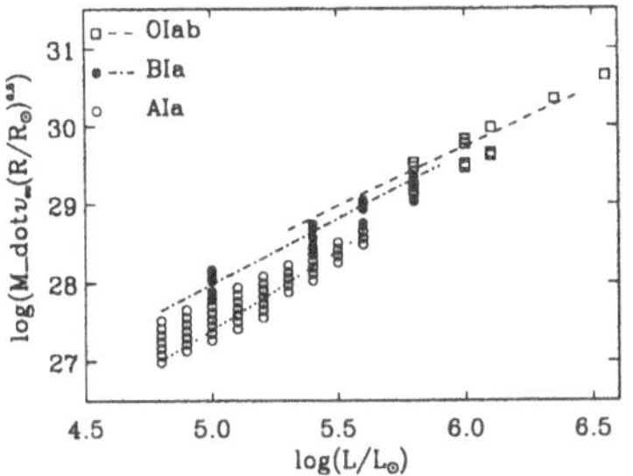

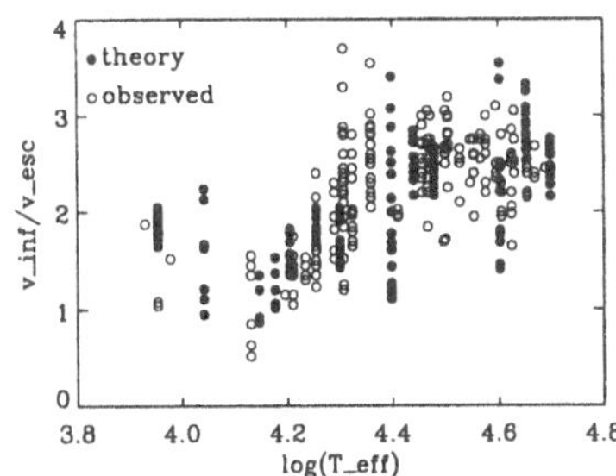

Fig. 3. New wind calculations compared with the observations. Left: Wind momenta as function of luminosity. The dotted, dashed-dotted and dashed curves correspond to the observed regression curves of Fig. 14 for A-, B- and O-supergiants, respectively. The (open and filled) circles and squares correspond to the models. Their scatter at a given luminosity is a result of the different masses or effective temperatures adopted. Right: Ratio of terminal velocity to photospheric escape velocity as function of effective temperature. Open circles represent observations taken from Prinja and Massa (1998). Filled circles correspond to the calculations and the scatter at a given temperature is a result of the different masses and luminosities which were adopted.

3 Spectral variability – the case of the A0Ia-supergiant HD 92207

The observed tight correlation of stellar wind momentum with luminosity seems to in variance with the variability of stellar wind lines observed in the spectra of many supergiants indicating stellar wind momenta to be a function of time. However, even strong changes in the observed line profiles of – for example – H_α do not neccessarily transform into large changes of stellar wind momentum because of the extremely strong dependence of H_α on mass-loss. Ruppersberg et al. 1998 (see also Ruppersberg 1998) have investigated this question by analyzing some hundred spectra of the the A0Ia-supergiant HD 92207 obtained within the Heidelberg Spectral Variability Survey by Kaufer et al. 1996 over a period of 150 days. For each spectrum effective temperature, gravity, radius, mass-loss rate and terminal velocity were carefully determined by line profile fits (see Fig. 4 for illustration). Mass-loss rates und terminal

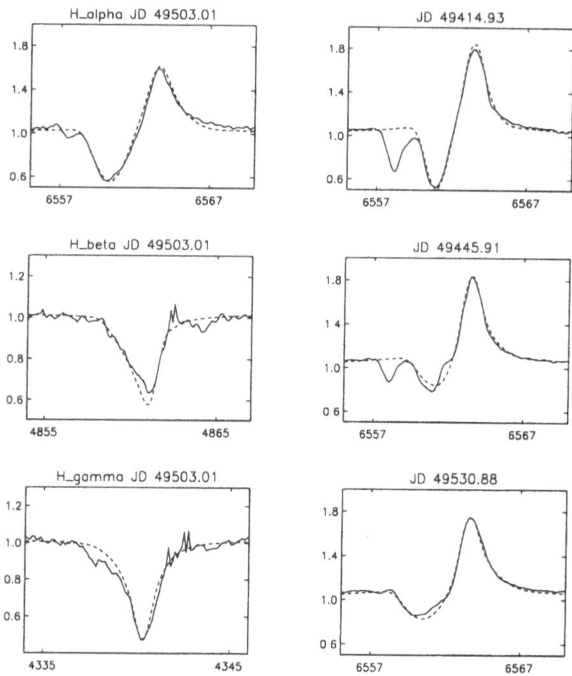

Fig. 4. Left: Simultaneous fit of the H_{α}, $_{\beta}$, $_{\gamma}$ profiles in one spectrum of HD 92207. Right: H_{α} fits at three different times. (From Ruppersberg *et al.*, 1999).

wind velocities obtained from this analysis did indeed show a significant variation as function of time. However, the variation in wind momentum turned out to be smaller than 0.2 dex (see Fig. 5) resulting only in a very small shift of this object in the wind momentum - luminosity plane typical for the general scatter within the WLR.

4 The Wind Momentum – Luminosity Relationship and the determination of extragalactic distances

It has long been a dream of stellar astronomers to use the spectra of the most luminous blue stars to determine the distances to other galaxies. Now, the WLR might provide the means to fullfill this dream.The basic technique is to derive the stellar parameters (temperature, gravity, metallicity) spectroscopically from optical absorption lines (see Kudritzki 1998 for references), and then to model the H_{α} profiles ($\rightarrow \dot{M}$ for O-, B- and A-supergiants, v_{∞} for A-supergiants) and the UV P-Cygni line profiles ($\rightarrow v_{\infty}$ for O-, B- and A-supergiants) to obtain the wind momentum. Application of the empirically calibrated WLR appropriate to the spectroscopically determined metallicity

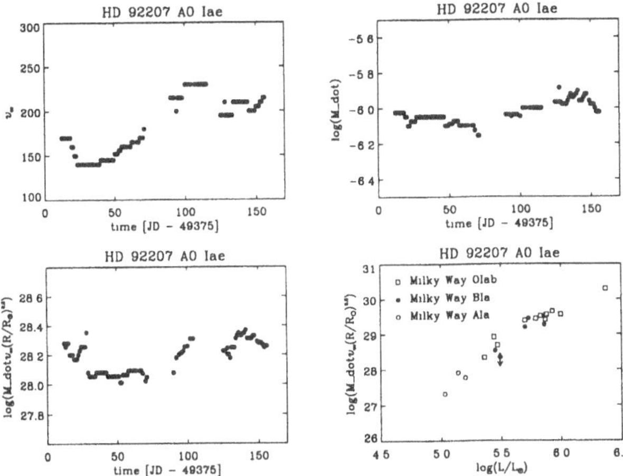

Fig. 5. Terminal velocity, mass-loss rate and stellar wind momentum of HD 92207 as function of time (upper left and right, lower left, respectively). The location of HD 92207 in the wind momentum – luminosity plane superimposed by the arrow indicating the maximum amount of momentum variability measured. (From Ruppersberg *et al.*, 1999).

(for a discussion how to deal with different metallicities, see again Kudritzki 1998) then allows one to determine the intrinsic luminosity from which the distance follows once the stellar apparent magnitude and colours(→ reddening and extinction in conjunction with the stellar parameters determined spectroscopically) are known. In practice, one will adopt a distance for all stars in a galaxy to locate them in the momentum/luminosity plane and will then iterate the distance until the distribution coincides with the calibrated sample.

4.1 A first investigation in M31 and M33

A crucial test of the WLR method is the investigation of wind momenta of blue supergiants in M31. In an extensive collaboration (see acknowledgements), we have started a systematic study using the WHT and Keck telescopes for optical spectroscopy and the HST for multicolour photometry and UV spectroscopy. While most of the Cycle 6 and 7 HST observations have still to be carried out, we are able to present the first analysis of two M31 A-supergiants (see Mc Carthy et al. 1997) based solely on Keck HIRES optical spectroscopy in Fig. 6 (see also Mc Carthy et al. 1995 for M33). The results are very encouraging. The H_α line profiles allow a precise determination of wind momenta. Adopting a distance modulus of 24.25 mag (Rozanski & Rowan-Robinson 1994) we find that both objects coincide very well with galactic WLR. Mc Carthy et al. 1997 estimate that with 10 to 20 objects it will be possible to obtain an independent M31 distance modules with an accuracy of 0.1 mag.

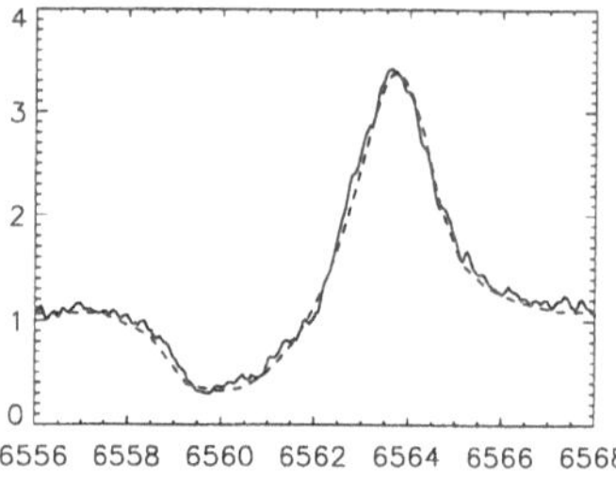

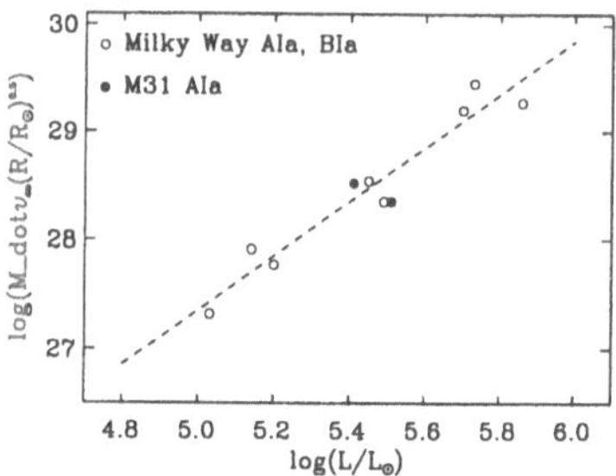

Fig. 6. Left: H_α fit of the Keck HIRES spectrum of the M31 A2 Ia-O supergiant 41-3654 yielding a precise determination of $\dot{M}$ and v_∞. Right: Wind momenta of two A-supergiants in M31 (obtained with Keck HIRES) compared with galactic A- and B-supergiants of the solar neighbourhood. Results from McCarthy *et al.* 1997.

4.2 The potential of the WLR-method

The application of the WLR method does not neccessarily require spectra of high S/N or high spectral resolution. As a very illustrative example we have used the H_α and H_β equivalent widths of LMC A-supergiants published by Tully and Wolff 1984 to derive gravity and – most importantly – wind momenta (adopting $v_\infty = 150$ km/s for all objects). Fig. 7 shows the resulting correlation between wind momentum and absolute magnitude for those objects where a good simultaneous determination of gravity and mass-loss rate was possible (for details, see Knoerndel et al. 1998). Despite the limited quality of the data, the tightness of the relation obtained is striking.

If one wants to use this observed relationship to estimate the possible accuracy of distance determinations, it is important to realize that determining a mass-loss rate from an H_α profile requires an assumption about the stellar radius and therefore the distance. Thus, both wind momentum and absolute magnitude depend on the radius adopted. However, as one can show analytically (Puls et al. 1996), the quantity

$$Q = \dot{M} v_\infty (\frac{R_*}{R_\odot})^{-1.5} \tag{2}$$

is an invariant of the profile fitting, i.e. the strength of H_α as a wind line depends exactly on this quantity. As a consequence, we also plot Q as a function of M_V in Fig. 7 to quantify the distance independent scatter. The standard deviation in absolute magnitude from the mean relation $Q = f(M_V)$

is on the order of 0.3 mag. This confirms the estimate by Mc Carthy et al. 1997 that with ten to twenty objects per galaxy, distance moduli as accurate as 0.1 mag should be achievable.

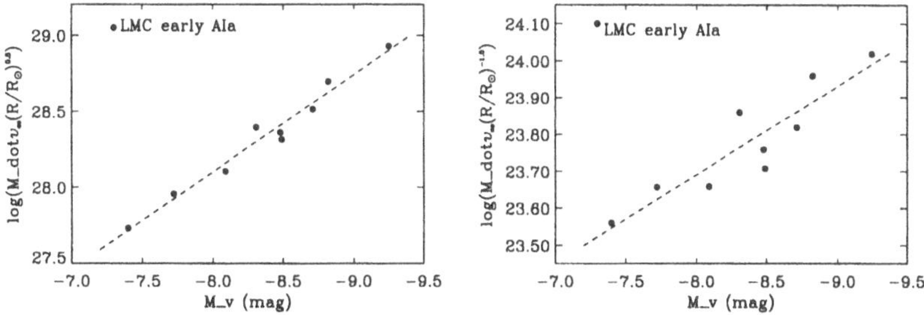

Fig. 7. Left: Wind momentum determined from the Balmer line equivalent widths of LMC AO Ia - supergiants published by Tully and Wolff (1984) as function of absolute magnitude (adopting a distance modulus of 18.5 mag). From Knoerndel et al., 1998. Right: Same as the left figure, but instead of wind momentum the invariant Q of the H_α fitting (see text) is plotted.

The uncertainties in WLR distance moduli might, therefore, be comparable to those obtainable from Cepheids in galaxies. The advantage of the WLR-method, however, is that individual reddening (and therefore extinction) as well as metallicity can be derived directly from the spectrum of every object. Moreover, it is a new independent primary method for distance determination and can contribute to the investigation of systematic errors of extragalactic distances. However, the crucial question is to what distances will the method be applicable.

The best spectroscopic targets at large distances are A-supergiants. Since massive stars evolve at almost constant luminosity towards the red, A-supergiants are the optically brightest "normal" stellar objects because of the effects of Wien's law on the bolometric correction. In addition, for these objects we can determine the wind momentum solely by optical spectroscopy at H_α (without need of the UV). This means that we can use ground-based telescopes of the 8m class for spectroscopy rather than the tiny HST (which is then needed for accurate photometry only).

From Fig. 7 we see that the brightest A-supergiants have absolute magnitudes between −9 and −8 mag (objects like 41–3654 in M31 are another example). Such objects would be of apparent magnitude between 20 and 21 in galaxies 6 Mpc away, certainly not a problem for medium (2 Å) resolution spectroscopy with 8 m class telescopes. Even in a galaxy like M100 at a distance of 16 Mpc (Freedman *et al.* 1994b; Ferrarese et al. 1996) these objects would still be accessible at magnitudes around 22.5 and would yield wind momentum distances if medium resolution spectroscopy on a 8 m class telescope were combined with HST photometry. Indeed, the HST colour magnitude diagram published by Freedman et al. 1994a may show the presence of such objects in M100.

One might ask, of course, whether a medium resolution of 2 Å would still allow a sufficiently accurate determination of effective temperature, surface gravity, abundances, and wind momentum. We note that the rotational velocities of A-supergiants are typically on the order of $40\,\mathrm{km\,s^{-1}}$, a broadening which is matched by a spectral resolution of roughly 1.4 Å at $\mathrm{H}\alpha$. Therefore we expect that the medium resolution situation will not be dramatically worse, provided sufficient S/N can be achieved and accurate sky-, galaxy-, and H II region-background subtraction can be performed at very faint magnitudes. Experiments with observed spectra degraded with regard to S/N, resolution and sky emission indicate that such observations are challenging but feasible.

In summary, the WLR results obtained so far in the Local Group galaxies are very encouraging. We are optimistic that after further tests and calibration steps, **we will have a new distance determination method capable of reaching out as far as to the Virgo and Fornax clusters of galaxies.**

Acknowledgements

It is a pleasure to thank Artemio Herrero and Ilu Monteverde from the IAC, Jim McCarthy (Caltech), Kim Venn (Minnesota), Stephen Smartt (La Palma) and the Munich crowd Rudi Gabler, Stefan Haser, Oliver Knoerndel, Margie and Danny Lennon (now at La Palma), Paco Najarro (now Madrid), Adi Pauldrach, Joachim Puls, Uwe Springmann and Gudrun Taresch for their contributions.

References

Ferrarese, L. et al. 1996, ApJ 464, 568

Freedman, W. et al. 1994a, ApJ 435, L31

Freedman, W. et al. 1994b, Nature 371, 757

Kaufer, A., Stahl, O., Wolf, B., Gäng, Th., Gummersbach, C.A., Kovacs, J., Mandel, H., Szeifert, Th. 1996, A&A 305, 887

Knoerndel, O., Kudritzki, R.P., Puls, J., Lennon, D.J. 1998, A&A, in preparation

Kudritzki, R.P., Pauldrach, A.W.A., Puls,J., & Abbott, D.C. 1989, A&A 219, 205

Kudritzki, R.P., Lennon, D.J., Haser, S.M., Puls, J., Pauldrach, A., Venn, K., & Voels, S.A. 1996a, in " Science with the Hubble Space Telescope II", eds. P. Benvenuti et al., 285–296

Kudritzki, R.P., Mendez, R.H., McCarthy, J.K., Puls, J. 1997, Proc. IAU Symp. 180, "Planetary Nebulae", eds. H.J. Habing and H.J.G.L. Lamers, page 64 -74, invited paper, (Kluwer Academic Publishers)

Kudritzki, R.P., Springmann, U., Puls, J., Pauldrach, A.W.A., Lennon, M. 1998a, ASP Conf. Series Vol. 131, p. 299 - 308

Kudritzki, R.P., Puls, J., Lennon, D.J., Venn, K.A., McCarthy, J.K., & Herrero, A. 1998b, A&A , in preparation.

Kudritzki, R.P., Springmann, U., Puls, J., Pauldrach, A.W.A., Lennon, M. 1998c, A&A , in preparation.

Kudritzki, R.P. 1998, "Quantitative Spectroscopy of the Brightest Blue Supergiant Stars in Galaxies" in Proc. of "Stellar Astrophysics for the Local Group", VIII Canary Island Winterschool for Astrophysics, eds. A. Aparicio et al., Cambridge University Press, p. 149-262, invited lectures

McCarthy, J.K., Lennon, D.J., Venn, K.A., Kudritzki, R.P., Puls, J., & Najarro, F. 1995, ApJ 455, L35

McCarthy, J.K., Kudritzki, R.P., Lennon, D.J., Venn, K.A., Puls, J. 1997, ApJ 482, 757

Monteverde, I., Herrero, A., Lennon, D.J., & Kudritzki, R.P. 1997, ApJ Letters 474, L107

Pauldrach, A.W.A., Lennon, M., Hoffmann, T.L., Sellmaier,F., Kudritzki, R.P., Puls, J. 1998, ASP Conference Series Vol. 131, page 258 - 277

Prinja, R.K., Massa, D.L. 1998, ASP Conf. Series, Vol. 131, p. 218 - 227

Puls, J., Kudritzki, R.P., Herrero, A., Pauldrach, A., Haser, S.M., Lennon, D.J., Gabler, R., Voels, S.A., Vilchez, J.M., Wachter, S., & Feldmeier, A. 1996, A&A 305, 171

Rozanski, R., & Rowan-Robinson, M. 1994, MNRAS 271, 530

Ruppersberg, R., Kudritzki, R.P., Puls, J., Kaufer, A. 1998, in preparation for A&A

Ruppersberg, R. 1998, Diplomarbeit, University of Munich

Santolaya-Rey, E., Puls, J., & Herrero, A. 1997, A&A 323, 488

Springmann, U. 1997, doctoral thesis, University of Munich

Springmann, U., Puls, J. 1998, ASP Conf. Series, Vol. 131, p. 286 - 298

Tully, R.B., Wolff, S.C. 1984, ApJ 281, 67

Discussion

S. Owocki: One remarkable implication of the observed wind-momentum luminosity relation is that the line parameters are nearly universal for O, B, and A stars. But if you were to add WR stars to your observational comparison, they would not fit by a factor of ten. This suggests that there is something about WR winds that makes line opacity a factor of ten more efficient in driving mass. All this is quite independent of a single scattering momentum limit, but represents a kind of "opacity problem".

Conference Summary: The Demise of Spherical and Stationary Winds

Immo Appenzeller[1,2]

[1] Landessternwarte, Königstuhl, D-69117 Heidelberg, Germany
[2] Max-Planck-Institut für Astronomie, Königstuhl, D 69117 Heidelberg, Germany

Abstract. The observational evidence and the theoretical work presented at this colloquium demonstrate that significant non-spherical effects and time variations are very common phenomena in the winds from hot stars. Hence, the assumptions of spherical symmetry and stationarity, while highly successful for the theory of stellar structure, appear to be inadequate for describing and understanding the winds of hot stars. This conclusion has significant consequences for the appearance and evolution of hot stars.

1 Introduction

In view of the huge distances of our research objects stellar astrophysics has been a surprisingly successful branch of science. Today the interiors of the Sun (and of main-sequence stars about anywhere in the Universe) are in many aspects better understood than the deep interior of our Earth, and many details of stellar atmospheres can be modeled more reliably than the behavior of the atmosphere in which we live.

Much of this success has been due to the fact that the basic physics of stars is relatively simple. In most cases their matter can be described by the equation of state of an ideal gas. Stellar structure can be approximated very well assuming spherical symmetry. And although all stars evolve with time, the evolutionary time scale is so much longer than the dynamic time scale that static or stationary models provide excellent approximations.

Since about 70 years it has been known that luminous hot stars tend to lose mass. Later it became clear that late-type stars like our Sun also have stellar winds and that stellar mass loss is a rather common phenomenon occurring in all regions of the HR diagram. In view of the simplicity and the success of the spherical and static stellar structure models it was natural to start modeling stellar winds with spherical and stationary configurations too. Although direct observations of Planetary Nebulae, Nova and Supernova remnants, wind-blown IS bubbles, and X-ray observations of the solar corona showed conspicuous deviations from spherical symmetry, the notion of spherically symmetric and stationary winds survived, mainly due to our lack of knowledge on the real wind geometries.

At this conference much additional evidence for non-spherical and non-stationary winds was reported. Moreover, we learned that major deviations from spherical symmetry and stationarity follow naturally from realistic wind theories. Hence, IAU Colloquium 169 may one day be remembered as the meeting where it was realized that the assumptions of spherical symmetry and stationarity (while very successful for stellar interiors) is fundamentally inadequate for stellar winds. In the following I shall try to summarize the results which were presented at this Colloquium and which lead us to this conclusion.

2 Probing the Wind Geometry

2.1 Observational Progress

The occurrence of stellar winds was initially inferred from spectroscopic data (Beals, 1930). Spectroscopy also provided the first hints for deviations from spherically symmetric flows, and this technique continues to be one of the main information sources on the wind geometries. Perhaps the most impressive data sets shown at this meeting were the extensive spectroscopic time series of O stars, BA supergiants and Be stars obtained with IUE and with dedicated ground-based instruments, such as HEROS, and described elsewhere in this volume by A.W. Fullerton, Andreas Kaufer, Geraldine Peters, Stanislav Štefl, and Thomas Rivinius. All these observations resulted in convincing evidence for rotational modulations of the wind spectra, proving conclusively that the observed winds are not spherically symmetric and, in fact, not even azimuthally symmetric. Various plausible models for wind geometries have been derived and presented at this meeting. However, the discussion also showed that, while spectroscopic time series can prove the presence of asymmetries conclusively, the derivation of the exact wind structure from spectroscopic data is complex and much more difficult.

A more direct method to prove the presence of deviations from spherical symmetry is polarimetry. This technique has been used to investigate flattened objects and circumstellar disks since more than three decades (cf. e.g. Appenzeller 1965, Appenzeller and Hiltner 1967). But major progress in this field was achieved only during the past decade with the development of improved instrumentation, better models and the introduction of spectropolarimetry. This progress was demonstrated impressively at this meeting by the reports and posters presented by Karin Bjorkman, Mário Magalhães, Cláudia Vilega Rodrigues, and R. V. Yudin. But, as analyzed beautifully in John Brown's talk, inferring the geometric details from polarimetry and spectropolarimetry alone requires caution, and a successful fit to a simple model does not guarantee a correct derivation for the underlying geometry. Therefore, polarimetric observations cannot fully replace direct high angular resolution imaging.

As expected some of the key results in the field of high-resolution direct imaging of winds presented at this meeting came from the HST (cf. the corresponding contributions by Antonella Nota, Regina Schulte-Ladbeck, and others to this volume). There were also interesting new radio data (S. White) and - more surprisingly - FIR data from ISO (N. R. Trams et al.). That ISO (with its angular resolution strongly limited by the diffraction disk of its, expressed in wavelength, rather small aperture) should allow us to image the wind zones of stars appears unexpected on first glance. On the other hand, ISO records radiation from the cold matter. Hence ISO allows us to see the cool and dusty outermost regions of wind flows, which are sufficiently extended to be resolved with infrared satellites.

Among the most exciting results presented at this meeting were data obtained from optical interferometric imaging (F. Vakili and others). The data produced so far show clearly the wavelength dependence of the shape, orientation and size of the wind flows on a sub-arcsec scale. The next step obviously must be to obtain a more complete coverage of the u-v plane and to reconstruct real images of the observed stellar winds. While this will require much additional development work, it was very encouraging to see real progress in this field. As a precaution it must be noted however, that for direct and interferometric imaging the length scales which can actually be resolved will always depend on the objects distance. In this respect indirect methods like polarimetry and spectroscopy have the advantage of working equally well for any distance. Hence all these techniques will remain important for future work.

2.2 A Choice of Theories

Non-spherical mass loss was inferred first for Be stars. Very soon a relation with the rotational velocity was discovered and rotation became the prime suspect for causing non-spherical winds (Sletteback 1949). That other physical mechanisms may (also) be involved became clear when strongly nonsymmetric winds were found in the (slowly rotating) B[e] supergiants (Zickgraf et al. 1985). Initially it appeared difficult to reconcile this result with the established theories of (radiation driven) winds of hot stars. Today there is no longer any doubt that through various mechanisms even a small amount of rotation can cause large asymmetries in radiation driven winds. (For details see the contributions by Jon Bjorkman, Joachim Puls, Henny Lamers and others to this volume). However, the lively discussion at this colloquium could not clarify whether rotationally induced compression of the gas in the equatorial plane or bi-stable winds (or a combination) are the main source of the observed asymmetries. The theory of wind compressed disks appears better developed at present, but seem to have problems reproducing the required wind parameters. No such problems seem to exist in the case of rotationally induced bi-stability, but this could simply be due to a lower stage of development of this theory. Finally, one has to keep in mind that both concepts are approximations. And magnetic fields (invoked by various authors) may complicate things further.

Fortunately, the different theories predict significantly different wind geometries. Rotationally induced wind compression will result in geometrically thinner disks than bi-stable winds or magnetically supported disks. Hence future interferometric images of the wind geometry may well help us to identify the dominant mechanism unambiguously.

3 Variable Winds

Variability of luminous hot stars and of their winds has been observed since several decades. (Note that for the hot star where mass loss has been inferred first, P Cyg, variability had been discovered already in 1600, making P Cyg one of the first two stars – apart from Novae and SNe – for which variations were established). But only during the past ten years it became clear that wind variations is an omnipresent feature of these stars.

The variations may be minor (as in θ^1 Ori C) or dramatic (as in the LBVs). The small amplitude variations seem to be reasonably well understood as effects of rotation (as already discussed in Section 2.1), pulsations (see Section 4), or wind instabilities (see the contributions of A. Feldmeier and Stan Owocki to this volume). As pointed out by W.-R. Hamann and by Gloria Koenigsberger the formation of clumps seems to be a major cause of the variations in the dense winds of the Wolf-Rayet stars. On the other hand, the true physical origin of the dramatic *large amplitude* variations of the LBVs remains unclear. Although our empirical knowledge on the properties is fairly extensive by now (cf. Otmar Stahl's contribution) little progress has been reported concerning the theory of these large amplitude LBV variations. There exist since many years various suggestions for mechanisms (discussed already in the proceedings of IAU meetings Symposium 116 and IAU Colloquium 113). But these ideas have not been followed up by detailed model computations. Moreover, as pointed out by Roberta Humphreys at this meeting, the LBVs may well consist of at least two distinct classes with different physical variability mechanisms.

4 Pulsations

For many decades investigations of hot pulsating stars were restricted essentially to the β Cep (or β CMa) stars, although the theorists had predicted since a long time that pulsational instability may be present in a much larger fraction of the blue part of the HR-diagram (see e.g. Osaki 1987). At this meeting pulsational instability of various modes (radial, non-radial, strange mode) was reported for about any effective temperature and luminosity where hot stars can be found (for details see the contributions to this volume by Henrichs, Baade, Osaki, Glatzel, Guzik, and Ødegaard). Also new at this meeting was the finding that beating effects between the many different non-radial modes could have important astrophysical consequences and perhaps play a major role for the mass loss from Be stars (as discussed in a contribution by Thomas Rivinius).

5 Consequences of Non-radial and Non-stationary Winds

5.1 Evolutionary Effects

An obvious consequence of winds from hot stars is the formation of wind blown bubbles, circumstellar nebulae, and sometimes cool and dusty disks. Progress in our understanding of these phenomena has been reported at this meeting (and is described in this volume) in the contributions of Mordecai Mac Low, Linda Smith et al. and R. Waters.

It is well known that stellar winds have a profound influence on the evolutionary tracks of massive stars in the HR diagram (cf. A. Maeder, G. Meynet, and N. Langer in this volume). Therefore, in all modern evolutionary calculations for massive stars mass loss has been taken into account in some semi-empirical way (see e.g. Meynet et al. 1994). However, all these calculations assumed for simplicity spherical and stationary mass loss. As pointed out at this meeting, more realistic assumptions may affect the evolutionary time scales and the computed luminosities significantly. New tracks will have to be calculated before a quantitative assessment of the expected differences can be carried out.

5.2 Luminous Hot Stars as Distance Indicators

With luminosities often surpassing those of globular clusters and of whole small galaxies LBVs can be observed and identified easily in many resolved extragalactic systems. Hence they are potentially very important extragalactic distance indicators, provided their brightness can be calibrated with sufficient accuracy.

Various methods to obtain such calibrations have been suggested in the literature. At this meeting two such methods have been discussed in some detail. R.-P. Kudritzki reported new results on the Wind Momentum Luminosity Relation (WMLR) method (Kudritzki et al. 1995), and Berhard Wolf and others discussed the use of the Amplitude-Luminosity Relation (ALR) of the LBVs (Wolf (1989)).

On first glance one might suspect that time variations and deviations spherical symmetry could compromise the accuracy of the WMLR method. However, as demonstrated in R.-P. Kudritzki's contribution, fortunately these effects influence the WMLR only to a minor extent. Hence the WMLR should remain one of the potentially most accurate and powerful procedure to determine the distance of resolved galaxies.

Wolf's ALR method, of course, is based on time variations, and in view of the statistical calibration should not be influenced much by non-spherical winds. Hence its potential should not be affected by the results discussed at this meeting. Comparing the WMLR and the ALR, the WMLR has the obvious advantage of giving accurate results from a single epoch observation, while the ALR requires observations at several observing epochs at uncomfortably large time intervals. On the other hand, being a purely photometric method the ALR in principle can reach much larger distances. It is somewhat surprising that not more use is being made of this method.

6 Outlook

Because of their short life expectancy luminous hot stars are relatively rare objects. Moreover, most galactic hot luminous stars are hidden from our view as they are strongly concentrated to the galactic plane and are obscured by the galactic dust clouds. Therefore, much of our knowledge about these objects is based on observations in other nearby galaxies, notably the Magellanic Clouds. On the other hand large telescopes and efficient instruments are needed to study extragalactic stars. Just at the time of the final preparations of IAU Colloquium 169 the ESO-VLT UT1, the first very large telescope in the southern hemisphere started its operations. Together with the already routinely operating two Keck telescopes and several comparable instruments close to completion or under construction these new facilities will make observations of extragalactic stars much easier and, with powerful new focal plane instruments such as FORS, much more economic. There is no doubt that these new generation of telescopes will greatly stimulate research in the area of hot luminous stars with all the techniques discussed at this colloquium.

One of unique features of the ESO VLT will be the possibility to combine the four unit telescopes and several smaller auxiliary instruments to a powerful optical interferometer, which promises major progress in the interferometric imaging of stellar envelopes.

I am somewhat sorry to note that in view of these new technical opportunities which will become available during the next months and years, it seems safe to predict that much of the observational data reported at IAU Colloquium 169 and much of what has been written down in this volume will soon become completely obsolete and replaced by much superior results. On the other hand, I am also convinced that the high scientific standard of this colloquium and the very lively discussions of this meeting will certainly have a significant impact on the research in our field during next years and that in this way our meeting may have contributed much to a productive and efficient use of the new large telescopes for stellar astrophysics.

References

Appenzeller, I. (1965): ApJ **141**, 1390

Appenzeller, I. and Hiltner, W.A. (1967): ApJ **149**, 353
Beals, C.S. (1930): Pub. Dom. Ap. Obs. Victoria **4**, 288
Kudritzki, R.-P., Lennon, Puls, J., (1995), in "Science with the VLT", J.R. Walsh and I.J. Danziger eds., Springer, Heidelberg 1994, p. 246
Meynet, G., Maeder, A., Schaller, G., Schaerer, D., Charbonnel, C. (1994): Astron. Astrophys. Suppl.**103**, 97
Osaki, Y. (1987), in "Instabilities in Luminous Early Type Stars", H.J.G.L.M. Lamers and C.W.H. de Loore eds., D. Reidel Publ. Co. , Dordrecht 1987, p. 39
Slettebak, A. (1949): ApJ **110**, 498
Wolf, B., (1989), in Proc. IAU Coll. **113** on *Physics of Luminous Blue Variables*, K. Davidson, A.F.J. Moffat, H.J.G.L.M. Lamers eds., Kluwer, Dordrecht, p. 91.
Zickgraf F.-J., Wolf, B., Stahl, O., Leitherer, C., Klare, G. (1985): Astron. Astrophys. **143**, 421

Immo Appenzeller, Ulrich Bastian and Bernhard Wolf

Object Index

Springer and the environment

At Springer we firmly believe that an international science publisher has a special obligation to the environment, and our corporate policies consistently reflect this conviction.

We also expect our business partners – paper mills, printers, packaging manufacturers, etc. – to commit themselves to using materials and production processes that do not harm the environment. The paper in this book is made from low- or no-chlorine pulp and is acid free, in conformance with international standards for paper permanency.

Lecture Notes in Physics

For information about Vols. 1–483
please contact your bookseller or Springer-Verlag

Vol. 484: L. Schimansky-Geier, T. Pöschel (Eds.), Stochastic Dynamics. XVIII, 386 pages. 1997.

Vol. 485: H. Friedrich, B. Eckhardt (Eds.), Classical, Semi-classical and Quantum Dynamics in Atoms. VIII, 341 pages. 1997.

Vol. 486: G. Chavent, P. C. Sabatier (Eds.), Inverse Problems of Wave Propagation and Diffraction. Proceedings, 1996. XV, 379 pages. 1997.

Vol. 487: E. Meyer-Hofmeister, H. Spruik (Eds.), Accretion Disks – New Aspects. Proceedings, 1996. XIII, 356 pages. 1997.

Vol. 488: B. Apagyi, G. Endrédi, P. Lévay (Eds.), Inverse and Algebraic Quantum Scattering Theory. Proceedings, 1996. XV, 385 pages. 1997.

Vol. 489: G. M. Simnett, C. E. Alissandrakis, L. Vlahos (Eds.), Solar and Heliospheric Plasma Physics. Proceedings, 1996. VIII, 278 pages. 1997.

Vol. 490: P. Kutler, J. Flores, J.-J. Chattot (Eds.), Fifteenth International Conference on Numerical Methods in Fluid Dynamics. Proceedings, 1996. XIV, 655 pages. 1997.

Vol. 491: O. Boratav, A. Eden, A. Erzan (Eds.), Turbulence Modeling and Vortex Dynamics. Proceedings, 1996. XII, 245 pages. 1997.

Vol. 492: M. Rubí, C. Pérez-Vicente (Eds.), Complex Behaviour of Glassy Systems. Proceedings, 1996. IX, 467 pages. 1997.

Vol. 493: P. L. Garrido, J. Marro (Eds.), Fourth Granada Lectures in Computational Physics. XIV, 316 pages. 1997.

Vol. 494: J. W. Clark, M. L. Ristig (Eds.), Theory of Spin Lattices and Lattice Gauge Models. Proceedings, 1996. XI, 194 pages. 1997.

Vol. 495: Y. Kosmann-Schwarzbach, B. Grammaticos, K.M. Tamizhmani (Eds.), Integrability of Nonlinear Systems. VII, 380 pages. 1997.

Vol. 496: F. Lenz, H. Grießhammer, D. Stoll (Eds.), Lectures on QCD. VII, 483 pages. 1997.

Vol. 497: J. P. Greve, R. Blomme, H. Hensberge (Eds.), Stellar Atmospheres: Theory and Observations. VIII, 352 pages. 1997

Vol. 498: Z. Horváth, L. Palla (Eds.), Conformal Field Theories and Integrable Models. Proceedings, 1996. X, 251 pages. 1997.

Vol. 499: K. Jungmann, J. Kowalski, I. Reinhard, F. Träger (Eds.), Atomic Physics Methods in Modern Research, IX, 448 pages. 1997.

Vol. 500: D. Joubert (Ed.), Density Functionals: Theory and Applications, XVI, 194 pages. 1998.

Vol. 501: J. Kertész, I. Kondor (Eds.), Advances in Computer Simulation. VIII, 166 pages. 1998.

Vol. 502: H. Aratyn, T. D. Imbo, W.-Y. Keung, U. Sukhatme (Eds.), Supersymmetry and Integrable Models. Proceedings, 1997. XI, 379 pages. 1998.

Vol. 503: J. Parisi, S. C. Müller, W. Zimmermann (Eds.), A Perspective Look at Nonlinear Media. From Physics to Biology and Social Sciences. VIII, 372 pages. 1998.

Vol. 504: A. Bohm, H.-D. Doebner, P. Kielanowski (Eds.), Irreversibility and Causality. Semigroups and Rigged Hilbert Spaces. XIX, 385 pages. 1998.

Vol. 505: D. Benest, C. Froeschlé (Eds.), Impacts on Earth. XVII, 223 pages. 1998.

Vol. 506: D. Breitschwerdt, M. J. Freyberg, J. Trümper (Eds.), The Local Bubble and Beyond. Proceedings, 1997. XXVIII, 603 pages. 1998.

Vol. 507: J. C. Vial, K. Bocchialini, P. Boumier (Eds.), Space Solar Physics. Proceedings, 1997. XIII, 296 pages. 1998.

Vol. 508: H. Meyer-Ortmanns, A. Klümper (Eds.), Field Theoretical Tools for Polymer and Particle Physics. XVI, 258 pages. 1998.

Vol. 509: J. Wess, V. P. Akulov (Eds.), Supersymmetry and Quantum Field Theory. Proceedings, 1997. XV, 405 pages. 1998.

Vol. 510: J. Navarro, A. Polls (Eds.), Microscopic Quantum Many-Body Theories and Their Applications. Proceedings, 1997. XIII, 379 pages. 1998.

Vol. 511: S. Benkadda, G. M. Zaslavsky (Eds.), Chaos, Kinetics and Nonlinear Dynamics in Fluids and Plasmas. Proceedings, 1997. VIII, 438 pages. 1998.

Vol. 512: H. Gausterer, C. Lang (Eds.), Computing Particle Properties. Proceedings, 1997. VII, 335 pages. 1998.

Vol. 513: A. Bernstein, D. Drechsel, T. Walcher (Eds.), Chiral Dynamics: Theory and Experiment. Proceedings, 1997. IX, 394 pages. 1998.

Vol. 514: F. W. Hehl, C. Kiefer, R.J.K. Metzler, Black Holes: Theory and Observation. Proceedings, 1997. XV, 519 pages. 1998.

Vol. 515: C.-H. Bruneau (Ed.), Sixteenth International Conference on Numerical Methods in Fluid Dynamics. Proceedings. XV, 568 pages. 1998.

Vol. 516: J. Cleymans, H. B. Geyer, F. G. Scholtz (Eds.), Hadrons in Dense Matter and Hadrosynthesis. Proceedings, 1998. XII, 253 pages. 1999.

Vol. 517: Ph. Blanchard, A. Jadczyk (Eds.), Quantum Future. Proceedings, 1997. X, 244 pages. 1999.

Vol. 518: P. G. L. Leach, S. E. Bouquet, J.-L. Rouet, E. Fijalkow (Eds.), Dynamical Systems, Plasmas and Gravitation. Proceedings, 1997. XII, 397 pages. 1999.

Vol. 519: A. Pękalski, K. Sznajd-Weron (Eds.), Anomalous Diffusion. From Basics to Applications. Proceedings, 1998. XVIII, 378 pages. 1999.

Vol. 520: J. A. van Paradijs, J. A. M. Bleeker (Eds.), X-Ray Spectroscopy in Astrophysics. EADN School X. Proceedings, 1997. XV, 530 pages. 1999.

Vol. 521: L. Mathelitsch, W. Plessas (Eds.), Broken Symmetries. Proceedings, 1998. VII, 299 pages. 1999.

Vol. 523: B. Wolf, O. Stahl, A.W. Fullerton (Eds.), Variable and Non-spherical Stellar Winds in Luminous Hot Stars. Proceedings, 1998. XX, 424 pages. 1999.

Monographs

For information about Vols. 1–10 please contact your bookseller or Springer-Verlag

Vol. m 11: A. D. Yaghjian, Relativistic Dynamics of a Charged Sphere. XII, 115 pages. 1992.

Vol. m 12: G. Esposito, Quantum Gravity, Quantum Cosmology and Lorentzian Geometries. Second Corrected and Enlarged Edition. XVIII, 349 pages. 1994.

Vol. m 13: M. Klein, A. Knauf, Classical Planar Scattering by Coulombic Potentials. V, 142 pages. 1992.

Vol. m 14: A. Lerda, Anyons. XI, 138 pages. 1992.

Vol. m 15: N. Peters, B. Rogg (Eds.), Reduced Kinetic Mechanisms for Applications in Combustion Systems. X, 360 pages. 1993.

Vol. m 16: P. Christe, M. Henkel, Introduction to Conformal Invariance and Its Applications to Critical Phenomena. XV, 260 pages. 1993.

Vol. m 17: M. Schoen, Computer Simulation of Condensed Phases in Complex Geometries. X, 136 pages. 1993.

Vol. m 18: H. Carmichael, An Open Systems Approach to Quantum Optics. X, 179 pages. 1993.

Vol. m 19: S. D. Bogan, M. K. Hinders, Interface Effects in Elastic Wave Scattering. XII, 182 pages. 1994.

Vol. m 20: E. Abdalla, M. C. B. Abdalla, D. Dalmazi, A. Zadra, 2D-Gravity in Non-Critical Strings. IX, 319 pages. 1994.

Vol. m 21: G. P. Berman, E. N. Bulgakov, D. D. Holm, Crossover-Time in Quantum Boson and Spin Systems. XI, 268 pages. 1994.

Vol. m 22: M.-O. Hongler, Chaotic and Stochastic Behaviour in Automatic Production Lines. V, 85 pages. 1994.

Vol. m 23: V. S. Viswanath, G. Müller, The Recursion Method. X, 259 pages. 1994.

Vol. m 24: A. Ern, V. Giovangigli, Multicomponent Transport Algorithms. XIV, 427 pages. 1994.

Vol. m 25: A. V. Bogdanov, G. V. Dubrovskiy, M. P. Krutikov, D. V. Kulginov, V. M. Strelchenya, Interaction of Gases with Surfaces. XIV, 132 pages. 1995.

Vol. m 26: M. Dineykhan, G. V. Efimov, G. Ganbold, S. N. Nedelko, Oscillator Representation in Quantum Physics. IX, 279 pages. 1995.

Vol. m 27: J. T. Ottesen, Infinite Dimensional Groups and Algebras in Quantum Physics. IX, 218 pages. 1995.

Vol. m 28: O. Piguet, S. P. Sorella, Algebraic Renormalization. IX, 134 pages. 1995.

Vol. m 29: C. Bendjaballah, Introduction to Photon Communication. VII, 193 pages. 1995.

Vol. m 30: A. J. Greer, W. J. Kossler, Low Magnetic Fields in Anisotropic Superconductors. VII, 161 pages. 1995.

Vol. m 31 (Corr. Second Printing): P. Busch, M. Grabowski, P.J. Lahti, Operational Quantum Physics. XII, 230 pages. 1997.

Vol. m 32: L. de Broglie, Diverses questions de mécanique et de thermodynamique classiques et relativistes. XII, 198 pages. 1995.

Vol. m 33: R. Alkofer, H. Reinhardt, Chiral Quark Dynamics. VIII, 115 pages. 1995.

Vol. m 34: R. Jost, Das Märchen vom Elfenbeinernen Turm. VIII, 286 pages. 1995.

Vol. m 35: E. Elizalde, Ten Physical Applications of Spectral Zeta Functions. XIV, 224 pages. 1995.

Vol. m 36: G. Dunne, Self-Dual Chern-Simons Theories. X, 217 pages. 1995.

Vol. m 37: S. Childress, A.D. Gilbert, Stretch, Twist, Fold: The Fast Dynamo. XI, 406 pages. 1995.

Vol. m 38: J. González, M. A. Martín-Delgado, G. Sierra, A. H. Vozmediano, Quantum Electron Liquids and High-Tc Superconductivity. X, 299 pages. 1995.

Vol. m 39: L. Pittner, Algebraic Foundations of Non-Com-mutative Differential Geometry and Quantum Groups. XII, 469 pages. 1996.

Vol. m 40: H.-J. Borchers, Translation Group and Particle Representations in Quantum Field Theory. VII, 131 pages. 1996.

Vol. m 41: B. K. Chakrabarti, A. Dutta, P. Sen, Quantum Ising Phases and Transitions in Transverse Ising Models. X, 204 pages. 1996.

Vol. m 42: P. Bouwknegt, J. McCarthy, K. Pilch, The W3 Algebra. Modules, Semi-infinite Cohomology and BV Algebras. XI, 204 pages. 1996.

Vol. m 43: M. Schottenloher, A Mathematical Introduction to Conformal Field Theory. VIII, 142 pages. 1997.

Vol. m 44: A. Bach, Indistinguishable Classical Particles. VIII, 157 pages. 1997.

Vol. m 45: M. Ferrari, V. T. Granik, A. Imam, J. C. Nadeau (Eds.), Advances in Doublet Mechanics. XVI, 214 pages. 1997.

Vol. m 46: M. Camenzind, Les noyaux actifs de galaxies. XVIII, 218 pages. 1997.

Vol. m 47: L. M. Zubov, Nonlinear Theory of Dislocations and Disclinations in Elastic Body. VI, 205 pages. 1997.

Vol. m 48: P. Kopietz, Bosonization of Interacting Fermions in Arbitrary Dimensions. XII, 259 pages. 1997.

Vol. m 49: M. Zak, J. B. Zbilut, R. E. Meyers, From Instability to Intelligence. Complexity and Predictability in Nonlinear Dynamics. XIV, 552 pages. 1997.

Vol. m 50: J. Ambjørn, M. Carfora, A. Marzuoli, The Geometry of Dynamical Triangulations. VI, 197 pages. 1997.

Vol. m 51: G. Landi, An Introduction to Noncommutative Spaces and Their Geometries. XI, 200 pages. 1997.

Vol. m 52: M. Hénon, Generating Families in the Restricted Three-Body Problem. XI, 278 pages. 1997.

Vol. m 53: M. Gad-el-Hak, A. Pollard, J.-P. Bonnet (Eds.), Flow Control. Fundamentals and Practices. XII, 527 pages. 1998.

Vol. m 54: Y. Suzuki, K. Varga, Stochastic Variational Approach to Quantum-Mechanical Few-Body Problems. XIV, 324 pages. 1998.

Vol. m 55: F. Busse, S. C. Müller, Evolution of Spontaneous Structures in Dissipative Continuous Systems. X, 559 pages. 1998.